Springer Series in Optical Sciences

Volume 229

Springer Series in Optical Sciences is led by Editor-in-Chief William T. Rhodes, Florida Atlantic University, USA, and provides an expanding selection of research monographs in all major areas of optics:

- lasers and quantum optics
- ultrafast phenomena
- optical spectroscopy techniques
- optoelectronics
- information optics
- applied laser technology
- industrial applications and
- other topics of contemporary interest.

With this broad coverage of topics the series is useful to research scientists and engineers who need up-to-date reference books.

More information about this series at http://www.springer.com/series/624

Arash Rahimi-Iman

Polariton Physics

From Dynamic Bose–Einstein Condensates in Strongly-Coupled Light–Matter Systems to Polariton Lasers

Springer

Arash Rahimi-Iman
Physics Department
Philipps-Universität Marburg
Marburg, Germany

ISSN 0342-4111 ISSN 1556-1534 (electronic)
Springer Series in Optical Sciences
ISBN 978-3-030-39331-1 ISBN 978-3-030-39333-5 (eBook)
https://doi.org/10.1007/978-3-030-39333-5

This Springer imprint is published by the registered company Springer Nature Switzerland AG
The registered company address is: Gewerbestrasse 11, 6330 Cham, Switzerland

Foreword

The world of polaritons is truly rich of wonderful physics. It is full of phase transitions, superfluids and Bose–Einstein condensation studies. A magical realm opens up to the interested reader that is filled with fascinating phenomena and manifestations such as long-range order, repulsive expansion of the fluid, nonlinearity inversion and condensation in real space, solitons and shock waves, quantum vortex molecules, precession and vortex transmutations, and even polariton supernovas, black and white holes as well as other cosmological parallels such as polariton Hawking radiation. A variety of futuristic polariton platforms are in the line for being utilized, ranging from polariton lasing for the generation of coherent light up to the use of polariton fluids for neuronal machine learning. We are now also at the verge of entering the true quantum regimes of single or few polariton particles.

All these features and promises were and are here, still disclosing new wonders to us and to the courageous discoverer, who is necessarily exploring at times with hectic pace and at times with serendipity. The very first time we ourselves—including the author—started wandering inside the world of microcavity polaritons, we really did not know much about them, but the place where this term was coming from: the merging of polar vibrations and photons, created in words by the pioneering scientist Hopfield and his works. He did so really hopping between two fields, and linking two realms, that of photons of light and that of electric charge of matter. Likely, most of us, at the time we were new to the polariton subject, were thinking that we had always studied and worked prevalently with light, before realizing that most of the time what we had been already doing—and would have continued to do prevalently later on—was to study the interactions of light with matter. Was it a nonlinear medium, a couple of nanostructured mirrors or a microcavity, a photovoltaic cell, a sensing device or new kind of laser source, everything was dealing with the interactions, and hybridization, between both of these two components: Photons and electrons.

Among the discoveries in the scientific yet magic world of polaritons, I also met Arash, the author of this book. Our first encounter was in fact driven by the first demonstration of an electrically-driven polariton laser, which the very same author highlighted at an international conference in 2013 in his talk. His mood of freshness and kindness I have kept in good memory, and not less his scientific vigor, commitment and enthusiasm. I am convinced that this is well reflected in his knowledge-sharing efforts and the way this book introduces the fascinating and complex subject of polariton physics to the broad readership. With the same strong drive which he had to bring us together in a research project (in an effort to hybridize Italian–German polariton research), he also remained dedicated to strong light–matter interactions and gave life to this book.

Because polaritons are not only science, but also technology, and even if it is hard to believe, it can be though easily foreseen that one way or another they will be present in our future lives. Be it a kind of laser-like device, an effective quantum source or optical computer, a superfluid gyroscope, or an unpredictable device, the evolution of technology—and in the specific case that of nanophotonics—will bring polaritons into our houses. That are microcavity polaritons for now: A so-called platform embedding a quantum fluid which is the mirror in which to project the future. But first of all, with this book you have brought the fundamentals and the technology behind polariton systems into your house, together with the crystalline and crystallized vision of their current status of research.

So, were you an interested wandering reader, a university or Ph.D. student wishing to understand some polariton effects, looking for a related experimental scheme or device, a researcher or a professor tackling a specific problem or searching for a reference, a vast pool of information would be found inside this book. Never mind a bit of science fiction that is already in the polariton world, which is a melangé itself of old school and solid physics together with most recent discoveries and futuristic promises. And you will forgive the author, Arash, if—in bringing you across this realm—he couldn't bring up all the works and effects relevant to polariton fluids: They are simply too many, and evolving.

I wish a pleasant reading.

Lecce, Italy
November 2019

Lorenzo Dominici

Preface

What an exciton–polariton condensate is—or a *polariton laser*? Where the similarities and differences between polariton and conventional lasers lie? How optical microcavities gave access to the study of condensate physics? These questions and topics will be subject of this work, which will introduce the reader to the rich physics of polariton systems. Insights shall be provided into theoretical basics, technological aspects and experimental studies in this fascinating field of science.

A new kind of light source is on the verge of being utilized, which is based on pronounced light–matter interaction. In such a light source, new quasi-particles named (exciton–)polaritons are formed via the strong coupling of matter excitation and confined light. At very low threshold densities, they are supposed to act as laser-like coherent light sources. Therefore, these particles have to experience a Bose condensation effect. Ultimately, condensates on demand are promised by polariton devices which can be harnessed for future applications.

Polaritonic light sources can serve as an ideal test bed for various studies of (dynamical) Bose–Einstein condensation as well as related effects and phenomena, i.e. bosonic many-particle effects and quantum degeneracy, in solids. Moreover, they are believed to pave the way for alternative and energy-efficient sources of coherent light in the low-emission-intensity regime, as was recently demonstrated in publications on electrically-driven polariton lasers and further encouraged by room-temperature condensate reports.

On the one hand, technological advances in recent years and decades in the fields of growth, processing and characterization of semiconductor optoelectronic systems facilitated the realization of appropriate polariton devices, comprising quantum structures and high-quality optical microresonators. On the other hand, novel material systems employed in polariton research enabled experiments at elevated temperatures, where future practical devices are expected to operate. These developments clearly motivate the discussion of polariton physics in this work.

This book offers an overview on polariton *Bose–Einstein condensation* and the emerging field of *polaritonics*. The reader will get acquainted with both fundamental and device-oriented research. Following a summary of theoretical considerations (Chaps. 2 and 3), light is shed on the concept of the polariton laser (Chap. 4), polariton microcavities (Chap. 5), the technical realization of optoelectronic devices with polaritonic emission (Chap. 6), their characterization (Chaps. 7 and 8) and the role of external fields which can be utilized for manipulation and control of exciton–polaritons (Chap. 9). A glossary provides simplified summaries of the most frequently discussed topics, allowing readers to quickly familiarize themselves with the content.

The book pursues an uncomplicated and intuitive approach to the topics covered, while also providing a brief outlook on current and future work. It is my aim that its straight-forward content will make it accessible to a broad readership, ranging from research fellows, lecturers and students to interested science and engineering professionals in the interdisciplinary domains of nanotechnology, photonics, materials sciences and quantum physics.

Truly, polariton physics has become a rapidly developing field of solid-state physics. In fact, such a book cannot fully cover the rich physics of exciton–polaritons and all related studies. It might display and discuss an arbitrary selection of experimental and theoretical works known in the literature, affected by my own experiences, perception and interests in this research field. I am aware of the plethora of exciting research work which has been concluded on this subject and which is still being pursued by various expert groups. Accidentally or unintentionally, some important works might not be covered in the here provided overview.

Nevertheless, I would like to say that this book was composed by me with an as-balanced-as-possible view, with great appreciation for the community members' efforts and achievements, and with care. It can be said with confidence that any updated version of such textbook would be more complete, more up-to-date and possibly also more diverse.

A variety of books and review articles exist on this subject and they might serve as an excellent source of both historic and up-to-date knowledge on this matter, perfectly allowing to fill gaps in the topical coverage. Many of them are found as references throughout this book, and the interested reader is encouraged to get familiar with their content. One should also note that the scope of this book would hardly allow me to cover many of these related subjects and some aspects of discussed work in more detail. The aim has been to provide a wide and rich platform for a broad audience to get acquainted with polariton physics.

Suggestions for improvements are very welcome, as such work can only become a sustainable project and of great value to the research community with the help of expert input and steady adjustments. Thus, I will be thankful for constructive feedback and helpful remarks. Moreover, even thorough proofreading cannot prevent the occurrence of mistakes. Therefore, correction hints are appreciated.

Chapter icons used throughout the book are meant to visually guide the reader through the content; image material is courtesy of the author. The use of various figures from Springer-Nature journals is acknowledged. Similarly, I am grateful for the use of images from journals of the Optical Society of America. Additional images and artworks provided are a courtesy of the author, with the aim of depicting concepts and techniques, of sketching physical processes, structures or theoretical models in a simplified manner, and of giving insight into the behaviour of polariton systems.

Marburg, Germany Arash Rahimi-Iman

Acknowledgements

Every book and research project is the result of enduring endeavours, helpful people and fruitful collaborations. Thus, I would like to express my gratitude to all those who positively affected this work, and the following list is an attempt to summarize that.

I am grateful to Claus Ascheron for having enabled, supported and encouraged this book project, and the editors of the Springer Verlag for their continuous support and patience.

Similarly, I am grateful to former co-workers at Würzburg University, A. Forchel, S. Höfling, C. Schneider, M. Kamp and their team members, among others S. Brodbeck, K. Winkler, M. Lermer, T. Braun, and their technical assistants, for the excellent samples (grown and processed in house in a world-class facility based on long-term experience) with which the subject of polariton physics could be explored by me and co-workers, and for valuable discussions, and S. Reitzenstein, S. Höfling and L. Worschech, for their guidance and support on the quantum optics and polariton work.

I also would like to thank former students in Würzburg, K. McLeod, M. Amthor, J. Fischer, S. Holzinger, A. Schade, J. Senkpiel, P. Rieder, A. Cavers and many more, who had been supervised by me over the years and involved in polariton studies, many of whom became helpful co-workers as postgraduates.

Special thanks are devoted to former co-workers at Stanford University, Y. Yamamoto, G. Roumpos, N.Y. Kim, N. Na, S. Utsunomiya and team members, among others for hosting me at various occasions for joint research activities at the Ginzton Laboratory, for all the fruitful and important discussions at a very early stage, and for the successful collaboration.

Similarly, special thanks are devoted to (former) co-workers in the field of polariton physics from other institutions worldwide, A. V. Chernenko and V. Kulakovskii (Chernogolovka, Russia), I. Shelykh and I. Savenko (Reykjavik, Iceland; NTU, Singapore), M. Bayer, M. Aßmann, J.-S. Tempel, and co-workers (Dortmund, Germany), M. Durnev (St. Petersburg, Russia) and A. Kavokin (St. Petersburg, Russia; Southampton, UK), L. Dominici and D. Sanvitto (Lecce, Italy), and many others for the fruitful discussions, invaluable collaborations and lasting support. It shall be mentioned that helpful discussions with many other researchers in the polariton community have been always appreciated.

I would like to thank my team and project members, who have supported and accompanied my research endeavours in the past years, particularly L. M. Schneider, F. Wall, O. Mey, C. Palekar for their commitment to the polariton and light–matter interaction subject, and my ‘Semiconductor Laser and Spectroscopy’ team members (including former members) who eagerly contributed to my other activities such as M. Wichmann, M. A. Gaafar, F. Zhang, K. A. Fedorova, C. Möller, C. Kriso, H. Guoyu, S. Lippert, J. Kuhnert, O. M. Abdulmunem, Q. Ngo, M. Shah and many more. Moreover, many thanks are devoted to the head of the semiconductor photonics group, M. Koch, for his support over the last years which provided an outstanding research opportunity as well as a fruitful platform for the preparation of this work and for independent explorations, and also to faculty members G. Weiser and W. Heimbrodt for enriching discussions on semiconductor optics and support in the field of semiconductor spectroscopy, respectively.

I also would like to thank those with an important impact on my early scientific endeavours before reaching the polariton subject, among others S. L. Tait and K. Kern (Max-Planck Institute for Solid State Research, Germany), and S. Albrecht, J. Geurts, C. Kistner, S. Reitzenstein and A. Forchel (Würzburg, Germany).

Similarly, multiple fellowships by the German Academic Exchange Service (Deutscher Akademischer Austausch Dienst, DAAD) and the German National Academic Foundation (Studienstiftung des Deutschen Volkes) during all phases of my personal academic development helped to obtain the necessary flexibility, mobility and freedoms for continuous progress in the domain of science.

Last but not least, I am grateful for the continuous support from my family, which guaranteed a successful work on this topic and the accomplishment of all set goals.

All together, the very motivating environment, the outstanding research possibilities and the wide support received enabled all the successful studies on my side and have become the driving force behind this book project.

Financial support by the Deutsche Forschungsgemeinschaft for my current polariton project via Grant No. RA2841/9-1 is acknowledged, which provided even a non-financial additional boost and positive impact with regard to the completion of this book project.

Marburg, Germany Arash Rahimi-Iman

Contents

1 Towards Polariton Condensates and Devices 1
1.1 Introduction 1
1.2 Bose–Einstein Condensation of Polaritons 5
1.3 Endeavours to Achieve Polariton Lasers 11
1.4 More Exciton-Polariton Physics 15
1.5 Further Polaritons Not Detailed in This Book 19
References 22

2 Fundamentals of Polariton Physics 33
2.1 The Origin of Polaritons 34
2.1.1 Fundamental Light–Matter Interaction 34
2.1.2 Cavity–Polaritons 36
2.2 Building-Blocks for Polariton Formation 37
2.2.1 Excitons in Quantum Wells 37
2.2.2 Confined Photons 45
2.3 Light–Matter Coupling 50
2.3.1 Exciton–Polaritons 52
2.3.2 Detuning Dependencies of Polariton Modes 55
References 62

3 On the Condensation of Polaritons 65
3.1 Bosonic Many-Particle Features 66
3.1.1 Condensation of a Bose Gase 66
3.1.2 Criteria for Condensation 67
3.1.3 Dynamical Bose–Einstein Condensation of Polaritons 71
3.2 Excitation and Relaxation Dynamics 74
3.2.1 Excitation of Polaritons 74
3.2.2 Relaxation Towards the Energy Minimum 75

3.2.3 The Bottleneck Effect 77
3.2.4 Stimulated Ground-State Scattering 78
References 81

4 The Concept of Polariton Lasing 87
4.1 Polariton Lasers—Electrically-Driven, Please! 88
4.1.1 What Is It About? 88
4.1.2 The Stimulated Scattering Process 90
4.2 Comparison with Photon Lasing (Lasing in the Weak-Coupling Regime) 95
4.2.1 What Is a Laser? 95
4.2.2 Stimulated Emission, Laser Conditions and Coherence Properties 98
4.2.3 Bernard–Duraffourg Condition in Semiconductors 103
4.2.4 Similarities and Differences Between Polariton and Photon Lasers 104
4.3 Identification of Polariton Lasing 107
4.3.1 Prerequisites and the Signatures of a Polariton Condensate 107
4.3.2 Overview on the Typical Experimental Procedure 112
References 113

5 Optical Microcavities for Polariton Studies 119
5.1 Fabry–Pérot Microcavities 120
5.1.1 Distributed Bragg Reflectors 120
5.1.2 Planar Microresonator Structures 122
5.2 Implementation of Quantum Wells 125
5.2.1 Distribution of Quantum Wells 126
5.2.2 Number of Quantum Wells 127
5.2.3 Excitation Schemes 127
5.3 Optical Properties of Resonators 128
5.3.1 Free-Spectral Range, Cavity Finesse, Photonic Density of States 129
5.3.2 Resonator Quality 131
References 134

6 Technological Realization of Polariton Systems 139
6.1 Growth and Processing of Microcavity Devices 140
6.1.1 Epitaxy of Multilayered Structures 140
6.1.2 Potential Landscapes and Polariton Boxes 141
6.1.3 Doped Microresonators 143
6.1.4 Polariton Diodes 144

6.2 Microcavities for Different Material Systems 147
6.2.1 II/VI Microresonators 149
6.2.2 Inorganic Room-Temperature Polariton Systems 150
6.2.3 Organic Materials 152
6.2.4 Perovskite-Based Exciton–Polariton Systems 154
6.2.5 Monolayer Transition-Metal Dichalcogenides 155
References 159

7 Spectroscopy Techniques for Polariton Research 167
7.1 Optical Spectroscopy 168
7.1.1 Reflection and Transmission Measurements 168
7.1.2 Micro-Photoluminescence Experiments 170
7.1.3 Micro-Electroluminescence Studies 171
7.2 Imaging and Real-Space Spectroscopy 172
7.2.1 Sample Imaging for Position Monitoring or Interferometry 173
7.2.2 Spatially-Resolved Spectra 175
7.3 Fourier-Space-Resolved Spectroscopy 176
7.3.1 Goniometer-Like Technique 177
7.3.2 Pinhole Translation Method 179
7.3.3 Single-Shot Angle-Resolved Acquisition 181
7.4 Time-Resolved Spectroscopy 184
7.4.1 Streak-Camera Measurements 184
7.4.2 Pump–Probe Techniques 186
References 191

8 Optically-Excited Polariton Condensates 195
8.1 The Observation of Polariton Condensation 196
8.1.1 Condensate Studies in the Literature 196
8.1.2 Optical Pumping Schemes 198
8.1.3 Spectral Features of Polaritons 201
8.2 Condensation Experimentally Characterized 203
8.2.1 Real-Space and Momentum-Space Distribution of Condensate Emission 203
8.2.2 Stimulated Scattering and Macroscopic Ground-State Occupation 205
8.2.3 Link to BEC via Spatial Coherence Measurements 213
8.2.4 Photon Statistics 218
8.3 Special Condensate Features 223
8.3.1 Polaritons at Their Extremes 226
8.3.2 Coherent Polariton Lasers 229
8.3.3 Superfluidity and Vortices in Condensates 230
References 232

9 Polaritons in External Fields 241
9.1 Effects of External Fields on Quantum-Well Excitons 242
9.1.1 Electro-Optical Tuning 242
9.1.2 Coupling to Strong Transient Electric Fields 245
9.1.3 Magneto-Optics with Excitons 246
9.2 Magneto-Polaritons in Microcavity Systems 250
9.2.1 Manipulating the Excitonic Component of Polaritons 250
9.2.2 Spinor Condensates in External Magnetic Fields 253
9.3 Interaction with Transient Fields 257
9.3.1 Terahertz Radiation and Polaritons 257
9.3.2 Addressing the Dark Side of Polaritons 259
References 259

Glossary 263

Index 269

Acronyms

Here, lists of commonly and specifically used abbreviations, symbols and the like are provided.

Abbreviations

0D	Zero-dimensional
1D	One-dimensional
2D	Two-dimensional
3D	Three-dimensional
II-VI	Group II/VI semiconductors
III-V	Group III-V semiconductors
arb.u.	Arbitrary unit
Al	Aluminium
As	Arsenic
ASE	Amplified spontaneous emission
APD	Avalanche photodiode
BCS	Bardeen–Cooper–Schrieffer
BEC	Bose–Einstein condensation/condensate
BKT	Berezinski–Kosterlitz–Thouless (theory)
C	Cavity mode
cav	Cavity
CB	Conduction band
Cd	Cadmium
cQED	Cavity quantum electrodynamics
CCD	Charge-coupled device (e.g. CCD camera)
CMOS	Complementary metal-oxide-semiconductor (e.g. CMOS camera)
crit	Critical
cw	Continuous wave
DBR	Distributed Bragg reflector

DOS	Density of states
e^-	Electron
EID	Excitation-induced dephasing
EL	Electroluminescence
EP	Exceptional point
FF	Far-field
FSR	Free spectral range
FWHM	Full width at half maximum
Ga	Gallium
h^+	Hole, defect electron
HBT	Hanbury–Brown-and-Twiss (experimental configuration)
high-Q	High-quality
In	Indium
IR	Infrared
KT	Kosterlitz–Thouless (phase transition)
LA	Longitudinal acoustic
LASER	Light amplification by stimulated emission of radiation
LED	Light-emitting diode
ℓN_2	Liquid Nitrogen
ℓHe	Liquid Helium
LO	Longitudinal optical
LP	Lower polariton, referring to the lower energy mode/branch of a polaritonic system
μCav	Microcavity
μEL	Micro-electroluminescence
μPL	Micro-photoluminescence
MBE	Molecular beam epitaxy
MeLPPP	Methyl-substituted ladder-type poly(para-phenylene)
ML	Monolayer
MoS_2	Molybdenum disulfide
$MoSe_2$	Molybdenum diselenide
MOVPE	Metal-organic vapour-phase epitaxy
NA	Numerical aperture
NF	Near-field
NIR	Near-infrared
ODLRO	Off-diagonal long-range order
OPO	Optical parametric oscillator
PL	Photoluminescence
pn	Positive-negative (…junction/diode)
p-i-n	Positive-intrinsic-negative, referring to a structure's material doping
Q-factor	Quality factor
QCSE	Quantum-confined Stark effect
QD	Quantum dot
QW	Quantum well
SE	Spontaneous emission

SEM	Scanning electron microscopy
TA	Transverse acoustic
TDAF	2,7-bis[9,9-di(4-methylphenyl)-fluoren-2-yl]-9,9-di(4-methylphenyl) fluorene
Te	Tellurium
th	Threshold
THz	Terahertz
TMDC	Transition-metal dichalcogenide
TO	Transverse optical
UP	Upper polariton (…mode/branch)
UV	Ultraviolet
vac	Vacuum
VB	Valence band
VCSEL	Vertical-cavity surface-emitting laser
VIS	Visible (spectral range)
WS_2	Tungsten disulfide
WSe_2	Tungsten diselenide
X	Exciton
XP	Exciton–polariton
XX	Biexciton

Symbols

h	Planck's quantum of action
$\hbar$	$= \frac{h}{2\pi} \approx 1.054572 \cdot 10^{-34}\,\mathrm{J\,s}$
k_B	$\approx 8.617333 \cdot 10^{-5}$ eV/K, Boltzmann constant
c	$= 2.99792458 \cdot 10^{8}\,\frac{\mathrm{m}}{\mathrm{s}}$, speed of light
π	≈ 3.141593
$\mathbf{k}$	Wavevector
$k_{\|\|}$	Transversal component of the wavevector, corresponding to an in-plane momentum, for instance, in the plane of a quantum well, of unit $[k_{\|\|}] = \mu\mathrm{m}^{-1}$
$k_\perp$	Longitudinal component of the wavevector, corresponding to an out-of-plane momentum, for instance, vertical to the plane of a quantum well
e	$\approx 1.602177 \cdot 10^{-19}$ C, elementary charge
m_e	$\approx 9.109384 \cdot 10^{-31}$ kg, free electron mass
ε_0	$\approx 8.854188 \cdot 10^{-12}\,\frac{\mathrm{A\,s}}{\mathrm{V\,m}}$, dielectric constant, vacuum permittivity

ε_r	Relative dielectric constant (vacuum: 1 per definition); high frequency constant: $\varepsilon_r(\infty)$, background constant ε_b; in general, frequency-dependent complex dielectric function $\varepsilon_r(\omega)$
ε	$= \varepsilon_0 \cdot \varepsilon_r$, permittivity of a medium
n	Refractive index
E	$= \hbar\omega = h\nu$, Energy
ω	Angular frequency
ν	Frequency
λ	$= hc/E$, wavelength
E_0	Energy of a mode's (or system's) ground state
E_g	Band gap energy
γ	Linewidth, FWHM
θ	Angle
α	Absorption coefficient
ΔE	$= E_{C,0} - E_{X,0}$, spectral detuning of two resonances (here of the ground-state photonic (C) and excitonic (X) modes, respectively)
Q	$= E/\gamma$, quality factor
F	Finesse
g	Coupling constant (coupling strength)
$2\hbar\Omega$	$= 2g = E_{\mathrm{Rabi}}$, (vacuum) Rabi splitting, with 2Ω the Rabi oscillation frequency
M	Transition-matrix element
f	Oscillator strength
$\lvert X\rvert^2$	Excitonic fraction of a polariton (one of the Hopfield coefficients)
$\lvert C\rvert^2$	Photonic fraction of a polariton (one of the Hopfield coefficients)
τ	Lifetime (also used as symbol representing the delay time as a parameter in some experiments)
υ	$= N/V$, particle density (number N divided by volume V)
$\lvert\varphi\rangle$	Wavefunction (quantum-mechanical state)
$\mathbf{E}$	Electric field (vector)
$\mathbf{B}$	Magnetic flux (vector)
T	Temperature
P	Pump power/rate
j	Current density

Chapter 1
Towards Polariton Condensates and Devices

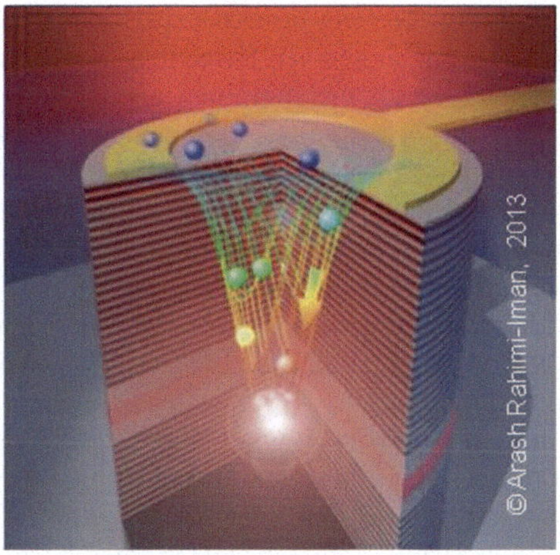

Abstract Strong light–matter coupling has given rise to fascinating phenomena and observations in the past decades. The emergence of the field of polaritonics has been particularly evidenced since sophisticated microcavity experiments allowed the realization and investigation of Bose–Einstein-like condensation of exciton–polaritons and the superfluid nature of degenerate Bose gases in solids. Thus, this chapter summarizes the history and the developments regarding polariton research and provides an introduction to the following chapters. Hereby, the concept of Bose–Einstein condensation of polaritons and endeavors to achieve polariton lasers within the wider framework of polariton studies are also briefly highlighted.

1.1 Introduction

The interaction of light and matter has fascinated human kind for aeons, not to mention the very nature of both. With centuries of intense work having focused on the description of light and matter in the modern ages, the possibilities peaked in the era of quantum physics, where quantum optical properties govern the principles

A. Rahimi-Iman, *Polariton Physics*, Springer Series in Optical Sciences 229,
https://doi.org/10.1007/978-3-030-39333-5_1

of energy exchange in light–matter systems and open up new pathways to a better understanding of both light, and matter. This naturally leads to the excitement directed to the exploration of the physics of light–matter coupling and the attractiveness of *polaritons*, hybrid quasi-particles being composed of light and of matter.

When polariton formation was first discussed as a phenomenon in bulk systems as linear superposition of light and matter excitation in 1958 [1, 2], neither a practical device based on polaritons was in the scientists' minds at that time, nor any technology was at hand to realize such. However, in the early days of *microcavity polaritons* following their discovery [3], a clear strive was evidenced towards the exploitation of the beneficial properties of those light quasi-particles in matter in order to achieve the long desired experimental demonstration of the distinguished many-particle phenomenon referred to as *Bose–Einstein condensation (BEC)*. Quickly, a new testbed for the study of various many-particle effects and quantum phenomena such as superfluidity, superconductivity and collective coherence was praised owing to the experimental system's features, most importantly its particle properties and its facilitated characterization via optical spectroscopy. Typical examples of polariton systems based on planar microresonator structures for both (a) inorganic semiconductors and (b) organic materials are sketched in Fig. 1.1 and displayed together with a characteristic in-plane polariton dispersion in a schematic energy–momentum diagram originating from the strong coupling of excitonic resonance and cavity-photon mode (c).

Yet, it is not only the many-particle domain that has drawn great attention from scientists in the field of quantum optics, but also the single-particle regime, where coupling takes place between merely one quantum of matter excitation and the vacuum field of an optical microcavity,[1] allowing one to enter the field of *cavity quantum electrodynamics (cQED)* [4, 5].

This can be mainly accomplished thanks to astonishing advances in the spheres of semiconductor and nano technology with respect to fabrication and characterization of high-quality structures with nanoscale precision and low dimensionality in the last decades. With various technological tools having been developed to push the boundaries to ever more compact, cost-effective and energy-efficient devices, integrated circuits, photo-detectors and novel light emitters have been enabled—all of them featuring very prominent representatives and applications. These have unleashed a big impact on our lives (not only via consumables and gadgets) during the same period of time. While the LED[2] is a device often not fully recognized or specified as a product of quantum physics, a more prominent result of quantum technologies can be found in the laser,[3] and a rather exotic example is given by non-classical light sources such as single-photon sources.

Many of these concepts for novel light sources owe a lot to the development and availability of state-of-the-art optical microcavities [5, 7–10]. Without these microresonator structures, neither low-threshold laser operation, efficient single-photon

[1]A structure that confines light in one or more directions in the microscale is considered an optical microcavity.

[2]Acronym for **l**ight-**e**mitting **d**iode.

[3]Acronym for **l**ight **a**mplification by **s**timulated **e**mission of **r**adiation.

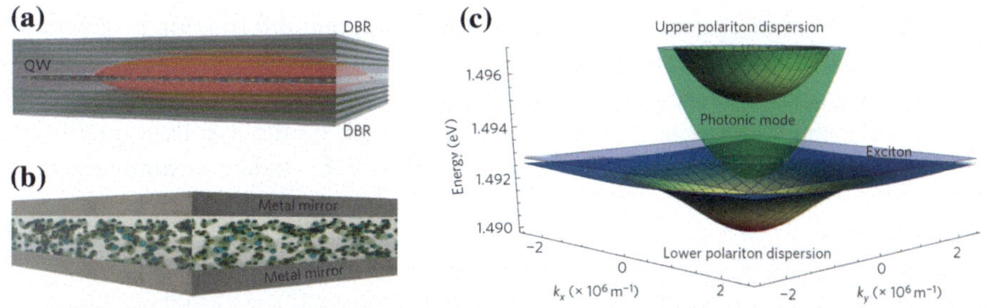

Fig. 1.1 **a** Typical inorganic semiconductor microcavity based on highly-reflective distributed Bragg mirrors (DBRs) and an active region consisting of a few quantum wells (QWs). **b** Alternative structure design employing a uniformly filled organic layer of randomly distributed molecules between to mirrors. Owing to the typically high binding energy of molecule excitons in organic semiconductors, a very high coupling strength and correspondingly large normal-mode splitting in the hundreds-of-meV range can be obtained. Such polaritons remain robust quasi-particles even at room temperature. In contrast, the relatively low oscillator strength of QW free excitons in inorganic crystal lattices results in a comparably small coupling constant on the order of a few meV up to tens of meV for large-band-gap materials. Therefore, large exciton binding energies are preferred for polariton research at elevated temperatures. A schematic representation of typical energy–momentum dispersions for a cavity–polariton system is displayed in **c**. Both, the uncoupled bare exciton (flat, transparent blue) and photon (curved, transparent green) modes are indicated with their characteristic dependence on the in-plane momentum, here projected in the x–y plane of wave-vectors k. Due to strong light–matter coupling, the system's modes hybridize and two new eigen states are formed, represented by the lower and upper polariton dispersion branches (coloured opaque meshes). At zero momentum, both branches feature their energy minimum. At resonance detuning (indicated by a white ring, where the photon dispersion dips into the exciton counterpart here), the energy splitting is referred to as Rabi splitting. Of particular attractiveness is the naturally formed ground state in the lower branch which can serve as a final state for stimulated scattering into a polariton condensate. Reused with permission. Reference [6] Copyright 2016, Springer Nature

emission nor localized light–matter coupling would have been feasible. Additionally, the realization of a novel source of coherent light would have been out of reach which was proposed to be an energy-efficient alternative to the conventional (photon) laser. Owing to its similarities with the photon laser regarding its emission features [11, 12], it was referred to as *polariton laser*.

Technological advances finally enabled researchers to systematically place quantum emitters into optical cavities in order to bring together spatially confined electrons, or more precisely bound electron–hole pairs forming *excitons*, and photons for the sake of a well controlled light–matter interaction in semiconductor microresonator structures. This has given rise to the studies of both weak and strong coupling of matter excitation—the exciton—with the cavity field.[4] The former situation (weak light–matter coupling, short *weak coupling*) is known by the description of the Purcell effect [13], and deals with the enhancement or suppression of spontaneous emission by quantum emitters. Together with non-classical emission from quantum states, they serve as basis for efficient single-photon sources and enable amazing appli-

[4]In the single particle regime, this constitutes the vacuum field of the resonator.

cations in the field of optical quantum communication and information processing. In contrast to the weak-coupling regime, strong light–matter coupling (short *strong coupling*) describes the case of coherent energy exchange between emitter and cavity light field and—as key to cQED—relies very much on the excellent quality of optical microresonator structures. The fundamentals of this field are summarized in Chap. 2.

To achieve strong coupling, a two-level-system-like exciton is placed in a high-quality optical microcavity. Via light absorption, an exciton is created, and its decay feeds the cavity field. This process can become reversible if the light–matter interaction is strong enough. Then, back-and-forth coherent exchange of energy can take place between emitter and cavity light field within the lifetime of both entities. Here, the dissipation channels of the coupled system are spontaneous emission as well as non-radiative decay for excitons, and decay as well as absorption for photons (see Fig. 1.2). Since the quality of the optical cavity drastically limits the photon lifetime, the quality factor of the optical resonator is an important figure of merit for the realization of strong-coupling systems. If the coupling rate exceeds the loss rates, a normal-mode splitting can be observed in the spectral domain, whereas damped population oscillations (between the exciton and photon state) occur in the time domain.

In the strong-coupling regime, the system can neither be described by a pure photon, nor exciton. The frequency of the periodic oscillation of energy between the exciton and photon state is determined by the Rabi frequency (in reference to the original work on Rabi oscillations [14]) and corresponds directly to the coupling strength.

Since the strong light–matter coupling regime gives birth to the aforementioned hybrid quasi-particles, many investigations of the last two decades aimed at the study of this regime and the characterization of the thereby obtained polaritons. One goal was to utilize the bosonic nature of these light-weight particles—as they are excitons dressed with photons, hence also referred to as *exciton–polaritons*—for the experimental demonstration of Bose–Einstein condensation in solids [15, 16] (expected in the early hours of polariton research to be one of the very first demonstrations of BEC, if not the first one due to the favourable features of polaritons). This motivates at this point to ask what BEC refers to and what has made it so prominent among the theoretical predictions of a phenomenon in the past century. Indeed, as we will learn in a short summary of its history, its importance has to do with the quantum nature of our world. Moreover, for polariton physicists, it is the basis of polariton condensation and polaritonic lasing. More details on condensation, polariton lasing and polariton microcavities are provided in Chaps. 3, 4 and 5, respectively.

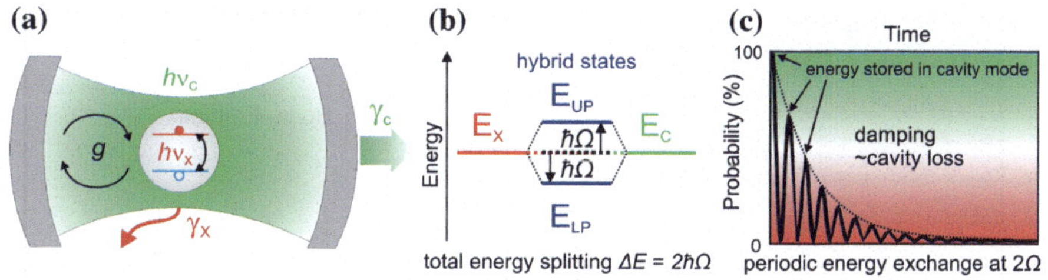

Fig. 1.2 **a** Sketch of a coupled cavity–emitter system with coupling strength g, with resonances $h\nu_{X/C}$ for exciton/photon, respectively. If spatial and spectral resonance is provided, the emitter (or an ensemble of emitters) can interact considerably with the resonant photon mode of the cavity. This can either lead to the Purcell effect in the weak-coupling regime, or to the hybridization of states (Rabi splitting) in the strong-coupling limit, provided that the possible decay rates $\gamma_{X/C}$ for the coupled emitter mode and light state, respectively, are sufficiently low. **b** If two harmonic oscillators in resonance couple to each other strongly, the interaction leads to new eigen-states of the coupled system. In fact, such mode splittings occur whenever two resonances with sufficiently strong coupling cross each other, evidenced by a mode anti-crossing in the energy (i.e. spectral) domain. **c** In the time domain, strong light–matter coupling exhibits a periodic oscillation between the probabilities to find the system in the photonic or excitonic state. This back-and-forth coherent energy exchange between two modes (oscillators) occurs at the Rabi frequency and is a well understood feature of coupled systems (here an excitation of the light field and of the emitter). It decays with a certain damping rate proportional to the lifetimes of the involved excitations

1.2 Bose–Einstein Condensation of Polaritons

BEC is indeed a fascinating phenomenon, the experimental demonstration of which in the light of theoretical predictions followed a remarkable journey, on which scientists encountered many interesting aspects of many-particle physics, overcame various experimental challenges and targeted several bosonic particles for its first achievement.

A short summary of the history of BEC and the phenomenon itself may be given at first, before light is shed on the hunt for an experimental demonstration of polariton condensates.

Bose–Einstein condensation is an effect common for all bosons and characterized by its long-range order in a fluid of condensed particles. It describes the macroscopic occupation of a single coherent quantum state [17, 18] and was first demonstrated experimentally for a diluted gas of atoms in 1995 after great efforts were undertaken to provide ultra-cold conditions in the nano-Kelvin range inside a magnetic trap for atoms [19–21]—70 years later than the theoretical predictions for an ideal Bose gas made by Albert Einstein in 1925 based on Satyendranath Bose's considerations for ideal bosons in 1924 [22, 23].

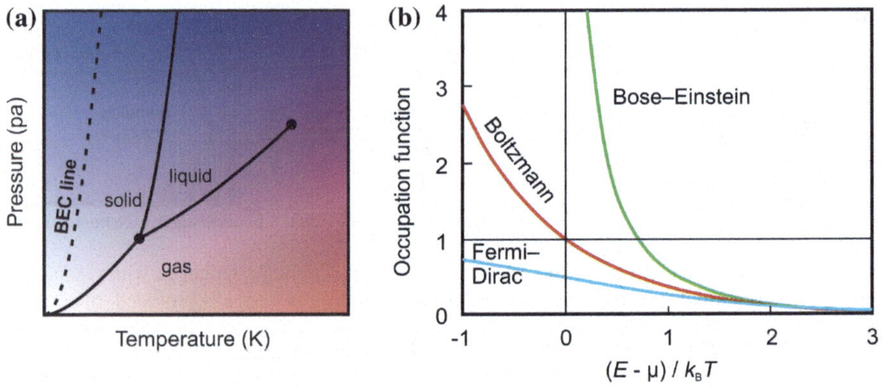

Fig. 1.3 **a** Schematic representation of a general phase diagram with gas, liquid and solid phases, together with the BEC line, after [18]. See Sect. 3.1 for further explanations. **b** The occupation functions for classical particles, fermions and bosons, which are the Boltzmann, the Fermi–Dirac, and the Bose–Einstein distribution functions, respectively, plotted on a common scale against $(E - \mu)/k_{\mathrm{B}}T$, i.e. with respect to the chemical potential μ as the reference energy, after [24]

Since that time, the term BEC refers to an extreme aggregate state of a bosonic system governed by quantum-mechanical interaction effects between bosons which features indistinguishable particles and the spontaneous formation of collective coherence on a macroscopic scale. In such a system, if the temperature T undercuts a critical temperature T_{crit} or the particle density $\upsilon_{\mathrm{particle}}$[5] exceeds a critical value (υ_{crit}), the major fraction of particles accumulates in one and the same quantum mechanical state and can be described by a macroscopic wave-function. However, if the temperature of the system was 0 K, all particles would occupy a single state and thereby form a giant matter wave, manifesting a pure Bose condensate.

Because all bosons, even composite ones, obey Bose–Einstein statistics (Fig. 1.3), they can in principle undergo the phase transition from a (dilute) gaseous state to a quantum fluid. Among well known Bose particles are photons, phonons, excitons, polaritons, Cooper pairs (responsible for superconductivity), Helium-4 (well known from the historic study of superfluidity), and the Higgs bosons (understood to be in a condensed state and being responsible for the elementary particles' mass). A wide range of background information about this amazing subject and the properties of BECs can be obtained for instance through outstanding literature dedicated to this remarkable phenomenon such as [17] with its compilation of numerous review papers. Owing to the vacuum being considered in the literature to contain a condensate of Higgs bosons and quark condensates, and owing to mesonic condensates—possibly important for characteristics of neutron stars—being considered nuclear analogues of exciton condensates [17], fundamental studies on exciton and exciton-–polariton condensates in solids become highly attractive.

[5] Often, the letter n is used in the literature for particle densities.

The prerequisites for condensation can be easily derived from the well-known formula for BEC in an ideal 3D environment [18]:

$$k_B T_{\text{critical}} = \frac{2\pi \hbar^2}{m_{\text{particle}}} \left(\frac{\upsilon_{\text{particle}}}{2.612} \right)^{2/3}, \tag{1.1}$$

with k_B the Boltzmann constant, and m_{particle} the particle effective mass. It should be noted early on that the application of this equation actually is not totally correct in the case of 2D confined particles (see literature with a focus on microcavity systems such as [5, 9, 16]).

Figure 1.4 illustrates what BEC refers to and how the parameters of the Bose gas matter to the formation of a macroscopically-occupied coherent quantum state. With the temperature of a Bose gas decreasing, the particles' de-Broglie wavelength $\lambda_{\text{dB}} \approx (h^2/mk_B T)^{1/2}$, which is inversely proportional to the square root of the temperature, increases and the mean particle distance is reduced, allowing next-neighbour wave-functions to overlap and to facilitate relaxation into a common ground state. Similarly, the particle density has an impact on the mean separation of two bosons, which is essential for particle interaction.

Another crucial parameter for BEC to set in for a bosonic system is the (effective) mass of its particles m_{particle}, which drastically alters the critical temperature for the condensation effect. Thus, a strive for ever lighter particles was huge in order to facilitate the observation of BEC in experimental systems available.

Later in history, the discovery of superfluidity of Helium in 1938 [26, 27] was linked to BEC [28–30] and a quantum field theory of BEC of interacting particles was introduced by N.N. Bogoliubov in 1947 [31]. For the following decades, further research was conducted on superfluidity as the only experimental research related to the phenomenon of BEC, until studies of excitons in solids, composite bosons consisting of fermionic electrons and holes,[6] emerged. Yet, long-lasting intensive research after the predictions of exciton condensation [32, 33] and pioneering work on collective properties of such quasi-particles [34] did not bring up the desired results on this matter. On the one hand, success was impeded by a high decoherence rate via phonon scattering in appropriate structures. On the other hand, the Mott transition [35] from an insulator state of matter in the bosonic state to a metallic (electron–hole) plasma in the fermionic regime above a critical particle density [36] prevented scientists for decades to claim the achievement of excitonic BEC. Moreover, as excitons in 3D semiconductors are rather considered polaritons [2], their cooling would result in continuous photon leakage. Because bulk polaritons—described by the light cone at low momenta—do not exhibit a non-zero energy minimum on the lower branch (in contrast to cavity–polaritons), they are not supposed to undergo a condensation process [10]. Later attempts focussed on excitons in coupled quantum wells, and even indirect excitons in double quantum wells were targeted [37]. Only within the last two decades, such a phase transition to a condensate was claimed to have been observed for specially prepared excitons at very low temperatures in

[6]Defect electrons in the valence band of semiconductors with a charge opposite to that of electrons.

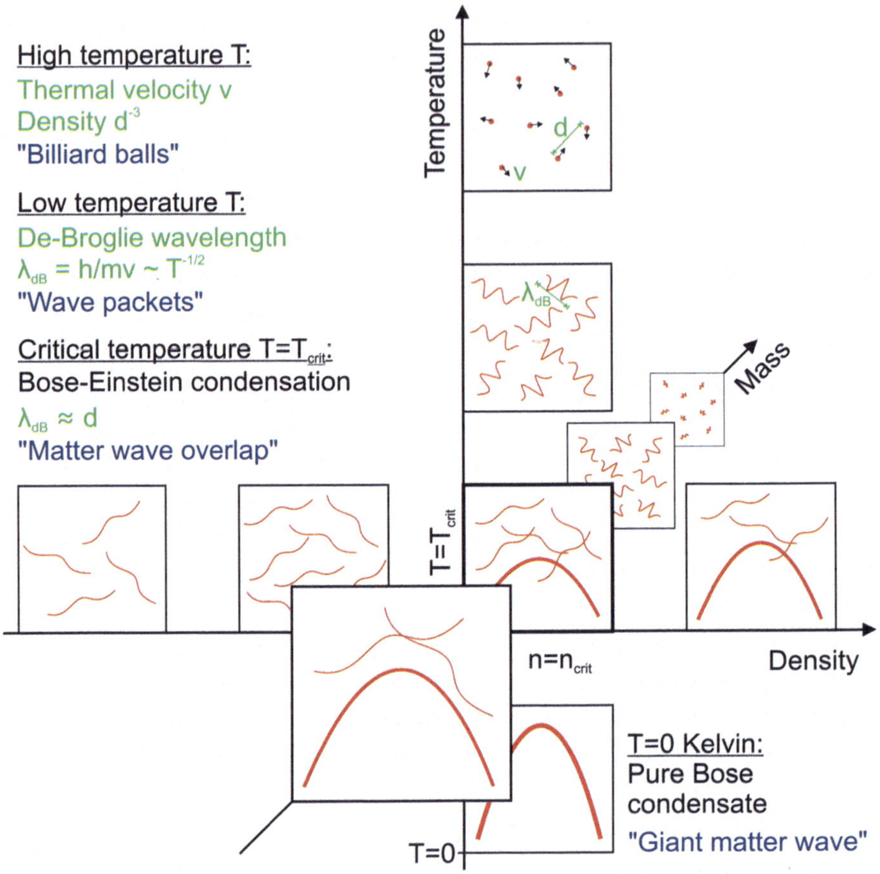

Fig. 1.4 Sketch of the physics behind Bose–Einstein condensation (BEC): The formation of a condensate is illustrated in a three-axes diagram with the fundamental column representing the decrease of temperature in order to achieve matter wave overlap. At high temperatures, the gas of particles obeys the Boltzmann statistics. Below a critical temperature T_{crit}, a BEC is formed. The ideal gas of bosons is described by the Bose–Einstein distribution function. While the majority of particles is in the same quantum degenerate state, a minority of uncondensed particles coexists. This is expressed by the condensate fraction $\upsilon_0/\upsilon_{\mathrm{tot}}$, with υ_{tot} the total density and υ_0 the ground-state occupation. From left to right, the effect of a density increase is indicated. At low enough temperatures, the critical particle density υ_{crit} needs to be exceeded in order to obtain a condensate. A third dimension pointing into the viewers direction reveals the impact of the particle mass. While in the background of the image, heavy particles at given temperature hardly overlap due to their short de-Broglie wavelength λ_{dB}, by reducing the mass, at given temperature, the critical conditions can be met. In other words, lighter particles allow one to observe BEC at elevated temperatures. This scheme has been prepared by the author in analogy to the very pedagogical (one-dimensional) chart created by W. Ketterle and used on his research group's online representation (see [25]), intentionally conceptualized as a practical extension of the appreciated "original" for clarity

Fig. 1.5 Typical signatures of a macroscopic build-up of ground-state occupation, obtained in a measurement of the momentum space distribution of polariton emission in a microcavity system, here for a CdTe-based structure. The measurement at cryogenic temperatures shows the transition from a thermally distributed gas of polaritons to a Bose–Einstein-like condensate. Above threshold, the majority of particles accumulates at $k = 0$ where the polariton system has its energy minimum. Reused with permission. Reference [45] Copyright 2006, Springer Nature

the milli-Kelvin range, with its experimental signatures obtained after an evident struggle [38–42].

With cavity–polaritons on stage since the 1990s, a new system was available with an effective particle mass of its bosons so light, that the critical temperature for condensation was expected at elevated temperatures in the range of a few K up to room temperature. Additionally, polaritons formed in semiconductor microresonator structures confining photons in 2D naturally exhibit a finite energy ground state, which originates from the cavity-photon dispersion with fundamental cavity resonance at zero momentum. Major efforts were put into the investigation of polariton condensation in the following decade. Yet, the desired condensation effect for polaritons (Fig. 1.5) was only achieved after the first demonstration of BEC. Having got affirmation by the atomic BEC results, the race towards exciton–polariton condensates led to demonstrations of dynamic condensation of polaritons in optically excited GaAs as well as CdTe microresonator structures [43, 44]. Shortly after, those achievements were complemented by further studies and set in relation to Bose–Einstein condensation based on studies of the particles' occupation distribution and the spontaneous build-up of spatial coherence [45] (CdTe), [46, 47] (GaAs).

These achievements were a direct consequence of the elevated critical temperature for polaritons, which exhibit an effective mass that is reduced by four orders of magnitude in comparison to that of pure excitons in solids. Moreover, owing to the strong light–matter interaction, excitons dressed with photons in the cavity are much more delocalised than their uncoupled pendants and, thus, are less prone to trapping, dephasing and non-radiative recombination due to inhomogeneities in solids. Eventually, condensation has been obtained even at room temperature both for inorganic [48–51] (GaN), [52–55] (ZnO), and organic semiconductor structures [56]. Recently, even BEC of polaritons in an organic semiconductor microcavity based on spatial coherence studies at room temperature was claimed [57]. Systems for polariton studies based on different materials are highlighted in Chap. 6.

Yet, polaritons are not isolated from their surroundings, unlike atoms in an atomic BEC maintained in a magnetic trap, and will be coupled out of the microcavity as photons owing to the finite reflectivity of the resonator mirrors. As a consequence, polaritons feature a limited lifetime and a polariton gas in such system cannot establish thermodynamic equilibrium. Thus, one generally refers to a dynamical BEC of polaritons when describing experimental systems which undergo the phase transition from a Boltzmann-distributed gas to a macroscopically occupied ground state with coherence properties known for BEC [16, 58]. Nevertheless, from the supposed weakness of the polariton gas its peculiar strength arises. Without the out-coupling losses naturally provided by the microcavity, no direct experimental access to the quasi-particles' dynamics and distribution would have been given via spectral analysis of angle-resolved emission [59, 60].

In conclusion, these are the properties that distinguish polaritons from other bosonic particles as an ideal testbed for the investigation of many-particle effects and collective systems in solids.

However, according to theoretical considerations by Mermin, Wagner [61] and Hohenberg [62], an infinitely large 2D system cannot give rise to conventional BEC, because long-range order cannot set in. Yet, the existence of condensates in 2D systems of finite size or with local potential variations are explainable by localization effects [9]. This strongly motivates studies on natural and artificial confinement potentials for polaritons, such as the BEC experiment of Balili et al. demonstrated [46]. 2D condensation effects are also linked to vortex–antivortex pair formation, which establishes finite spatial coherence by screening local phase defects (see Sects. 3.1.2 and 8.3.3).

It is worth noting that the observation of BEC in optical microcavities has not been only confined to light–matter systems in the strong-coupling regime. Indeed, Bose–Einstein condensation of photons in an optical microcavity was reported by Klaers et al. in 2010 [63]. In previous theoretical works, thermalization processes had been considered that conserve photon number (a prerequisite for BEC), involving Compton scattering with a gas of thermal electrons [64] or photon–photon scattering in a nonlinear resonator configuration [65, 66]. In contrast, black-body radiation does not show a phase transition to BEC, since cooling of the photon gas would mean losing photons to the walls of the black body. However, it was reported that for a two-dimensional photon gas in a dye-filled optical microcavity, number-conserving thermalization was experimentally observed [67], referring to that cavity system as a white-wall box [63]. For such system, it was shown that by providing both a confining potential and a non-vanishing effective photon mass, a 2D gas of trapped, massive bosons was obtained, which could undergo a phase transition to a BEC mediated by the weak interactions (multiple scatterings) with the dye molecules at room temperature.

1.3 Endeavours to Achieve Polariton Lasers

At a time, when cavity–polariton studies were at their initial stage and the demonstration of a polariton condensate still due, a novel type of laser based on polariton condensation was proposed in 1996 by Imamoglu et al. [11].

A polariton-condensation device would in stark contrast to real lasers, i.e. conventional "photonic" lasers, not employ *stimulated emission of photons*, but would utilize *stimulated scattering of polaritons* into a common quantum state [11]—an effect strongly related to BEC. From the resulting coherent ground-state emission, one can derive laser-like operation with orders of magnitude lower threshold powers [68, 69]. Owing to its similar emission properties, such device was referred to as *polariton laser* in analogy to its photonic pendant [12].

Corresponding polariton laser operation was demonstrated several times and differences as well as similarities between polaritonic lasing (condensate emission) and photonic lasing were studied [68, 70–73]. Indeed, early observations of bosonic final-state stimulation, macroscopic ground-state occupation and, hence, polariton lasing go back to the end of the 1990s [74, 75], with the most-recognized first polariton condensation and BEC reports being from the year 2002 [43] and 2006 [45], respectively. Many characteristic measurements for condensates are summarized in Chap. 8, mainly focusing on optically-pumped systems. Thereby, measurements regarding special features such as superfluidity, spatial coherence and photon statistics will be highlighted as well.

Until 2013, only optical excitation, and not electrical current injection, was used to drive such "lasers" in the regime of strong coupling. Thus, electrical pumping of condensates has become a major goal of scientists world wide in the field of polariton research, while incoherent electroluminescence from polariton LEDs was presented multiple times in the last decade [76–78]. Most important for the first-time demonstration of an electrically-driven polariton laser was considered to be the unambiguous distinction between lasing in the regime of strong light–matter coupling and lasing in the regime of weak coupling [58, 79]. The latter scenario, in which no polaritons are present, corresponds to the operation of a VCSEL,[7] with current densities high enough to achieve an electron–hole plasma, i.e. transparency, in laser terminology equivalent to the term *population inversion*, and more accurately referred to as Bernard–Duraffourg condition in semiconductor lasers [80].

In this context, it is natural to recapitulate what a polariton laser actually is about. A polariton laser is indeed a unique device which involves light–matter interaction and its system employs strongly-coupled excitons and photons—typically in a microcavity with quantum-well active medium, in which spatial and spectral resonance

[7] Acronym for ***v**ertical-**c**avity **s**urface-**e**mitting **l**aser*.

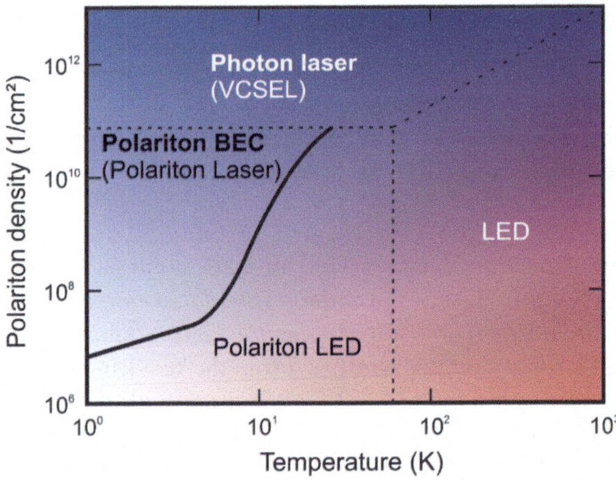

Fig. 1.6 Schematic phase diagram for a GaAs-based polaritonic system after [12]. In contrast to atoms, cavity–polaritons can condense at orders of magnitude higher temperatures, rendering them a useful platform for condensation studies in solids. The parameters that can be tuned in order to obtain polariton condensates are the density of particles and the temperature. At cryogenic temperatures, the transition from a polariton LED to a polariton laser (condensate) may occur in high-quality QW-microcavities, as indicated by a solid curve. At too high temperatures, polaritons in GaAs break up due to thermally-induced exciton linewidth broadening (dephasing) and the structure acts as a simple LED, whereas at too high densities, Coulomb interactions break up excitons and a transition from a Mott insulator state to an electron–hole plasma takes place. In the corresponding phase diagram, the limits of the strong-coupling regime with regards to temperature and density are indicated by vertically and horizontally drawn dashed lines, respectively. Beyond these boundaries, the system operates in the weak-coupling regime, in which the device acts as an ordinary photon laser (at high particle densities), i.e. VCSEL, or LED (below the conventional lasing threshold), with the threshold condition indicated by a dotted line

of emitter and light field are established. The quasi-particles that are formed in the semiconductor device, called polaritons, are utilized to form a coherent bath of polaritons via dynamic Bose–Einstein condensation. Thus, a polariton laser uses bosonic many-particle effects to achieve coherent ground state emission, for which a stimulated scattering process between polaritons is involved—in contrast to stimulated emission in conventional lasers.

A polariton laser is promised to be an alternative and energy-efficient source of coherent radiation due to the quantum degeneracy of polaritons. Owing to the lower excitation densities needed for polaritons to exist in usual VCSEL-like structures before an electron–hole plasma occurs above the Mott density of charge carriers, the threshold for condensation naturally undercuts the conventional lasing threshold in such devices (Fig. 1.6). When macroscopic occupation of the polaritonic energy ground-state is obtained, a beam of photons is emitted simply due to polaritonic decay from a coherent state. Further information about the concept behind polariton lasers, differences to conventional lasers, and its identification is provided in Chap. 4.

Since the polariton is a composite particle consisting of exciton and photon in optical microcavities, after its lifetime, i.e. when it decays, the energy bound in the particle is set free. Due to the fact that photonic decay from the optical cavity is much more probable than excitonic dephasing, out-coupling of light frequently takes place from the polaritonic device with the information of the decayed polaritonic particle carried by the released photon. This provides a unique access to bosons in solids, with the information of the particles' energy and momentum simply being extracted by optical spectroscopy of the emitted light in connection with the emission angle.

Given the continuous loss of particles intrinsically occurring in a polariton device, the dynamic condensate state is continuously repopulated by external pumping to compensate for optical losses in the case of a steady-state operation scheme. For a practical device, this scheme is preferably achieved via electrical current injection. A typical structure design for electrical pumping is summarized in Chap. 6.

Eventually, a real polariton laser resembles one device with three possible operation regimes (Fig. 1.7): that of a polariton LED (incoherent emission), of a polariton laser (condensate emission), and of a conventional laser/VCSEL (stimulated emission). What has to be always experimentally proven for a functioning polariton device, is that a matter component exists in the polaritonic regime of the system in operation.

Why all the effort, one could ask. As big economic and industrial advances have been achieved by semiconductor lasers, which have had a strong impact on our daily lives via all technologies enabled by lasers, energy consumption and optimization of efficiency has become of great significance. Obviously, polariton lasers do not target the high-power device market, as the operation regime is very sensitive to excitation densities and naturally requires a density lower than laser operation in such structures. In contrast, the concept of coherent light from polariton lasers gains importance when it comes to applications where coherent light from low-power lasers is used and power saving is an issue. With the threshold being up to orders of magnitudes less than for conventional lasers, yet, polariton lasers are not expected to reach the power bottom of ultra-low-threshold lasers such as quantum-dot nanolasers [81–83]. Still, the exploitation of many-particle effects in solids under electrical excitation for more energy-efficient sources of coherent light remains an attractive idea. Nevertheless, one major motivation for the work on polariton lasers directly feeds from the desire to investigate fascinating effects such as Bose–Einstein condensation of quasi-particles, which arise from strong light–matter interaction. In this context, it becomes clear that with an electrically pumped polariton laser at hand, one can achieve polariton condensates on demand for various studies and purposes.

In recent years, several realizations of an electrically-driven polariton laser have been presented to the scientific community. With the pioneering work of scientists from the University of Würzburg, Germany, in collaboration with their partners from

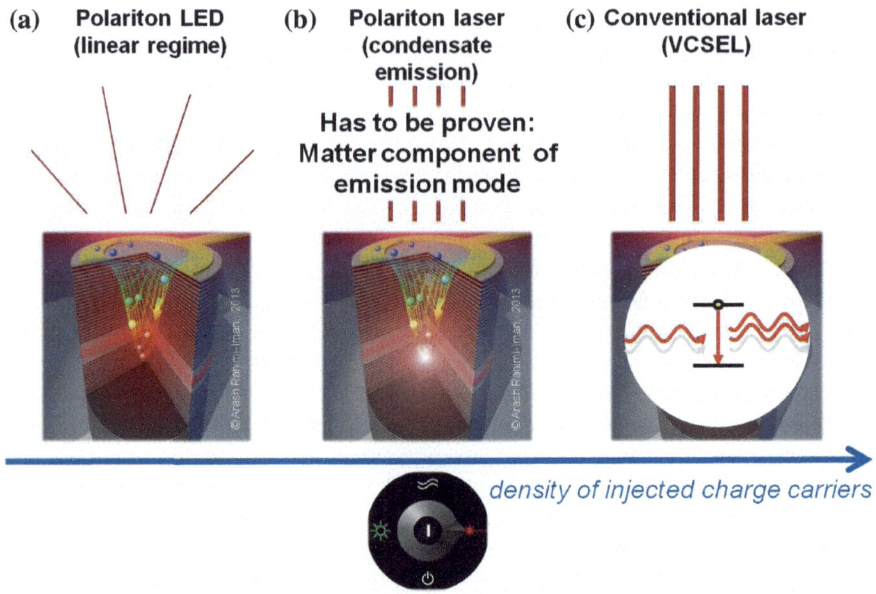

Fig. 1.7 Three operation regimes of an electrically-driven polariton-laser device. **a** Below condensation threshold, the device operates as a polariton LED in the so-called linear regime underlying Boltzmann distribution. This is indicated by the low density of polaritons (balls) in the polariton system (schematic), with divergent output beam corresponding to the emission from a wide range of states along the polariton dispersion (coloured mesh displayed on top of the figure, blueish/reddish colours represent higher/lower energies). Image courtesy of the author. **b** Above the first nonlinearity threshold, for which the matter component in the emitting resonance is still preserved, the device acts as a polariton laser. Courtesy of the author: The schematic drawing of a polariton laser diode based on the QW-microcavity. It indicates the emission of a directed, coherent light beam that originates from radiative decay of polaritons out of the condensate state, in which the bosonic quasi-particles are described by the macroscopic wave-function of the quantum degeneracy in the ground state. In contrast to stimulated scattering of quasi-particles from higher states (blue) into a lower-energetic final state (red), conventional lasing relies on stimulated photon emission from an active medium in its excited state (**c**). Above the second nonlinearity threshold, where excitons are dissociated into an electron–hole plasma, resembling population inversion in the semiconductor diode laser, the system operates in the weak-coupling regime as a conventional laser diode

Stanford, U.S.A., and institutes from Russia, Iceland and Singapore, they unambiguously demonstrated a device with two distinct nonlinearity thresholds corresponding to polariton lasing and photon lasing, respectively. Thereby, C. Schneider and A. Rahimi-Iman experimentally demonstrated the first electrically-pumped polariton laser in the group of S. Höfling with the help of their partners after years of research, having presented strong evidence of a matter component above the first "lasing" threshold [73]—i.e. the threshold for polariton condensation (Fig. 1.8). Simultaneously, and independent from this work, the group of P. Bhattacharya demonstrated a similar device with two-threshold behaviour [84]. While both prototype devices of different origin are operated at cryogenic temperatures and are based on the

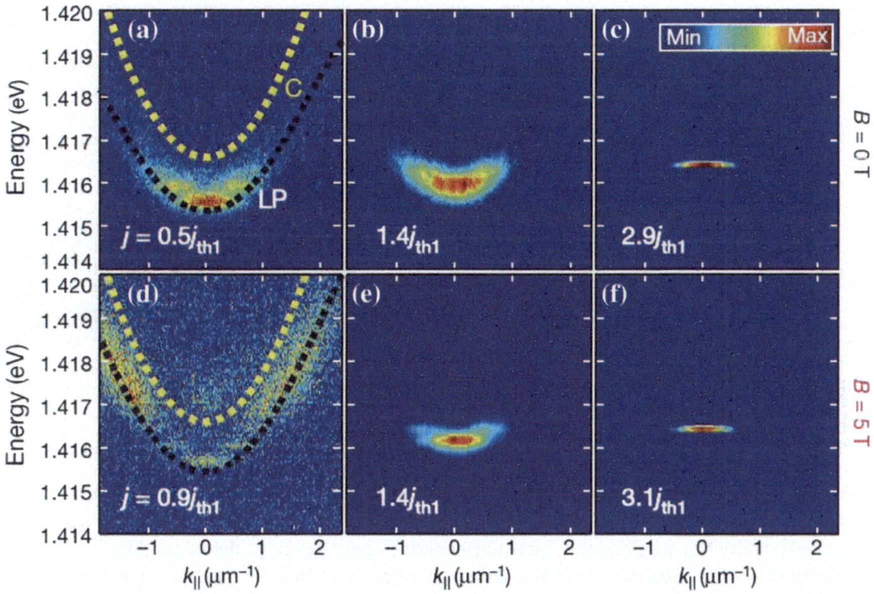

Fig. 1.8 Experimental signatures of polariton lasing evidenced in an excitation-density series that covers all the three operation regimes of an electrically-driven polariton-laser device. The left column presents the linear regime (polariton LED operation), the central and right columns the polariton-lasing and photon-lasing regimes, respectively. A clear feature of polariton-laser devices is their two-threshold behaviour. Above the second threshold, the operation regime resembles that of an ordinary VCSEL based on a QW-microcavity. In-between the two thresholds, the excitonic fraction of the emitting system with very narrow energy and momentum distribution is still preserved and the emission is attributed to polariton decay from the macroscopically occupied ground state, i.e. the condensate. Studies of the device's behaviour in a magnetic field (not displayed here) have not only revealed that the condensation effect becomes strongly pronounced at higher fields, they have also shown Zeeman splitting gradually decreasing with increasing excitation density until the excitonic fraction vanishes above the second threshold. Moreover, the magnetic-field dependence of the spinor-condensates' Zeeman splitting has indicated a spin-Meissner effect for polaritons. Reused with permission. Reference [73] Copyright 2013, Springer Nature

GaAs material system, another publication dealing with GaN microcavities recently claimed room-temperature operation of a polariton laser [85]. More details on polariton lasers and their conventional counterparts are found in Chap. 4.

1.4 More Exciton–Polariton Physics

Independent from the race for electrically-pumped condensates, a plethora of research has been performed in the field of optically-pumped polariton condensates, leading to novel insight into many aspects of microcavity polaritons. On the one

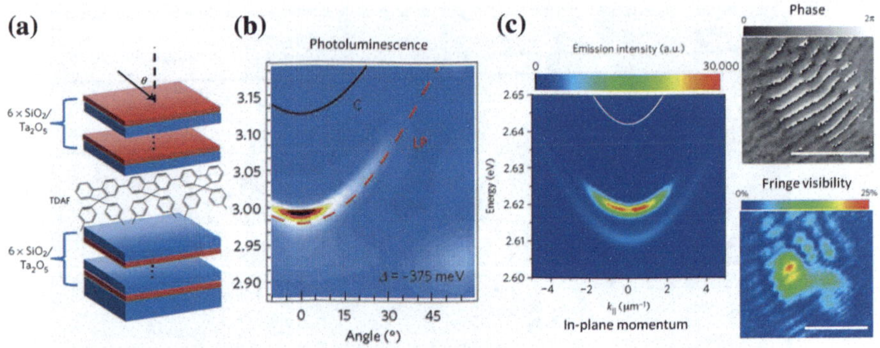

Fig. 1.9 Motivated by the possible observation of room-temperature BEC of polaritons, organic semiconductors were introduced into DBR microcavities comprising dielectric mirrors (**a**). An example material is the TDAF polymer, the chemical structure of which is displayed in **a**. Here, the emission from the optically excited microcavity (false-colour intensity contour plot) reveals the system's energy–momentum dispersion. The corresponding polariton dispersion is indicated for a low-density lower-polariton (LP) branch by a dashed curve that is well distanced from the bare (uncoupled) cavity (C) mode (solid curve) at negative cavity–exciton detuning. Above a critical particle density, a condensate is formed as indicated by the narrow energy–momentum distribution of photoluminescence. Reused with permission. Reference [86] Copyright 2014, Springer Nature. **c** Similar observation for a polymer-based system employing a MeLPPP film as active cavity region, with indicated bare cavity mode (white line). To support the claim of BEC observation, interferometric measurements were performed above the condensation threshold and spontaneous build-up of spatial coherence on a lateral length scale of 20 μm (corresponds to scale bar) demonstrated. Here, a typical phase map and fringe-visibility map are shown for the reported room-temperature polariton condensate. Reused with permission. Reference [57] Copyright 2013, Springer Nature

hand, the hunt for room temperature condensation has always prevailed. In this context, various claims of room-temperature polariton condensates have been made in different material systems [48, 56, 57, 85, 86] (see for instance organic microcavity condensates in Fig. 1.9), with the publication by Plumhof et al. indicating BEC in organic systems [57] similar to earlier work on inorganic structures [45]. On the other hand, the development of novel polariton LED structures of various kind including ones with buried potential landscapes [87] or Esaki diodes [88] were pursued and their properties studied.

Besides studies about dynamic condensation [43, 89, 90], the long-range coherence [45, 47, 91–93], the polarization [94–97] and the two-threshold behaviour in QW microcavities [71, 72], the characterization of the response of polaritons on electric [98–100] and magnetic fields [98, 101–104] was carried out. Even the fraction of uncondensed excitonic population in condensate systems—labelled the dark side of polariton condensates—has been recently analysed [105]. There are many more studies worth mentioning. However, for the sake of clarity, this will be subject of the consecutive chapters, mainly Chaps. 8 and 9.

In addition to standard condensation characterization, which also effectively dealt with the important question how to distinguish polariton lasing and photon lasing [58,

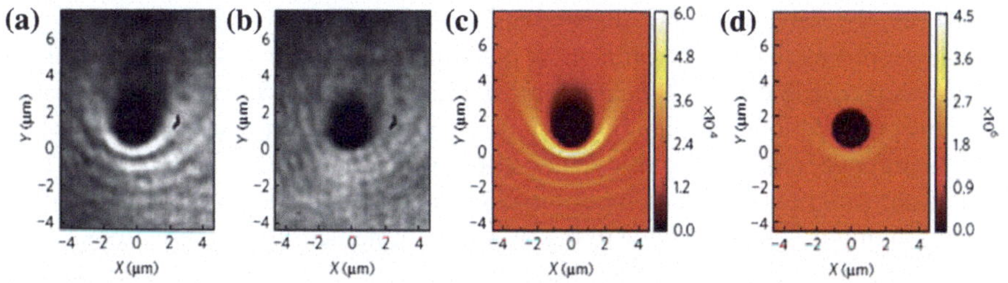

Fig. 1.10 Superfluidity of polaritons in a condensate at room temperature studied under pulsed excitation for an organic microcavity system: Real-space polariton distribution for flow with group velocity of 19 μm/ps from the bottom to the top of the image across an artificial defect. The 13 μm pump spot was centred 2 μm below a defect position to observe the fluid behaviour of polaritons at two polariton density regimes, 0.5×10^6 polaritons/square-μm and 10^7 polaritons/square-μm, respectively. It was shown that the interference fringes due to Rayleigh scattering of polaritons at the defect and the shadow cone beyond the defect, resulting from the reduced transmission, vanish almost perfectly at high densities, demonstrating a fluid with zero viscosity. The corresponding time-averaged calculation is shown for the supersonic (low density) (**c**) and superfluid (**d**) regimes at pump powers corresponding to peak polariton densities comparable to those in **a** and **b**, respectively. The colour scale corresponds to the average polariton density during the time-integration interval of 800 fs. Reused with permission. Reference [135] Copyright 2017, Springer Nature

73, 79, 106], advanced research led to several significant findings. For instance, the achievement of a single-mode polariton laser enabled the demonstration of a very high temporal coherence [107], whereas the exploration of spinor-condensates of exciton–polaritons in external magnetic fields [95, 108–110] led to the observation of the spin-Meissner effect of polariton condensates [95, 104, 111, 112] (see Sect. 9.2). Technologically of great interest are also the realization and investigation of lattices [113–120] and molecules [121] of polariton condensates, to name but a few examples (see Sect. 6.1.2).

In the field of polariton fluids, the condensation effect was rigorously exploited to study superfluidity, i.e. frictionless/lossless motion, of polaritons [122–127] (Fig. 1.10), vortices [123, 128, 129] and even vortex-antivortex pair formation [130–132] (see Sect. 8.3). Other recent investigations targeted backjets in polariton fluids [133] and polariton solitons [116, 134] and their characterization.

Major attention has been drawn to potential landscapes in polariton systems after pioneering work in this field, which enabled both weak [90] or strong confinement [71, 136, 137] and the demonstration of condensates in polariton traps [87, 115, 138, 139] and lattices [113, 114]. Confinement potentials provide an attractive tool to localize polaritons and to define regions of light–matter interaction and polariton dissipation. While common cavity–polaritons are exhibited in 2D structures employing QWs, featuring the well-known polaritonic dispersion, 1D and 0D structures based on these 2D systems feature different spectral characteristics owing to their confinement potentials.

For instance, 1D structures in the form of etched wire-shaped microcavities [138] further reduce the freedom of motion of polaritons. Their confinement potential introduces several polaritonic dispersions with respective energy offset to one-another according to the 1D confinement potential's dimension, which has an impact on the spatial confinement of the photonic mode in such microcavities. Correspondingly, the unidirectional propagation of polaritons in a polariton gas or condensate and their spatial coherence along that axis can be straight-forwardly investigated [140, 141], as the polariton branches are not lost for the guided mode (energy–momentum dispersion in wire direction) due to the overall confinement in two directions.

However, in case of 3D confined polaritons, i.e. in 0D structures, discrete energy lines that arise as a consequence of the pronounced all-directional spatial confinement effect spoil the generally quite simple attribution of modes to polaritonic modes (i.e. strong-coupling regime) instead of cavity photon modes (i.e. weak-coupling regime). Both types of modes exhibit intriguing similarity due to the loss of the unambiguous in-plane (unidirectional) dispersion relation in 2D (1D) cavities. In order to target this drawback, magneto-optical studies were introduced as a means of characterization for polariton systems [103].

Another attractive technique to create potential landscapes is given by the generation of surface-acoustic waves (SAWs) [115, 116, 142], which can generate multiple trap condensates within the optical excitation spot and be used to study interaction between spatially separated and confined condensates [115]. It is worth noting that also the pump spot itself can imprint a potential landscape for polaritons [143]. Examples regarding polariton traps, molecules and lattices are further discussed in the literature, e.g. in [144], and will be highlighted in Sect. 6.1.2.

In general, one can observe an increasing amount of proposals and experiments regarding application-oriented polariton systems, as has been summarized in recent literature such as [6, 10]. For optoelectronic or quantum optical applications employing polariton systems, different approaches were suggested. Switches were proposed from the theoretical view point based on polaritons for use in all-optical circuits [145], later discussed on the level of integrated photonic circuits [146]. Also, functional all-optical elements for polaritonic logic circuits were proposed based on on-demand projection of circuits onto an unprocessed planar microcavity in a study using spatially-modulated non-resonant optical excitation [147]. Furthermore, spin switches [148] and transistors based on polariton spin [149] were suggested. One approach of a polariton transistor proposes ultrafast coherent switches with polariton condensates [150]. Others aimed at harnessing the control of a two-fluid polariton switch [151]. Another technique aims at manipulating and controlling the polariton state optically with ultrafast laser pulses directly targeting the polaritons' Bloch sphere [152]. Ultrafast optical switching is also proposed based on external electric bias applied to polariton diodes in the reverse direction to switch the photoluminescence between a polaritonic and photonic regime [100]. Experimental realization of all-optical polariton devices based on transistors [153], routers [154], resonant-tunneling diodes [155] and interferometers [156] is a rising topic in the domain of *polaritonics*. These few examples show the increasing interest in this subject and the interested reader is referred to the original articles and the discussions within.

Recently, also pioneering work on microcavities with embedded monolayers of 2D materials have brought polaritons into the spot light: on the one hand, 2D semiconductors in planar microresonator structures were employed to achieve 2D exciton–polaritons [157] (see Sect. 6.2); on the other hand, graphene islands were used to obtain strong coupling with light [158]. Owing to the fact that excitons in most of these new 2D materials feature a strong binding energy, these systems show a clear advantage with respect to high temperature operation and have become very promising testbeds for room-temperature polaritonics and BEC studies. Thus, a revival of polariton research is taking place in the sphere of monolayer and few-layer materials. Here, the ability to use valley physics—already discussed with regard to "*valleytronics*" with 2D materials—in combination with polarization-selective optical cavities, such as chiral microcavities (previously investigated for monolithically-grown QW and QD microcavity systems [159–161]), could open up new possibilities related to polarization- as well as spin-sensitive electro-optical devices, i.e. optical valleytronics.

With various tunable-cavity concepts at hand for polariton studies with organic or inorganic high-binding-energy semiconductor materials at room temperature, the material independent continuous adjustability of cavity resonances could be utilized for polariton-based chemistry allowing a precise adjustment of the polariton energy levels [162]. These platforms could also become useful for an efficient study of chemical reactions influenced by different light–matter coupling situations. Especially, the investigation of the transition between weak and strong coupling may become possible by smoothly tuning the coupling. This is attractive, since the two coupling regimes are linked by an exceptional point (EP). At this position, only one complex solution exists for the coupled system [163–165] which can be exploited for interesting phenomena.

1.5 Further Polaritons Not Detailed in This Book

Having its emphasis on polariton condensation and lasing, this work will not cover all aspects of polariton physics, but will at this point provide a brief overview on various polariton regimes achieved in solids by coupling electro-magnetic fields with matter states. For example, a plethora of polariton effects have been recently summarized for layered van-der-Waals materials [166], i.e. 2D materials. In this regard, also light–valley interactions in 2D semiconductors, e.g. in 2D-microcavity systems, have increasingly gained popularity with eyes towards optical "valleytronics" [167]. Here, one major aim lies in the harnessing of the new degree of freedom referred to as pseudo-spin, which is promoted by the spin–valley locking in 2D semiconductors.

Polaritons, which are coupled-oscillator systems, can be exhibited not only with quantum-well or organic-molecule excitons in microcavities (see Fig. 1.1), but also as exciton–polaritons in bulk crystals [1, 2, 168, 169], in systems with phonons [170–174], polarons [175, 176] as well as surface-plasmons [174, 177–179] (e.g. highlighted for 2D materials in [166]), and at wavelength ranges beyond the infrared in the form of terahertz (THz) polaritons in structures based on metamaterials nanocavities [180]. Moreover, beyond weak and strong coupling, one can obtain coupling strengths comparable to the transition frequencies in the light–matter system (or even larger), when entering the ultra-strong coupling (or even deep-strong coupling) regime [181, 182], for instance with strongly-confined light fields and organic molecules. Recently, even plasmon–exciton–polariton (PEP) lasers have been reported at room temperature using plasmonic structures with organic materials [183]. Such polariton-laser devices have the potential to enable on-chip photonic integration by harnessing active metamaterial planar technologies and lower thresholds than conventional lasers based on stimulated-scattering effects.

In the field of cavity quantum electrodynamics, exciton–polaritons are even obtained in the single particle limit when combining quantum dots with optical microcavities (e.g. micropillars or photonic crystals), where strong coupling can be investigated between the vacuum field and a single two-level matter system [184–186] (Fig. 1.11). On the one hand, from these single quantum emitters strongly coupled to the light field, photon anti-bunching can be obtained [187]. A similar sub-Poissonian photon statistics had been demonstrated for the one-atom laser [188], which also operates in the strong-coupling regime. On the other hand, with such QD-micropillar systems, also photon–photon coupling mediated by the exciton spin state [189] or the climbing of the Jaynes–Cummings ladder [190] can be investigated, to name but a few examples.

In the early years of semiconductor research concerning their optical properties and the interaction of light with excitons, scientists at that time encountered the bulk exciton–polariton. It was understood that the presence of an optically-excited electronic resonance is directly related to polariton formation in the semiconductor crystal due to the coupling of a polarization wave with the light field. What one observed, was not the propagation of the light-induced exciton through the lattice, but the propagation of a polariton, which is accompanied by the periodic emission and absorption of a photon by a (coherent) bulk exciton (macroscopic polarization), resembling a strongly-coupled light–matter system. This coherent energy oscillation between photon and exciton propagates through the crystal and has been characterized by the exciton–polariton dispersion in bulk, as introduced in Sect. 2.1 (for further details on the concept of polaritons, the interested reader is referred to [169]).

Similar to the formation of exciton–polaritons by photon-coupling with electronic excitations, optical waves in matter can also couple to phonon resonances, forming phonon–polaritons. These quasi-particles feature an anti-crossing behaviour of their dispersion similar to that of their excitonic counterparts, however, the characteristic splitting occurs in the phonon band instead. Phonon–polaritons were historically observed in ionic crystals [170, 191] and have become relevant for *polaritonics* [192]. They can help to improve high speed signal processing and to achieve integrating signal generation, guidance and readout in one platform that operates at frequencies between those of electronics and photonics.

When an incident light field couples to plasmonic resonances at metallic surfaces, this can lead to the formation of surface-plasmon–polaritons [193–195], which resonate and propagate as light–matter waves along the interface between the metal and dielectric (or air/vacuum). These electro-magnetic waves in the visible to infrared frequency range with strong spatial confinement and enhanced local field intensities

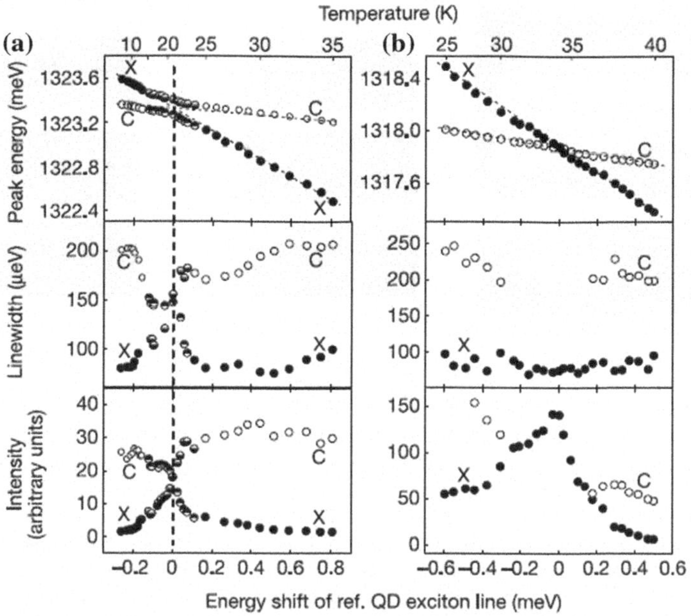

Fig. 1.11 Experimental demonstration of strong exciton–photon coupling in the single-emitter limit. The coupling is achieved in a high-quality Fabry–Pérot micropillar between the empty cavity field, i.e. the vacuum level of the 3D optical confinement potential, from which an atomistic density of states for photons results, and a single two-level emitter offered by a quantum dot, which resembles an artificial atom. Here, cryogenic temperatures were varied to achieve resonance tuning for two distinct cases: The left column (**a**) shows the behaviour of a strongly-coupled emitter–field system (featuring mode anti-crossing), whereas the right column (**b**) shows the weak-coupling regime counterpart (exhibiting a Purcell effect). From top row to bottom row, the peak energies, the linewidths and the intensities of the modes are displayed, respectively. The energy splitting of the modes under strong coupling is referred to as vacuum Rabi splitting. Reused with permission. Reference [184] Copyright 2004, Springer Nature

can be exploited in on-chip integrated light-based future information transmission schemes, whereas the waves can propagate faster than electric currents and are guided by surface modes. This opens up new possibilities in sub-wavelengths photonics technologies, enabling future nano-optics applications.

Indeed, polaritons can also occur at other energy ranges such as in the infrared and the THz spectral range, as polaritons represent strong coupling between electromagnetic radiation and excitation in matter. A prominent example of THz polaritons can be found for metamaterial systems comprising of metallic resonators on top of a semiconductor QW structure, in which intraband transitions of electrons can couple to the enhanced light field in the metamaterial nanocavity [180]. Artificial materials referred to as "metamaterials" sparked wide interest for applications such as optical cloaking and light-wave manipulation (see for instance [196]), and polaritons using metamaterials cavities can be attractive for new studies in the field of light–matter

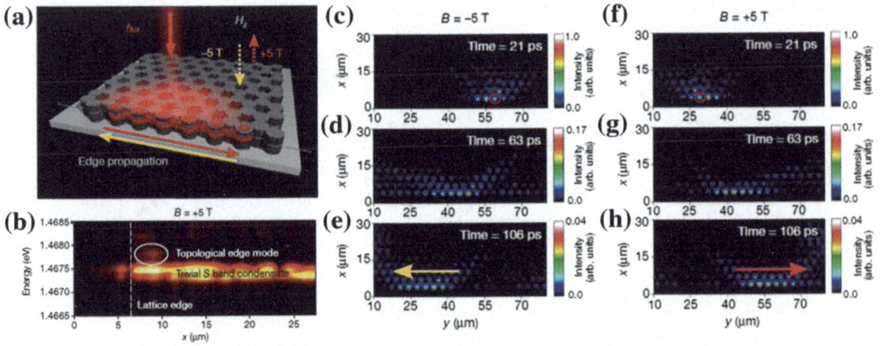

Fig. 1.12 Study of an exciton–polariton topological insulator. A scheme of the non-resonant laser excitation of left-moving (yellow) and right-moving (red) chiral topological polariton edge modes in an external magnetic field (dashed arrows) is illustrated in **a**. **b** A trivial S-band condensate throughout the structure and a topological edge mode well separated from the bulk were observed. This edge mode, which was well located at the zig-zag edge (dashed white line), clearly extended around the corner to the arm-chair configuration, not shown here. **c**–**h** The dominant propagation direction became inverted (yellow and red arrows) when the direction of the magnetic field was reversed. Furthermore, it was demonstrated that the polariton chiral edge mode propagating along the structure avoids a pre-installed defect site, not shown here. Reused with permission. Reference [120] Copyright 2018, Springer Nature

interaction, e.g. in the THz and infrared frequency bands, as well as for new opto-electronic device concepts, which could be used for optical filters and modulators.

Lately, the subject of topological properties also penetrated the polariton world. Exciton–polariton topological insulators have been recently characterized [197, 198] and a topological polariton condensate proposed and demonstrated [120] (Fig. 1.12). In such system, the polariton mode forward-propagating along the edge of a polariton-micropillar in-plane superlattice configuration can circumvent defects without losses owing to topologically protected states, similarly exploited in the so-called topological insulator laser [199, 200]. Topological edge states have been well under investigation since their recent demonstrations [201]. They can be attractive for future devices by harnessing the underlying features of topological insulators in 2D arrays. Such approach promises photonic integration involving low-loss components, whereas photonics in general may benefit from topological concepts, ultimately giving rise to robust, scatter-free light propagation in a new generation of optical structures and devices (see [199, 200]).

References

1. S.I. Pekar, The theory of electromagnetic waves in a crystal in which excitons are produced. Sov. Phys. JETP **6**, 785 (1958)
2. J.J. Hopfield, Theory of the contribution of excitons to the complex dielectric constant of crystals. Phys. Rev. **112**(5), 1555– (1958)

3. C. Weisbuch, M. Nishioka, A. Ishikawa, Y. Arakawa, Observation of the coupled exciton-photon mode splitting in a semiconductor quantum microcavity. Phys. Rev. Lett. **69**(23), 3314– (1992)
4. Y. Yamamoto, F. Tassone, H. Cao, *Semiconductor Cavity Quantum Electrodynamics* (Springer, Berlin, 2000)
5. A.V. Kavokin, J.J. Baumberg, G. Malpucch, F.P. Laussy, *Microcavities* (Oxford University Press, Oxford, 2007)
6. D. Sanvitto, S. Kéna-Cohen, The road towards polaritonic device. Nat. Mater. **15**, 1061 (2016)
7. K.J. Vahala, Optical microcavities. Nature **424**(6950), 839–846 (2003)
8. K. Vahala, *Optical Microcavities* (World Scientific, Singapore, 2004)
9. B. Deveaud, *The Physics of Semiconductor Microcavities* (WILEY-VCH, Weinheim, 2007)
10. A.V. Kavokin, J.J. Baumberg, G. Malpuech, F.P. Laussy. *Microcavities*, vol. 1 (Oxford University Press, Oxford, 2017)
11. A. Imamoglu, R.J. Ram, S. Pau, Y. Yamamoto, Nonequilibrium condensates and lasers without inversion: exciton-polariton lasers. Phys. Rev. A **53**(6), 4250 (1996)
12. A. Kavokin, G. Malpuech, F.P. Laussy, Polariton laser and polariton superfluidity in microcavities. Phys. Lett. A **306**(4), 187–199 (2003)
13. E.M. Purcell, Spontaneous emission probabilities at radio frequencies. Phys. Rev. **69**(11–12), 681 (1946)
14. I.I. Rabi, Space quantization in a gyrating magnetic field. Phys. Rev. **51**(8), 652–654 (1937)
15. A. Kavokin, G. Malpuech, *Cavity Polaritons* (Academic Press, Cambridge, 2003)
16. H. Deng, H. Haug, Y. Yamamoto, Exciton-polariton Bose-Einstein condensation. Rev. Mod. Phys. **82**(2), 1489 (2010)
17. A. Griffin, D.W. Snoke, S. Stringari, *Bose-Einstein Condensation* (Cambridge University Press, Cambridge, 1995)
18. L.P. Pitaevskii, S. Stringari, *Bose-Einstein Condensation* (Oxford University Press, Oxford, 2003)
19. M.H. Anderson, J.R. Ensher, M.R. Matthews, C.E. Wieman, E.A. Cornell, Observation of Bose-Einstein condensation in a dilute atomic vapor. Science **269**(5221), 198–201 (1995)
20. K.B. Davis, M.O. Mewes, M.R. Andrews, N.J. van Druten, D.S. Durfee, D.M. Kurn, W. Ketterle, Bose-Einstein condensation in a gas of sodium atoms. Phys. Rev. Lett. **75**(22), 3969– (1995)
21. C.C. Bradley, C.A. Sackett, J.J. Tollett, R.G. Hulet, Evidence of Bose-Einstein condensation in an atomic gas with attractive interactions. Phys. Rev. Lett. **75**(9), 1687 (1995)
22. S.N. Bose, Plancks Gesetz und Lichtquantenhypothese. Z. Phys. **26**, 178 (1924)
23. A. Einstein, Quantentheorie des einatomigen idealen Gases. 2. Abhandlung. Sitz. Ber. Preuss. Akad. Wiss. **22**, 261 (1925)
24. J.H. Davies, *The Physics of Low-Dimensional Semiconductors: An Introduction* (Cambridge University Press, Cambridge, 1998)
25. W. Ketterle, What is B.E.C.? (2014). (webpage: https://cua.mit.edu/ketterle_group/intro/whatbec/whtisbec.html)
26. P. Kapitza, Viscosity of liquid helium below the λ-point. Nature **141**, 74 (1938)
27. J.F. Allen, A.D. Misener, Flow of liquid helium II. Nature **141**, 75 (1938)
28. L. Tsitza, Transport phenomena in helium II. Nature **141**, 913 (1938)
29. F. London, The λ-phenomenon of liquid helium and the Bose-Einstein degeneracy. Nature **141**, 643–644 (1938)
30. L. Landau, Theory of the superfluidity of helium ii. PR **60**(4), 356–358 (1941)
31. N.N. Bogoliubov, On the theory of superfluidity. J. Phys. USSR **11**, 23 (1947)
32. S.A. Moskalenko, Reversible optico-hydrodynamic phenomena in a nonideal exciton gas. Sov. Phys. Solid State **4**(1), 199–204 (1962)
33. J.M. Blatt, K.W. Boer, W. Brandt, Bose-Einstein condensation of excitons. Phys. Rev. **126**(5), 1691 (1962)
34. L.V. Keldysh, A.N. Kozlov, Collective properties of excitons in semiconductors. Sov. Phys. JETP **27**, 521–528 (1968)

35. N.F. Mott, *Metal Insulator Transitions*, 2nd edn. (Taylor & Francis, Milton Park, 1990)
36. L. Kappei, J. Szczytko, F. Morier-Genoud, B. Deveaud, Direct observation of the mott transition in an optically excited semiconductor quantum well. Phys. Rev. Lett. **94**(14), 147403 (2005)
37. S.A. Moskalenko, D.W. Snoke, *Bose-Einstein Condensation of Excitons and Biexcitons: And Coherent Nonlinear Optics with Excitons* (Cambridge University Press, Cambridge, 2000)
38. L.V. Butov, A.C. Gossard, D.S. Chemla, Macroscopically ordered state in an exciton system. Nature **418**(6899), 751–754 (2002)
39. L.V. Butov, C.W. Lai, A.L. Ivanov, A.C. Gossard, D.S. Chemla, Towards Bose-Einstein condensation of excitons in potential traps. Nature **417**(6884), 47–52 (2002)
40. K. Yoshioka, E. Chae, M. Kuwata-Gonokami, Transition to a Bose-Einstein condensate and relaxation explosion of excitons at sub-Kelvin temperatures. Nat. Commun. **2**, 328 (2011)
41. A.A. High, J.R. Leonard, A.T. Hammack, M.M. Fogler, L.V. Butov, A.V. Kavokin, K.L. Campman, A.C. Gossard, Spontaneous coherence in a cold exciton gas. Nature **483**(7391), 584–588 (2012)
42. A.A. High, J.R. Leonard, M. Remeika, L.V. Butov, M. Hanson, A.C. Gossard, Condensation of excitons in a trap. Nano Lett. **12**(5), 2605–2609 (2012)
43. H. Deng, G. Weihs, C. Santori, J. Bloch, Y. Yamamoto, Condensation of semiconductor microcavity exciton polaritons. Science **298**(5591), 199–202 (2002)
44. M. Richard, J. Kasprzak, R. André, R. Romestain, L. Si Dang, G. Malpuech, A. Kavokin, Experimental evidence for nonequilibrium Bose condensation of exciton polaritons. Phys. Rev. B **72**(20), 201301 (2005)
45. J. Kasprzak, M. Richard, S. Kundermann, A. Baas, P. Jeambrun, J.M.J. Keeling, F.M. Marchetti, M.H. Szymańska, R. André, J.L. Staehli, V. Savona, P.B. Littlewood, B. Deveaud, L. Si Dang, Bose-Einstein condensation of exciton polaritons. Nature **443**(7110), 409–414 (2006)
46. R. Balili, V. Hartwell, D. Snoke, L. Pfeiffer, K. West, Bose-Einstein condensation of microcavity polaritons in a trap. Science **316**(5827), 1007–1010 (2007)
47. H. Deng, G.S. Solomon, R. Hey, K.H. Ploog, Y. Yamamoto, Spatial coherence of a polariton condensate. Phys. Rev. Lett. **99**(12), 126403– (2007)
48. S. Christopoulos, G. Baldassarri Höger von Högersthal, A.J. Grundy, P.G. Lagoudakis, A.V. Kavokin, J.J. Baumberg, G. Christmann, R. Butté, E. Feltin, J.-F. Carlin, N. Grandjean, Room-temperature polariton lasing in semiconductor microcavities. Phys. Rev. Lett. **98**(12), 126405 (2007)
49. G. Christmann, R. Butté, E. Feltin, J.-F. Carlin, N. Grandjean, Room temperature polariton lasing in a GaN/AlGaN multiple quantum well microcavity. Appl. Phys. Lett. **93**(5), 051102 (2008)
50. J.J. Baumberg, A.V. Kavokin, S. Christopoulos, A.J.D. Grundy, R. Butté, G. Christmann, D.D. Solnyshkov, G. Malpuech, G. Baldassarri Höger von Högersthal, E. Feltin, J.-F. Carlin, N. Grandjean, Spontaneous polarization buildup in a room-temperature polariton laser. Phys. Rev. Lett. **101**(13), 136409 (2008)
51. A. Das, J. Heo, M. Jankowski, W. Guo, L. Zhang, H. Deng, P. Bhattacharya, Room temperature ultralow threshold GaN nanowire polariton laser. Phys. Rev. Lett. **107**(6), 066405 (2011)
52. L. Sun, S. Sun, H. Dong, W. Xie, M. Richard, L. Zhou, L.S. Dang, X. Shen, Z. Chen, Room temperature one-dimensional polariton condensate in a ZnO microwire (2010), arXiv:1007.4686v1 (unpublished)
53. W. Xie, H. Dong, S. Zhang, L. Sun, W. Zhou, Y. Ling, L. Jian, X. Shen, Z. Chen, Room-temperature polariton parametric scattering driven by a one-dimensional polariton condensate. Phys. Rev. Lett. **108**(16), 166401 (2012)
54. H. Franke, C. Sturm, R. Schmidt-Grund, G. Wagner, M. Grundmann, Ballistic propagation of exciton-polariton condensates in a ZnO-based microcavity. New J. Phys. **14**, 013037 (2012)
55. F. Li, L. Orosz, O. Kamoun, S. Bouchoule, C. Brimont, P. Disseix, T. Guillet, X. Lafosse, M. Leroux, J. Leymarie, M. Mexis, M. Mihailovic, G. Patriarche, F. Réveret, D. Solnyshkov, J. Zuniga-Perez, G. Malpuech, From excitonic to photonic polariton condensate in a ZnO-based microcavity. Phys. Rev. Lett. **110**(19), 196406 (2013)

56. S. Kena-Cohen, S.R. Forrest, Room-temperature polariton lasing in an organic single-crystal microcavity. Nat. Phot. **4**(6), 371–375 (2010)
57. J.D. Plumhof, T. Stöferle, L. Mai, U. Scherf, R.F. Mahrt, Room-temperature Bose–Einstein condensation of cavity exciton-polaritons in a polymer. Nat. Mater. **13**, 247 (2014). Published online Dec. 2013
58. B. Deveaud-Plédran, On the condensation of polaritons. J. Opt. Soc. Am. B **29**(2), A138–A145 (2012)
59. R. Houdré, C. Weisbuch, R.P. Stanley, U. Oesterle, P. Pellandini, M. Ilegems, Measurement of cavity-polariton dispersion curve from angle-resolved photoluminescence experiments. Phys. Rev. Lett. **73**(15), 2043–2046 (1994)
60. M.S. Skolnick, T.A. Fisher, D.M. Whittaker, Strong coupling phenomena in quantum microcavity structures. Semicond. Sci. Technol. **13**, 645–669 (1998)
61. N.D. Mermin, H. Wagner, Absence of ferromagnetism or antiferromagnetism in one- or two-dimensional isotropic heisenberg models. Phys. Rev. Lett. **17**(22), 1133– (1966)
62. P.C. Hohenberg, Existence of long-range order in one and two dimensions. Phys. Rev. **158**(2), 383– (1967)
63. J. Klaers, J. Schmitt, F. Vewinger, M. Weitz, Bose-Einstein condensation of photons in an optical microcavity. Nature **468**(7323), 545–548 (2010)
64. Y.B. Zel'dovich, E.V. Levich, Bose condensation and shock waves in photon spectra. Sov. Phys. JETP **28**(6), 69 (1969)
65. R.Y. Chiao, Bogoliubov dispersion relation for a 'photon fluid': Is this a superfluid? Opt. Commun. **179**(1–6), 157–166 (2000)
66. E.L. Bolda, R.Y. Chiao, W.H. Zurek, Dissipative optical flow in a nonlinear Fabry-Perot cavity. Phys. Rev. Lett. **86**(3), 416–419 (2001)
67. J. Klaers, F. Vewinger, M. Weitz, Thermalization of a two-dimensional photonic gas in a white wall photon box. Nat. Phys. **6**(7), 512–515 (2010)
68. H. Deng, G. Weihs, D. Snoke, J. Bloch, Y. Yamamoto, Polariton lasing vs. photon lasing in a semiconductor microcavity. Proc. Natl. Acad. Sci. USA **100**(26), 15318–15323 (2003)
69. P. Tsotsis, P.S. Eldridge, T. Gao, S.I. Tsintzos, Z. Hatzopoulos, P.G. Savvidis, Lasing threshold doubling at the crossover from strong to weak coupling regime in GaAs microcavity. New J. Phys. **14**, 023060 (2012)
70. D. Bajoni, P. Senellart, A. Lemaître, J. Bloch, Photon lasing in GaAs microcavity: similarities with a polariton condensate. Phys. Rev. B **76**(20), 201305(R) (2007)
71. D. Bajoni, P. Senellart, E. Wertz, I. Sagnes, A. Miard, A. Lemaître, J. Bloch, Polariton laser using single micropillar GaAs-GaAlAs semiconductor cavities. Phys. Rev. Lett. **100**(4), 047401 (2008)
72. J.-S. Tempel, F. Veit, M. Aßmann, L.E. Kreilkamp, A. Rahimi-Iman, A. Löffler, S. Höfling, S. Reitzenstein, L. Worschech, A. Forchel, M. Bayer, Characterization of two-threshold behavior of the emission from a GaAs microcavity. Phys. Rev. B **85**(7), 075318 (2012)
73. C. Schneider, A. Rahimi-Iman, N.Y. Kim, J. Fischer, I.G. Savenko, M. Amthor, M. Lermer, A. Wolf, L. Worschech, V.D. Kulakovskii, I.A. Shelykh, M. Kamp, S. Reitzenstein, A. Forchel, Y. Yamamoto, S. Höfling, An electrically pumped polariton laser. Nature **497**, 348 (2013)
74. L. Si Dang, D. Heger, R. André, F. Bœuf, R. Romestain, Stimulation of polariton photoluminescence in semiconductor microcavity. Phys. Rev. Lett. **81**(18), 3920 (1998)
75. R.M. Stevenson, V.N. Astratov, M.S. Skolnick, D.M. Whittaker, M. Emam-Ismail, A.I. Tartakovskii, P.G. Savvidis, J.J. Baumberg, J.S. Roberts, Continuous wave observation of massive polariton redistribution by stimulated scattering in semiconductor microcavities. Phys. Rev. Lett. **85**(17), 3680–3683 (2000)
76. S.I. Tsintzos, N.T. Pelekanos, G. Konstantinidis, Z. Hatzopoulos, P.G. Savvidis, A GaAs polariton light-emitting diode operating near room temperature. Nature **453**(7193), 372–375 (2008)
77. D. Bajoni, E. Semenova, A. Lemaître, S. Bouchoule, E. Wertz, P. Senellart, J. Bloch, Polariton light-emitting diode in a GaAs-based microcavity. Phys. Rev. B **77**(11), 113303 (2008). March

78. A.A. Khalifa, A.P.D. Love, D.N. Krizhanovskii, M.S. Skolnick, J.S. Roberts, Electroluminescence emission from polariton states in GaAs-based semiconductor microcavities. Appl. Phys. Lett. **92**(6), 061107 (2008)
79. D. Bajoni, Polariton lasers. Hybrid light-matter lasers without inversion. J. Phys. D: Appl. Phys. **45**(31), 313001 (2012)
80. M.G.A. Bernard, G. Duraffourg, Laser conditions in semiconductors. Phys. Stat. Sol. (b) **1**(7), 699–703 (1961)
81. J.-M. Gérard, Solid-state cavity-quantum electrodynamics with self-assembled quantum dots, *Single Quantum Dots: Fundamentals, Applications, and New Concepts* (Springer, Berlin, 2003), pp. 269–314
82. S. Noda, Seeking the ultimate nanolaser. Science **314**(5797), 260 (2006)
83. S. Reitzenstein, A. Forchel, Quantum dot micropillars. J. Phys. D: Appl. Phys. **43**(3), 033001–(2010)
84. P. Bhattacharya, B. Xiao, A. Das, S. Bhowmick, J. Heo, Solid state electrically injected exciton-polariton laser. Phys. Rev. Lett. **110**(20), 206403 (2013)
85. P. Bhattacharya, T. Frost, S. Deshpande, M.Z. Baten, A. Hazari, A. Das, Room temperature electrically injected polariton laser. Phys. Rev. Lett. **112**, 236802 (2014)
86. K.S. Daskalakis, S.A. Maier, R. Murray, S. Kéna-Cohen, Nonlinear interactions in an organic polariton condensate. Nat. Mater. **13**(3), 271–278 (2014)
87. K. Winkler, C. Schneider, J. Fischer, A. Rahimi-Iman, M. Amthor, A. Forchel, S. Reitzenstein, S. Höfling, M. Kamp, Electroluminescence from spatially confined exciton polaritons in a textured microcavity. Appl. Phys. Lett. **102**, 041101 (2013)
88. S. Brodbeck, J.-P. Jahn, A. Rahimi-Iman, J. Fischer, M. Amthor, S. Reitzenstein, M. Kamp, C. Schneider, S. Höfling, Room temperature polariton light emitting diode with integrated tunnel junction. Opt. Express **21**(25), 31098–31104 (2013)
89. M. Richard, J. Kasprzak, R. Romestain, R. André, L. Si Dang, Spontaneous coherent phase transition of polaritons in CdTe microcavities. Phys. Rev. Lett. **94**(18), 187401 (2005)
90. S. Utsunomiya, L. Tian, G. Roumpos, C.W. Lai, N. Kumada, T. Fujisawa, M. Kuwata-Gonokami, A. Löffler, S. Höfling, A. Forchel, Y. Yamamoto, Observation of Bogoliubov excitations in exciton-polariton condensates. Nat. Phys. **4**(9), 700–705 (2008)
91. G. Roumpos, M. Lohse, W.H. Nitsche, J. Keeling, M. Hanna Szymańska, P.B. Littlewood, A. Löffler, S. Höfling, L. Worschech, A. Forchel, Y. Yamamoto, Power-law decay of the spatial correlation function in exciton-polariton condensates. Proc. Natl. Acad. Sci. USA (2012). https://doi.org/10.1073/pnas.1107970109
92. A. Rahimi-Iman, A.V. Chernenko, J. Fischer, S. Brodbeck, M. Amthor, C. Schneider, A. Forchel, S. Höfling, S. Reitzenstein, M. Kamp, Coherence signatures and density-dependent interaction in a dynamical exciton-polariton condensate. Phys. Rev. B **86**(15), 155308 (2012)
93. R. Spano, J. Cuadra, G. Tosi, C. Antón, C.A. Lingg, D. Sanvitto, M.D. Martín, L. Viña, P.R. Eastham, M. van der Poel, J.M. Hvam, Coherence properties of exciton polariton OPO condensates in one and two dimensions. New J. Phys. **14**(7), 075018 (2012)
94. G. Roumpos, C-W. Lai, T.C.H. Liew, Y.G. Rubo, A.V. Kavokin, Y. Yamamoto. Signature of the microcavity exciton-polariton relaxation mechanism in the polarization of emitted light. Phys. Rev. B **79**(19), 195310 (2009)
95. A.V. Larionov, V.D. Kulakovskii, S. Höfling, C. Schneider, L. Worschech, A. Forchel, Polarized nonequilibrium Bose-Einstein condensates of spinor exciton polaritons in a magnetic field. Phys. Rev. Lett. **105**(25), 256401 (2010)
96. A. Demenev, S. Gavrilov, V. Kulakovskii, Stimulated polariton-polariton scattering and dynamic Bose-Einstein condensation of polaritons in GaAs microcavities under excitation near the exciton resonance. JETP Lett. **95**(1), 38–43 (2012)
97. A. Bhattacharya, M. Zunaid Baten, I. Iorsh, T. Frost, A. Kavokin, P. Bhattacharya, Output polarization characteristics of a GaN microcavity diode polariton laser. Phys. Rev. B **94**(3), 5203 (2016)
98. D. Whittaker, T. Fisher, A. Afshar, M. Skolnick, P. Kinsler, J. Roberts, G. Hill, M. Pate, Vacuum rabi splitting in semiconductor microcavities with applied electric and magnetic fields. Il Nuovo Cimento D **17**(11), 1781–1786 (1995)

99. D. Bajoni, E. Semenova, A. Lemaître, S. Bouchoule, E. Wertz, P. Senellart, S. Barbay, R. Kuszelewicz, J. Bloch, Optical bistability in a GaAs-based polariton diode. Phys. Rev. Lett. **101**(26), 266402 (2008)
100. M. Amthor, S. Weißenseel, J. Fischer, M. Kamp, C. Schneider, S. Höfling, Electro-optical switching between polariton and cavity lasing in an InGaAs quantum well microcavity. Opt. Express **22**(25), 31146–31153 (2014)
101. T.A. Fisher, A.M. Afshar, M.S. Skolnick, D.M. Whittaker, J.S. Roberts, Vacuum Rabi coupling enhancement and Zeeman splitting in semiconductor quantum microcavity structures in a high magnetic field. Phys. Rev. B **53**(16), R10469 (1996)
102. S. Reitzenstein, S. Münch, P. Franeck, A. Rahimi-Iman, A. Löffler, S. Höfling, L. Worschech, A. Forchel, Control of the strong light-matter interaction between an elongated $In_{0.3}Ga_{0.7}As$ quantum dot and a micropillar cavity using external magnetic fields. Phys. Rev. Lett. **103**(12), 127401– (2009)
103. A. Rahimi-Iman, C. Schneider, J. Fischer, S. Holzinger, M. Amthor, S. Höfling, S. Reitzenstein, L. Worschech, M. Kamp, A. Forchel, Zeeman splitting and diamagnetic shift of spatially confined quantum-well exciton polaritons in an external magnetic field. Phys. Rev. B **84**(16), 165325 (2011)
104. J. Fischer, S. Brodbeck, A.V. Chernenko, I. Lederer, A. Rahimi-Iman, M. Amthor, V.D. Kulakovskii, L. Worschech, M. Kamp, M. Durnev, C. Schneider, A.V. Kavokin, S. Höfling, Anomalies of a nonequilibrium spinor polariton condensate in a magnetic field. Phys. Rev. Lett. **112**, 093902 (2014)
105. J.M. Ménard, C. Poellmann, M. Porer, U. Leierseder, E. Galopin, A. Lemaître, A. Amo, J. Bloch, R. Huber, Revealing the dark side of a bright exciton-polariton condensate. Nat. Commun. **5**, 4648 (2014)
106. V.B. Timoffeev, Bose condensation of exciton polaritons in microcavities. Fizika i Tekhnika Poluprovodnikov **46**(7), 865–883 (2012). engl. translation in Semiconductors **46**(7), S. 843, Springer (2012)
107. S. Kim, B. Zhang, Z. Wang, J. Fischer, S. Brodbeck, M. Kamp, C. Schneider, S. Höfling, H. Deng, Coherent polariton laser. Phys. Rev. X **6**(1), 011026 (2016)
108. I.A. Shelykh, Y.G. Rubo, A.V. Kavokin, Renormalized dispersion of elementary excitations in spinor polariton condensates. Superlattices Microstruct. **41**(5–6), 313–320 (2007)
109. D.D. Solnyshkov, H. Flayac, G. Malpuech, Black holes and wormholes in spinor polariton condensates. Phys. Rev. B **84**(23), 233405 (2011)
110. V.D. Kulakovskii, A.S. Brichkin, S.V. Novikov, C. Schneider, S. Höfling, M. Kamp, A. Forchel, N.A. Gippius, Magnetic field control of polarized polariton condensates in rectangular microcavity pillars. Phys. Rev. B **85**(15), 155322 (2012)
111. Y.G. Rubo, A.V. Kavokin, I.A. Shelykh, Suppression of superfluidity of exciton-polaritons by magnetic field. Phys. Lett. A **358**(3), 227–230 (2006)
112. P. Walker, T.C.H. Liew, D. Sarkar, M. Durska, A.P.D. Love, M.S. Skolnick, J.S. Roberts, I.A. Shelykh, A.V. Kavokin, D.N. Krizhanovskii, Suppression of Zeeman splitting of the energy levels of exciton-polariton condensates in semiconductor microcavities in an external magnetic field. Phys. Rev. Lett. **106**(25), 257401– (2011)
113. C.W. Lai, N.Y. Kim, S. Utsunomiya, G. Roumpos, H. Deng, M.D. Fraser, T. Byrnes, P. Recher, N. Kumada, T. Fujisawa, Y. Yamamoto, Coherent zero-state and π-state in an exciton-polariton condensate array. Nature **450**(7169), 529–532 (2007)
114. N. Young Kim, K. Kusudo, C. Wu, N. Masumoto, A. Löffler, S. Höfling, N. Kumada, L. Worschech, A. Forchel, Y. Yamamoto, Dynamical d-wave condensation of exciton-polaritons in a two-dimensional square-lattice potential. Nat. Phys. **7**(9), 681–686 (2011)
115. E.A. Cerda-Méndez, D.N. Krizhanovskii, K. Biermann, R. Hey, M.S. Skolnick, P.V. Santos, Wavefunction of polariton condensates in a tunable acoustic lattice. New J. Phys. **14**(7), 075011 (2012)
116. E.A. Cerda-Méndez, D. Sarkar, D.N. Krizhanovskii, S.S. Gavrilov, K. Biermann, M.S. Skolnick, P.V. Santos, Exciton-polariton gap solitons in two-dimensional lattices. Phys. Rev. Lett. **111**, 146401 (2013)

117. M. Abbarchi, A. Amo, V.G. Sala, D.D. Solnyshkov, H. Flayac, L. Ferrier, I. Sagnes, E. Galopin, A. Lemaître, G. Malpuech, J. Bloch, Macroscopic quantum self-trapping and Josephson oscillations of exciton polaritons. Nat. Phys. **9**, 275 (2013)
118. K. Winkler, J. Fischer, A. Schade, M. Amthor, R. Dall, J. Geßler, M. Emmerling, E.A. Ostrovskaya, M. Kamp, C. Schneider, S. Höfling, A polariton condensate in a photonic crystal potential landscape. New J. Phys. **17**, 023001 (2015)
119. F. Baboux, L. Ge, T. Jacqmin, M. Biondi, E. Galopin, A. Lemaître, L. Le Gratiet, I. Sagnes, S. Schmidt, H.E. Türeci, A. Amo, J. Bloch, Bosonic condensation and disorder-induced localization in a flat band. Phys. Rev. Lett. **116**(6), 066402 (2016)
120. S. Klembt, T.H. Harder, O.A. Egorov, K. Winkler, R. Ge, M.A. Bandres, M. Emmerling, L. Worschech, T.C.H. Liew, M. Segev, C. Schneider, S. Höfling, Exciton-polariton topological insulator. Nature **562**(7728), 552–556 (2018)
121. M. Galbiati, L. Ferrier, D.D. Solnyshkov, D. Tanese, E. Wertz, A. Amo, M. Abbarchi, P. Senellart, I. Sagnes, A. Lemaître, E. Galopin, G. Malpuech, J. Bloch, Polariton condensation in photonic molecules. Phys. Rev. Lett. **108**, 126403 (2012)
122. A. Amo, J. Lefrere, S. Pigeon, C. Adrados, C. Ciuti, I. Carusotto, R. Houdré, E. Giacobino, A. Bramati, Superfluidity of polaritons in semiconductor microcavities. Nat. Phys. **5**(11), 805–810 (2009)
123. D. Sanvitto, F.M. Marchetti, M.H. Szymanska, G. Tosi, M. Baudisch, F.P. Laussy, D.N. Krizhanovskii, M.S. Skolnick, L. Marrucci, A. Lemaître, J. Bloch, C. Tejedor, L. Viña, Persistent currents and quantized vortices in a polariton superfluid. Nat. Phys. **6**(7), 527–533 (2010)
124. D. Sanvitto, S. Pigeon, A. Amo, D. Ballarini, M. De Giorgi, I. Carusotto, R. Hivet, F. Pisanello, V.G. Sala, P.S.S. Guimaraes, R. Houdré, E. Giacobino, C. Ciuti, A. Bramati, G. Gigli, All-optical control of the quantum flow of a polariton condensate. Nat. Photon. **5**, 610–614 (2011)
125. A. Amo, S. Pigeon, D. Sanvitto, V.G. Sala, R. Hivet, I. Carusotto, F. Pisanello, G. Leménager, R. Houdré, E. Giacobino, C. Ciuti, A. Bramati, Polariton superfluids reveal quantum hydrodynamic solitons. Science **332**(6034), 1167–1170 (2011)
126. C. Antón, G. Tosi, M.D. Martín, L. Viña, A. Lemaître, J. Bloch, Role of supercurrents on vortices formation in polariton condensates. Opt. Express **20**(15), 16366–16373 (2012)
127. L. Dominici, G. Dagvadorj, J.M. Fellows, D. Ballarini, M. De Giorgi, F.M. Marchetti, B. Piccirillo, L. Marrucci, A. Bramati, G. Gigli, M.H. Szymañska, D. Sanvitto, Vortex and half-vortex dynamics in a nonlinear spinor quantum fluid. Sci. Adv. **1**(11), e1500807 (2015)
128. K.G. Lagoudakis, M. Wouters, M. Richard, A. Baas, I. Carusotto, R. Andre, L. Si Dang, B. Deveaud-Plédran, Quantized vortices in an exciton-polariton condensate. Nat. Phys. **4**(9), 706–710 (2008)
129. G. Nardin, G. Grosso, Y. Leger, B. Pietka, F. Morier-Genoud, B. Deveaud-Plédran, Hydrodynamic nucleation of quantized vortex pairs in a polariton quantum fluid. Nat. Phys. **7**(8), 635–641 (2011)
130. M.D. Fraser, G. Roumpos, Y. Yamamoto, Vortex-antivortex pair dynamics in an exciton-polariton condensate. New J. Phys. **11**, 113048 (2009)
131. G. Roumpos, M.D. Fraser, A. Löffler, S. Höfling, A. Forchel, Y. Yamamoto, Single vortex-antivortex pair in an exciton-polariton condensate. Nat. Phys. **7**(2), 129–133 (2011)
132. G. Tosi, F.M. Marchetti, D. Sanvitto, C. Antón, M.H. Szymańska, A. Berceanu, C. Tejedor, L. Marrucci, A. Lemaître, J. Bloch, L. Viña, Onset and dynamics of vortex-antivortex pairs in polariton optical parametric oscillator superfluids. Phys. Rev. Lett. **107**(3), 036401– (2011)
133. L. Dominici, M. Petrov, M. Matuszewski, D. Ballarini, M. De Giorgi, D. Colas, E. Cancellieri, B. Silva Fernández, A. Bramati, G. Gigli, A. Kavokin, F. Laussy, D. Sanvitto Real-space collapse of a polariton condensate. Nat. Commun. **6**, 8993 (2015)
134. M. Sich, D.N. Krizhanovskii, M.S. Skolnick, A.V. Gorbach, R. Hartley, D.V. Skryabin, E.A. Cerda-Méndez, K. Biermann, R. Hey, P.V. Santos, Observation of bright polariton solitons in a semiconductor microcavity. Nat. Photon. **6**, 50 (2011)
135. G. Lerario, A. Fieramosca, F. Barachati, D. Ballarini, K.S. Daskalakis, L. Dominici, M. De Giorgi, S.A. Maier, G. Gigli, S. Kéna-Cohen, D. Sanvitto, Room-temperature superfluidity in a polariton condensate. Nat. Phys. **13**, 837 (2017)

136. O. El Daïf, A. Baas, T. Guillet, J.-P. Brantut, R. Idrissi Kaitouni, J.L. Staehli, F. Morier-Genoud, B. Deveaud, Polariton quantum boxes in semiconductor microcavities. Appl. Phys. Lett. **88**(6), 061105–3 (2006)
137. R. Idrissi Kaitouni, O. El Daïf, A. Baas, M. Richard, T. Paraiso, P. Lugan, T. Guillet, F. Morier-Genoud, J.D. Ganiére, J.L. Staehli, V. Savona, B. Deveaud, Engineering the spatial confinement of exciton polaritons in semiconductors. Phys. Rev. B **74**(15), 155311– (2006)
138. E. Wertz, L. Ferrier, D.D. Solnyshkov, R. Johne, D. Sanvitto, A. Lemaître, I. Sagnes, R. Grousson, A.V. Kavokin, P. Senellart, G. Malpuech, J. Bloch, Spontaneous formation and optical manipulation of extended polariton condensates. Nat. Phys. **6**(11), 860–864 (2010)
139. G. Tosi, G. Christmann, N.G. Berloff, P. Tsotsis, T. Gao, Z. Hatzopoulos, P.G. Savvidis, J.J. Baumberg, Sculpting oscillators with light within a nonlinear quantum fluid. Nat. Phys. (2012). Advance online publication https://doi.org/10.1038/nphys2182
140. E. Wertz, A. Amo, D.D. Solnyshkov, L. Ferrier, T.C.H. Liew, D. Sanvitto, P. Senellart, I. Sagnes, A. Lemaître, A.V. Kavokin, G. Malpuech, J. Bloch, Propagation and amplification dynamics of 1d polariton condensates. Phys. Rev. Lett. **109**, 216404 (2012)
141. J. Fischer, I.G. Savenko, M.D. Fraser, S. Holzinger, S. Brodbeck, M. Kamp, I.A. Shelykh, C. Schneider, S. Höfling, Spatial coherence properties of one dimensional exciton-polariton condensates. Phys. Rev. Lett. **113**(20), 203902 (2014)
142. E.A. Cerda-Méndez, D. Krizhanovskii, K. Biermann, K. Guda, R. Bradley, R. Hey, P.V. Santos, M.S. Skolnick, One dimensional confinement of microcavity polaritons using non-piezoelectric surface acoustic waves. Phys. E: Low-dimensional Syst. Nanostruct. **42**(10), 2548–2551 (2010)
143. G. Roumpos, W.H. Nitsche, S. Höfling, A. Forchel, Y. Yamamoto, Gain-induced trapping of microcavity exciton polariton condensates. Phys. Rev. Lett. **104**(12), 126403 (2010)
144. A. Amo, J. Bloch, Exciton-polaritons in lattices: a non-linear photonic simulator. Comptes Rendus Physique **17**(8), 934–945 (2016)
145. T.C.H. Liew, A.V. Kavokin, I.A. Shelykh, Optical circuits based on polariton neurons in semiconductor microcavities. Phys. Rev. Lett. **101**(1), 016402 (2008)
146. T. Espinosa-Ortega, T.C.H. Liew, Complete architecture of integrated photonic circuits based on and and not logic gates of exciton polaritons in semiconductor microcavities. Phys. Rev. B **87**(19) (2013)
147. M. Aßmann, F. Veit, M. Bayer, A. Löffler, S. Höfling, M. Kamp, A. Forchel, All-optical control of quantized momenta on a polariton staircase. Phys. Rev. B **85**(15), 155320 (2012)
148. A. Amo, T.C.H. Liew, C. Adrados, R. Houdré, E. Giacobino, A.V. Kavokin, A. Bramati, Exciton-polariton spin switches. Nat. Photon. **4**, 361 (2010)
149. I.A. Shelykh, R. Johne, D.D. Solnyshkov, G. Malpuech. Optically and electrically controlled polariton spin transistor. Phys. Rev. B **82**(15) (2010)
150. T. Gao, P.S. Eldridge, T.C.H. Liew, S.I. Tsintzos, G. Stavrinidis, G. Deligeorgis, Z. Hatzopoulos, P.G. Savvidis, Polariton condensate transistor switch. Phys. Rev. B **85**, 235102 (2012)
151. M. De Giorgi, D. Ballarini, E. Cancellieri, F.M. Marchetti, M.H. Szymanska, C. Tejedor, R. Cingolani, E. Giacobino, A. Bramati, G. Gigli, D. Sanvitto, Control and ultrafast dynamics of a two-fluid polariton switch. Phys. Rev. Lett. **109**(26) (2012)
152. L. Dominici, D. Colas, S. Donati, J.P. Restrepo Cuartas, M. De Giorgi, D. Ballarini, G. Guirales, J.C. López Carreño, A. Bramati, G. Gigli, E. Del Valle, F.P. Laussy, D. Sanvitto, Ultrafast control and Rabi oscillations of polaritons. Phys. Rev. Lett. **113**(22), 226401 (2014)
153. D. Ballarini, M.D. Giorgi, E. Cancellieri, R. Houdré, E. Giacobino, R. Cingolani, A. Bramati, G. Gigli, D. Sanvitto, All-optical polariton transistor. Nat. Commun. **4**, 1778 (2013)
154. F. Marsault, H.S. Nguyen, D. Tanese, A. Lemaître, E. Galopin, I. Sagnes, A. Amo, J. Bloch, Single-wavelength, all-optical switching based on exciton-polaritons. Appl. Phys. Lett. **107**, 5263 (2015)
155. H.S. Nguyen, D. Vishnevsky, C. Sturm, D. Tanese, D. Solnyshkov, E. Galopin, A. Lemaitre, I. Sagnes, A. Amo, G. Malpuech, J. Bloch, Realization of a double-barrier resonant tunneling diode for cavity polaritons. Phys. Rev. Lett. **110**, 236601 (2013)

156. C. Sturm, D. Tanese, H. Nguyen, H. Flayac, E. Galopin, A. Lemaitre, A. Amo, G. Malpuech, J. Bloch, All-optical phase modulation in a cavity-polariton mach-zehnder interferometer. Nat. Commun. **5**, 3278 (2014)
157. X. Liu, T. Galfsky, Z. Sun, F. Xia, E.C. Lin, Y.H. Lee, S. Kéna-Cohen, V.M. Menon, Strong light–matter coupling in two-dimensional atomic crystals. Nat. Photon. **9**(1), 30–34 (2014)
158. F.H.L. Koppens, D.E. Chang, F. Javier García De Abajo, Graphene plasmonics: a platform for strong light-matter interactions. Nano Lett. **11**(8), 3370–3377 (2011)
159. A.A. Maksimov, I.I. Tartakovskii, E.V. Filatov, S.V. Lobanov, N.A. Gippius, S.G. Tikhodeev, C. Schneider, M. Kamp, S. Maier, S. Höfling, V.D. Kulakovskii, Circularly polarized light emission from chiral spatially-structured planar semiconductor microcavities. Phys. Rev. B **89**(4), 045316 (2014)
160. S.V. Lobanov, T. Weiss, N.A. Gippius, S.G. Tikhodeev, V.D. Kulakovskii, K. Konishi, M. Kuwata-Gonokami, Polarization control of quantum dot emission by chiral photonic crystal slabs. Opt. Lett. **40**(7), 1528 (2015)
161. A.A. Demenev, V.D. Kulakovskii, C. Schneider, S. Brodbeck, M. Kamp, S. Höfling, S.V. Lobanov, T. Weiss, N.A. Gippius, S.G. Tikhodeev, Circularly polarized lasing in chiral modulated semiconductor microcavity with GaAs quantum wells. Appl. Phys. Lett. **109**(17), 171106 (2016)
162. M. Hertzog, M. Wang, J. Mony, K. Börjesson, Strong light-matter interactions: a new direction within chemistry. Chem. Soc. Rev. **48**(3), 937–961 (2019)
163. S.R.K. Rodriguez, Classical and quantum distinctions between weak and strong coupling. Eur. J. Phys. **37**(2), 25802 (2016)
164. R. El-Ganainy, K.G. Makris, M. Khajavikhan, Z.H. Musslimani, S. Rotter, D.N. Christodoulides, Non-Hermitian physics and PT symmetry. Nat. Phys. **14**(1), 11–19 (2018)
165. W.D. Heiss, The physics of exceptional points. J. Phys. A: Math. Theor. **45**(44), 444016 (2012)
166. D.N. Basov, M.M. Fogler, F.J. García De Abajo, Polaritons in van der waals materials. Science **354**(6309) (2016)
167. K. Fai Mak, D. Xiao, J. Shan, Light–valley interactions in 2D semiconductors. Nat. Photon. **12**(8), 451–460 (2018)
168. C.H. Henry, J.J. Hopfield, Raman scattering by polaritons. Phys. Rev. Lett. **15**(25), 964–966 (1965)
169. C.F. Klingshirn, *Semiconductor Optics* (Springer, Berlin, 2012)
170. K. Huang, Lattice vibrations and optical waves in ionic crystals. Nature **167**(4254), 779–780 (1951)
171. D.W. Ward, J.D. Beers, T. Feurer, E.R. Statz, N.S. Stoyanov, K.A. Nelson, Coherent control of phonon-polaritons in a terahertz resonator fabricated with femtosecond laser machining. Opt. Lett. **29**(22), 2671–2673 (2004)
172. S. Dai, Q. Ma, T. Andersen, A.S. Mcleod, Z. Fei, M.K. Liu, M. Wagner, K. Watanabe, T. Taniguchi, M. Thiemens, F. Keilmann, P. Jarillo-Herrero, M.M. Fogler, D.N. Basov, Subdiffractional focusing and guiding of polaritonic rays in a natural hyperbolic material. Nat. Commun. **6**(1), 6963 (2015)
173. J.D. Caldwell, L. Lucas, G. Vincenzo, V. Igor, T.L. Reinecke, S.A. Maier, O.J. Glembocki, Low-loss, infrared and terahertz nanophotonics using surface phonon polaritons. Nanophotonics **4**(1), 44 (2015)
174. J-S. Wu, D.N. Basov, M.M. Fogler, Topological insulators are tunable waveguides for hyperbolic polaritons. Phys. Rev. B **92**(20), 205430 (2015)
175. M. Sidler, P. Back, O. Cotlet, A. Srivastava, T. Fink, M. Kroner, E. Demler, A. Imamoglu, Fermi polaron-polaritons in charge-tunable atomically thin semiconductors. Nat. Phys. **13**, 255 (2016)
176. W. Ning, J. Feist, F.J. Garcia-Vidal, When polarons meet polaritons: exciton-vibration interactions in organic molecules strongly coupled to confined light fields. Phys. Rev. B **94**(19), 195409 (2016)

177. M.I.S. Kawata, M. Ohtsu (eds.), *Near-Field Optics and Surface Plasmon Polaritons* (Springer, Berlin, 2001)
178. V.W. Brar, M.S. Jang, M. Sherrott, S. Kim, J.J. Lopez, L.B. Kim, M. Choi, H. Atwater, Hybrid surface-phonon-plasmon polariton modes in graphene/monolayer h-BN heterostructures. Nano Lett. **14**(7), 3876–3880 (2014)
179. V.A. Kosobukin, Plasmon-excitonic polaritons in superlattices. Phys. Solid State **59**(5), 999–1007 (2017)
180. A. Benz, S. Campione, S. Liu, I. Montaño, J.F. Klem, A. Allerman, J.R. Wendt, M.B. Sinclair, F. Capolino, I. Brener, Strong coupling in the sub-wavelength limit using metamaterial nanocavities. Nat. Commun. **4**(1), 2882 (2013)
181. A. Frisk Kockum, A. Miranowicz, S. De Liberato, S. Savasta, F. Nori, Ultrastrong coupling between light and matter. Nat. Rev. Phys. **1**, 19–40 (2019)
182. P. Forn-Diaz, L. Lamata, E. Rico, J. Kono, and E. Solano, Ultrastrong coupling regimes of light-matter interaction. Rev. Mod. Phys. **91**, 025005 (2019)
183. M. Ramezani, A. Halpin, A.I. Fernández-Domínguez, J. Feist, S.R.K. Rodriguez, F.J. Garcia-Vidal, J. Gómez Rivas. Plasmon-exciton-polariton lasing. Optica **4**(1), 31–37 (2017)
184. J.P. Reithmaier, G. Sek, A. Löffler, C. Hofmann, S. Kuhn, S. Reitzenstein, L.V. Keldysh, V.D. Kulakovskii, T.L. Reinecke, A. Forchel, Strong coupling in a single quantum dot-semiconductor microcavity system. Nature **432**(7014), 197–200 (2004)
185. T. Yoshie, A. Scherer, J. Hendrickson, G. Khitrova, H.M. Gibbs, G. Rupper, C. Ell, O.B. Shchekin, D.G. Deppe, Vacuum Rabi splitting with a single quantum dot in a photonic crystal nanocavity. Nature **432**(7014), 200–203 (2004)
186. E. Peter, P. Senellart, D. Martrou, A. Lemaître, J. Hours, J.M. Gérard, J. Bloch, Exciton-photon strong-coupling regime for a single quantum dot embedded in a microcavity. Phys. Rev. Lett. **95**(6), 067401– (2005)
187. D. Press, S. Götzinger, S. Reitzenstein, C. Hofmann, A. Löffler, M. Kamp, A. Forchel, Y. Yamamoto, Photon antibunching from a single quantum-dot-microcavity system in the strong coupling regime. Phys. Rev. Lett. **98**(11), 117402 (2007)
188. J. McKeever, A. Boca, A.D. Boozer, J.R. Buck, H.J. Kimble, Experimental realization of a one-atom laser in the regime of strong coupling. Nature **425**(6955), 268–271 (2003)
189. S. Reitzenstein, S. Münch, P. Franeck, A. Löffler, S. Höfling, L. Worschech, A. Forchel, I.V. Ponomarev, T.L. Reinecke, Exciton spin state mediated photon-photon coupling in a strongly coupled quantum dot microcavity system. Phys. Rev. B **82**(12), 121306– (2010)
190. J. Kasprzak, S. Reitzenstein, E.A. Muljarov, C. Kistner, C. Schneider, M. Strauss, S. Höfling, A. Forchel, W. Langbein, Up on the Jaynes-Cummings ladder of a quantum-dot/microcavity system. Nat. Mater. **9**, 304 (2010)
191. K.B. Tolpygo, Physical properties of a rock salt lattice made up of deformable ions. Zhurnal Eksperimentalnoi I Teoreticheskoi Fiziki (J. Exp. Theor. Phys.) **20**(6), 497–509 (1950). In Russian
192. T. Feurer, N.S. Stoyanov, D.W. Ward, J.C. Vaughan, E.R. Statz, K.A. Nelson. Terahertz polaritonics. Annu. Rev. Mater. Res. **37**(1), 317–350 (2019)
193. A.V. Zayats, I.I. Smolyaninov, A.A. Maradudin, Nano-optics of surface plasmon polaritons. Phys. Rep. **408**(3), 131–314 (2005)
194. W.L. Barnes, Surface plasmon-polariton length scales: a route to sub-wavelength optics. J. Opt. A: Pure Appl. Opt. **8**(4), S87–S93 (2006)
195. P. Berini, I. De Leon, Surface plasmon-polariton amplifiers and lasers. Nat. Photon. **6**(1), 16–24 (2012)
196. F. Capolino, *Applications of Metamaterials* (CRC Press, Boca Raton, 2009)
197. T. Karzig, C.E. Bardyn, N.H. Lindner, G. Refael, Topological polaritons. Phys. Rev. X **5**(3), 031001 (2015)
198. C.E. Bardyn, T. Karzig, G. Refael, T.C.H. Liew, Topological polaritons and excitons in garden-variety systems. Phys. Rev. B **91**(16), 161413(R) (2015)
199. M.A. Bandres, S. Wittek, G. Harari, M. Parto, J. Ren, M. Segev, D.N. Christodoulides, M. Khajavikhan, Topological insulator laser: experiments. Science **359**(6381) (2018)

200. G. Harari, M.A. Bandres, Y. Lumer, M.C. Rechtsman, Y.D. Chong, M. Khajavikhan, D.N. Christodoulides, M. Segev. Topological insulator laser: Theory. Science **359**, 6381 (2018)
201. M. Hafezi, S. Mittal, J. Fan, A. Migdall, J.M. Taylor, Imaging topological edge states in silicon photonics. Nat. Photon. **7**, 1001 (2013)

Chapter 2
Fundamentals of Polariton Physics

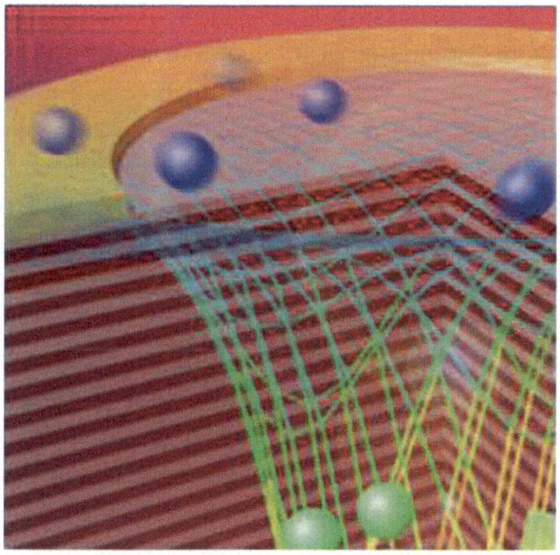

Abstract Light–matter interaction is a prerequisite for the observation of polaritons and for the utilization of excitonic particles in optical devices. It is particularly essential to deal with the recipe for strong coupling of excitons and photons in this early chapter to provide the necessary theoretical background for the dealings with the following content. Firstly, the principles of light–matter interaction will be introduced and the building blocks of polariton formation presented. Thereafter, the composite boson given by the quasi-particle polariton will be discussed and its general properties will be summarized. Here, the important role of the cavity–exciton detuning in optical microcavities with quantum-well excitons is sufficiently discussed in order to establish the necessary understanding of its impact on the quasi-particles' behaviour for later considerations of polariton condensates.

A. Rahimi-Iman, *Polariton Physics*, Springer Series in Optical Sciences 229,
https://doi.org/10.1007/978-3-030-39333-5_2

2.1 The Origin of Polaritons

2.1.1 Fundamental Light–Matter Interaction

Generally, when light is propagating through a medium, the situation is different from that in vacuum. In vacuum, photons are the particles of the propagating and oscillating electromagnetic field with angular frequency ω (2π times the frequency ν), the energy $E = \hbar\omega = h\nu$ (with energy corresponding to the wavelength in vacuum via $E = hc/\lambda$) and momentum $\mathbf{k} = (2\pi/\lambda)\hat{\mathbf{k}}$ (pointing in the direction defined by the unit vector $\hat{\mathbf{k}}$) with the magnitude $2\pi/\lambda$. The dispersion relation, which is the relation that links ω and $\mathbf{k}$, is given for photons in vacuum by $\omega(\mathbf{k}) = c\,|\mathbf{k}|$ or in terms of energy $E = \hbar c\,|\mathbf{k}| = \hbar\omega(\mathbf{k})$, whereas c is the speed of light in vacuum and defined as a natural constant. This dispersion is a straight line with slope $\hbar c$. Clearly, the vacuum light field appears unexcited at its dispersion minimum $\omega(0) = 0$ at $\mathbf{k} = 0$.

In contrast to propagation in vacuum, if light enters a polarizable medium, its propagation will be affected by the interaction with matter. Two scenarios can be then described. Firstly, the weak coupling case, in which the electromagnetic field and the excitation of the matter can be considered independent from each other. These quantities preserve their initial properties in the presence of each other, whereas the photon can cause an excitation of matter when resonant with an electronic oscillator in the medium and get "consumed". When it is far off from any resonances of a medium, light propagates similarly as in vacuum but with a modified velocity. In such case, $\mathbf{k} = \tilde{n}(\omega)\mathbf{k}_{\mathrm{vac}}$ resembles a "light-like" dispersion relation, with $\tilde{n}(\omega)$ the complex refractive index of the medium. This appears to be a useful description in many situations e.g. to describe optical excitation and refraction or propagation effects far away from resonances, but as the literature teaches, interaction with matter is more rich.

In fact, excitation is directly related to polarization. Since, in dipole approximation, light fields couple to optically allowed transitions through the dipole operator of matter resonances, they induce an oscillating polarization which in turn emits electromagnetic radiation that superimposes with the initial incident wave. This leads to the formation of polaritons in the strong light–matter coupling regime for resonances with pronounced oscillator strength. A result of such strong interactions is the hybridization of states and the obvious signature of coupled oscillators is an anti-crossing behaviour of the involved resonances (in the general classical picture that of two coupled harmonic oscillators). This general concept, which applies for instance to phonons, plasmons, and excitons interacting with light fields, is well described with numerous examples and illustrations in the literature (see for instance [1]). It is understood that the mixed state of polarization and light wave is quantized and exchanges energy with the environment as a quasi-particle by integer multiples of the corresponding polariton energies $\hbar\omega$. This gives the quanta of the polarization coupled to the photon its name *polariton*, quasi-particles of light in interacting matter. Polaritons in bulk systems are stationary states that only transform into photons at surfaces.

For bulk systems, the implicit representation of the dispersion relation $\omega(\mathbf{k})$ for polaritons can be found via

$$\frac{c^2k^2}{\omega^2} = \varepsilon(\omega), \tag{2.1}$$

whereas the relation $\mathbf{k}^2 = k^2 = \tilde{n}^2(\omega)k^2_{\text{vacuum}} = \varepsilon(\omega)(2\pi/\lambda)^2 = \varepsilon(\omega)(\omega/c)^2$ is used to obtain the so-called polariton equation [1]. Using the typical behaviour of the dielectric function $\varepsilon(\omega)$ in the vicinity of a single resonance according to the Lorentz model gives

$$\frac{c^2k^2}{\omega^2} = \varepsilon_\text{b} + \frac{f}{\omega_0^2 - \omega^2 + \text{i}\omega\gamma}, \tag{2.2}$$

whereas ω_0 is the frequency at the singularity of $\varepsilon(\omega)$, f the resonance's oscillator strength and γ the dephasing rate.

For zero damping, i.e. the ideal scenario of loss-free propagation, this dispersion relation results in two branches, the lower of which, coming from the lower energy side, behaves light-like before converging the energy dispersion of the matter state close to the resonance, thereby increasing its quasi-particle momentum (e.g. becoming exciton-like in an exciton–polariton system). Above resonance, the upper branch appears at low momentum within the light cone and becomes again light-like for higher energies. Characteristic for this strong-coupling situation is the anti-crossing behaviour of the two virtually crossing modes for no interaction, and a stop-band (Reststrahlenbande) evidenced in its frequency-dependent reflectivity between the transverse and longitudinal resonances ω_0 and ω_L, respectively. For finite damping due to absorption in the medium, the dispersion relation gets modified and a back-bending of the lower branch into the upper branch takes place, which leads to a reduction in reflectivity. These scenarios are indeed relevant for the characterization of semiconductor materials, but the occurrence of strong light–matter coupling is not confined to solids or crystals, although initially discussed in this light. An example graph of polariton modes in a three-dimensional system is shown in Fig. 2.1.

Here, the term anti-crossing is introduced. While the behaviour of a coupled-oscillator system can be derived quantum mechanically, the result is not different in the classical picture of two coupled identical harmonic oscillators experiencing a normal mode splitting of the eigen-frequencies, which will be explained in a more practical context later when discussing cavity–polaritons. Whenever two resonances in the strong-coupling regime tune into each other via a certain parameter, the expected crossing behaviour for negligible coupling turns into an anti-crossing one. This is schematically shown in Fig. 2.1. In the anti-crossing case, the repulsion of the resonances (mode splitting) scales with the coupling strength (later introduced as g and linearly proportional to the so-called Rabi frequency), which is proportional to the oscillator strength f of a resonance.

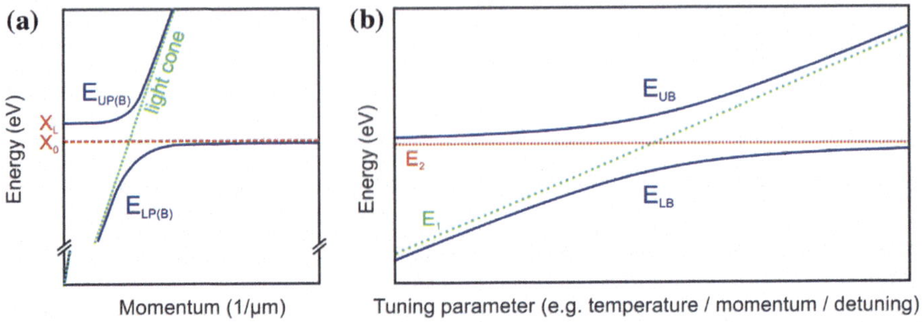

Fig. 2.1 Schematic representation of the bulk exciton–polariton dispersion (**a**) and the anti-crossing behaviour for two coupled oscillators tuned into resonance (**b**). In **a**, the resonances are the transverse (0 or T, corresponding to the singularity in the dielectric function for zero dephasing) and longitudinal (L) excitons (X), the vacuum light cone and the resulting lower/upper polariton branches (LP(B)/UP(B)), for $\varepsilon_B = 1$ (dielectric background constant). The chart in **b** displays the crossing of two resonances as a function of a parameter change. For a weakly coupled system, no deviation from the actual mode energies occurs. In contrast, coupled oscillators exhibit new eigen-states that are evidenced as a normal-mode splitting, the magnitude of which is proportional to the coupling strengths

2.1.2 *Cavity–Polaritons*

Exciton–polaritons, here simply referred to as polaritons in the following, are the result of strong light–matter interaction in semiconductor quantum-well microcavities, in which the strong coupling between the excitonic and the photonic modes gives rise to a new composite quasi-particle. Thus, they are also known as cavity–polaritons. Such quasi-particles, or in other words polaritonic modes, feature a hybrid nature and, thus, consist partly of light and partly of matter. In contrast to strong coupling, no new quasi-particles are formed in the weak-coupling regime of light–matter interaction; however, the emission rate of quantum emitters can be controlled in this regime as a consequence of the Purcell effect, which is very beneficial for optical devices such as low-threshold lasers and single photon sources. The principles of light–matter coupling will be briefly discussed in the following as one of the building blocks of polariton formation.

To achieve sustainable and strong-enough coupling between light and matter, both light, and matter have to be prepared in a well defined fashion, and brought together, of course. If spatial and spectral resonance of the required building blocks is guaranteed, and losses in the system are minimized, one can facilitate and maintain coherent energy exchange, which is the prerequisite for polariton formation. This process of periodic energy exchange is referred to as Rabi oscillations between exciton and photon states. The strength of light–matter coupling is reflected by the frequency, at which coherent energy exchange takes place, thus named Rabi frequency, in reference to the original studies on Rabi oscillations [2]. Particularly, since dissipation of matter excitation and photons out of the system cannot be avoided and the strength of coupling strongly depends on structural properties, special care has to be taken with regard to quality aspects and precision of structural design and fabrication. Owing to optical losses, which dominate the loss channels in high-quality polariton structures, Rabi oscillations can be found as damped oscillations in real systems. While this seems to be a major drawback, it has turned out to be a key advantage of polaritonic systems. Light coupled out of such strongly-coupled system after the polariton's lifetime can be measured by means of spectroscopy and bears direct information of the particle from which it was emitted, providing unique insight into excitonic processes in matter and bosonic many-body systems in solids.

2.2 Building-Blocks for Polariton Formation

This section is devoted to the building blocks that give rise to the existence of cavity–polaritons in semiconductor structures. In the first part, excitons in quantum-well (QW) structures are introduced as the necessary matter excitation for polariton formation, and the physics of QW exciton are wrapped up. In the second part, the confinement of photons in cavities is summarized before the last part covers the physics of light–matter interaction. As part of these considerations, light will be also shed on the basic properties of the involved optical and excitonic modes as well as the polaritonic features as a consequence of light–matter hybridization.

2.2.1 Excitons in Quantum Wells

Excitons in Semiconductors

In solids, Coulomb interaction between electrons and defect electrons, also known as holes, gives rise to a bound electron–hole pair—a bosonic quasi-particle with charge neutrality which is assembled of two fermions of opposite charge. This particle is named exciton (X) and represents a fundamental matter excitation (well discussed in literature on semiconductors, e.g. [1, 3–5]), since its electron (e^-) originally resulted from excitation from the valence band (VB) into the conduction band (CB) of the semiconductor (see Fig. 2.2a). The electron's charge amounts to $e = (-)1,602 \times 10^{-19}$ C and its energy can be described by

$$E_\mathrm{e}(k) = E_\mathrm{g} + \frac{\hbar^2 k^2}{2m_\mathrm{e,eff}}, \tag{2.3}$$

whereas E_g represents the energy band gap of the lattice material, $m_\mathrm{e,eff}$ is the effective mass of the excited electron in the crystal lattice, $\hbar$ Planck's constant (quantum of action) divided by 2π, and k the wave vector of the particle in the translation lattice, respectively.

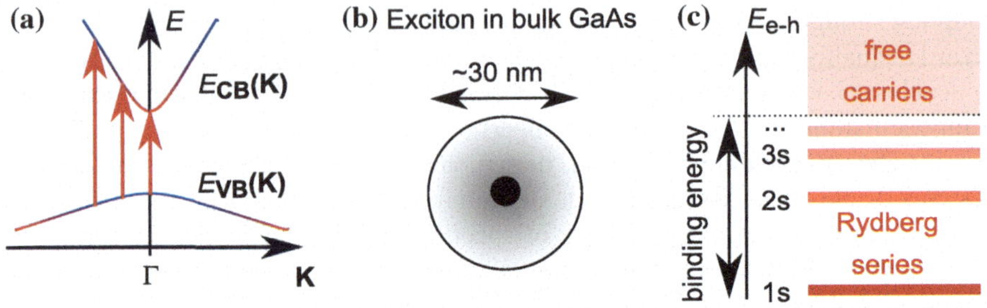

Fig. 2.2 **a** Sketch of matter excitation in a direct semiconductor as a consequence of photon-energy absorption (nearly-vertical arrows) with an electronic transition from the valence band (VB) to the conduction band (CB). Here, the bands around the Γ point for GaAs are oversimplified in the energy–momentum-space diagram. **b** Formation of bound electron–hole pairs, referred to as excitons (oversimplified sketch, with indicated size). This quasi-particle in its ground state represents the energetically-lowest matter excitation with centre-of-mass motion and corresponding effective mass. **c** Rydberg-series of energy states for the hydrogen-like electronic quasi-particle

As the electron is excited, an unoccupied state in the valence band is left behind, which remains as a hole (h^+) in the crystal lattice with properties similar to the electron, but inverse charge e^+ and slightly different effective mass $m_{\mathrm{h,eff}}$ as a defect electron in the valence band. The hole features the same momentum as the excited electron—yet opposite in sign—and the following energy

$$E_{\mathrm{h}}(k) = \frac{\hbar^2 k^2}{2m_{\mathrm{h,eff}}}. \tag{2.4}$$

If the particle momenta are low enough, i.e. $k \approx 0$, such Coulomb-bound system can form a free exciton, which is able to propagate in the host lattice of the solid and can be considered as an analogue to the hydrogen atom (or positronium atom)—provided that the valence and conduction bands are of simple form. For such hydrogenic system, a simple picture for its description is given by Bohr's atomic model, with Rydberg's formula describing it's excitation spectrum (see Fig. 2.2b and c). However, quantum mechanically, the exciton has to be described by matter waves with spatial distribution probabilities (see for instance [6–8]). Obviously, the higher dielectric screening in solids ($\varepsilon_{\mathrm{r}} \approx 10^1$) and the reduced ratio of effective masses between hole and electron ($m_{\mathrm{h,eff}}/m_{\mathrm{e,eff}} \approx 10^1$) compared to proton and electron in hydrogen lead to three orders of magnitude smaller corresponding binding energies of the exciton than the ionization energy of the hydrogen atom [9].

Using Bohr's model, the binding energy can be written as

$$E_{\mathrm{bind},i} = \frac{\hbar^2}{2m_{\mathrm{r}} a_{\mathrm{B,X}}^2 i^2}, \tag{2.5}$$

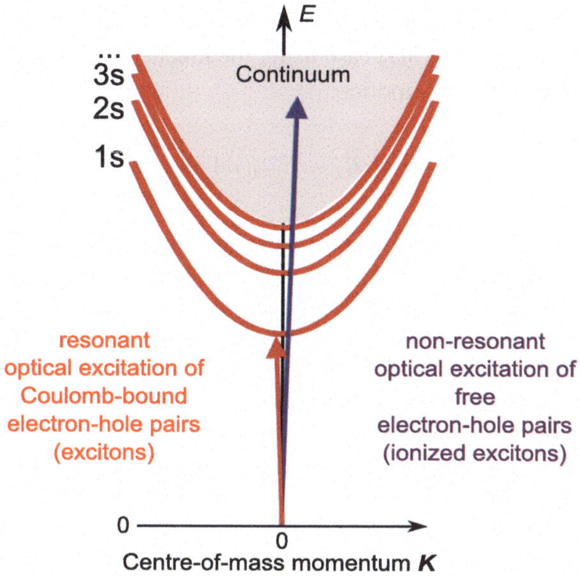

Fig. 2.3 Sketch of the excitonic bands. The dispersion relation of excitons in the two-particle picture shows the energy dependence for the ground-state excitation and the exciton's higher states as a function of the centre-of-mass momentum. The continuum band represents ionized excitons, i.e. free electron–hole pairs in the bands of the host lattice. Nevertheless, these high-energy excitations are still solutions of the exciton problem and the relaxing system will eventually return to its bound form in the absence of dissipative channels. Nearly-vertical arrows correspond to photons that can excite the system in either one of the exciton resonances (1s, 2s or higher, indicated by a short red arrow) or into the continuum states (long violet arrow). Note that the energies are not correctly scaled to display the relevant levels with respect to the far-off lying vacuum level ($E = 0$), i.e. the absolute ground state of the electronic excitation. Due to their finite lifetime, i.e. the spontaneous emission lifetime, excitons decay under emission of a photon under conservation of energy and momentum and, thus, the excitation in matter is so-to-say lost

whereas $m_\mathrm{r} = m_\mathrm{e,eff} m_\mathrm{h,eff} / m_\mathrm{X,eff}$ is the reduced effective mass with $m_\mathrm{X,eff} = m_\mathrm{e,eff} + m_\mathrm{h,eff}$ the effective total mass (relevant for the dispersion, Fig. 2.3) of the quasi-particle formed by e^- and h^+, and i the quantum number of the hydrogenic system. For the exciton in the ground state ($i = 1$), the Bohr radius is given as

$$a_\mathrm{B,X} = \frac{4\pi\hbar^2\varepsilon_\mathrm{r}\varepsilon_0}{m_\mathrm{r}e^2}. \tag{2.6}$$

As can be easily derived from a direct comparison to hydrogen, the Bohr radius of such a quasi-particle lies on the order of 10 nm (sketched for GaAs bulk and QW excitons in Fig. 2.4a), which indicates that such excitons can spread over many lattice sites in the host crystal (illustrated in Fig. 2.4c). Such weakly-bound and delocalized excitons are known as Wannier–Mott excitons [10, 11]. They are commonly found in inorganic semiconductors.

One can now provide an expression for the total energy of such an electron–hole pair in the three-dimensional (3D) case using the kinetic term $\frac{\hbar^2 \mathbf{K}^2}{2m_{\mathrm{X,eff}}}$ with the wave vector $\mathbf{K}$ of the centre-of-mass motion:

$$E_{\mathrm{X}}(\mathbf{k}, i) = E_{\mathrm{g}} - E_{\mathrm{bind},i} + \frac{\hbar^2 \mathbf{K}^2}{2m_{\mathrm{X,eff}}}. \tag{2.7}$$

This is the dispersion relation for free excitons and schematically depicted in Fig. 2.3 as parabolic bands ($i = 1, 2, 3, \ldots,$ and continuum $i = \infty$) in the two-particle picture. While the charged components of the neutral bosonic composite particle are both fermions and feature a non-integer spin, thus are subject to Pauli's principle known from quantum mechanics, the exciton has an integer spin and underlies boson statistics. This peculiarity of excitons, together with its electrical and optical properties, make it so attractive for bosonic many-particle physics. This quasi-particle has become very important for the study of matter excitation and light–matter interaction in solids [1, 7, 12, 13]. Some relevant features of exciton systems are analysed for instance in [14–16].

Quantum-Well Exciton

Quantum wells are a widely used form of semiconductor heterostructure and its realization gave birth to many important opto-electronic achievements in the past decades. They represent a finite quantum-mechanical potential well for electronic particles in solids and reduce their degree of freedom for motion in space from 3D in bulk matter to 2D in the case of quantum wells. Here excitons shall be significantly confined spatially in one direction which requires the employment of a type-I heterostructure in growth direction of the quantum-well structure [1, 4, 7, 17, 18]. Such confinement potential realized by a thin layer of semiconductor material sandwiched by barrier layers allows in-plane propagation of excitons, while the electrons and holes are both prohibited to penetrate the surrounding barrier layers, which are composed of a material with higher band gap.

The thickness of the thin potential well is chosen to match the de-Broglie wavelength of electrons and holes in the quantum-well material (see Fig. 2.4d), and typically lies on the order of the exciton Bohr radius. However, the Bohr radius is not the figure of merit for confinement, but the electron and hole wave-functions "size". It is understood that the spatial expansion and the exciton properties will adapt to and strongly depend on the thickness of the confinement potential and that the exciton will lose its bulk and isotropic character it features in 3D. In short, being squeezed into the quantum well plane, the overlap of electron and hole wave-functions increase and with it the exciton oscillator strength and binding energy. It shall be also noted here that the density of states (DOS) for the electronic system (in the effective-mass approximation) changes from a quasi-continuous square-root-like increase with energy for 3D to a step-like increase of a constant number of states per QW sub-band. These features are represented by the schematically-drawn

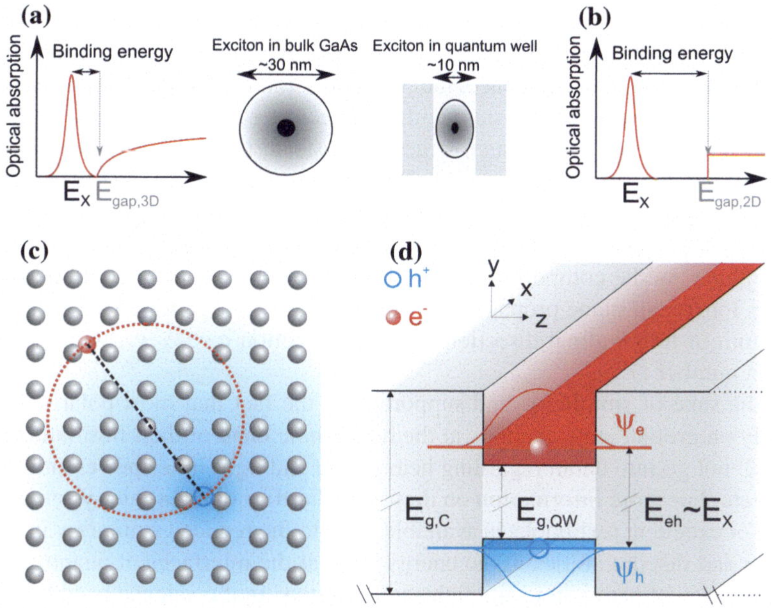

Fig. 2.4 Sketch of the optical absorption spectrum in bulk (**a**), with 3D exciton resonance, and in a quantum well (**b**), with "lens-shaped" one-dimensionally confined exciton (i.e. in a 2D system). The corresponding spherical and ellipsoid quasi-particle shapes are indicated schematically next to the spectra. Also shown are the schematic representation of a free exciton in its host crystal's lattice (**c**) and the band structure of a quantum well with quantized energy levels for electron and hole (**d**). Here, the corresponding energy band gaps $E_{g,QW}$ and $E_{g,C}$ of the quantum well and the surrounding (barrier) materials, respectively, and the recombination energy E_X of its quantum-well excitons are displayed, neglecting the energy-reducing influence of the exciton binding energy for simplicity. The indicated electron–hole pair is confined in growth direction of the quantum-well structure, but can freely propagate in the plane of confinement, i.e. the quantum well. In fact, the barrier material of quantum wells in microcavities typically corresponds to the cavity spacer material

optical-absorption spectra as well as exciton shapes for the 3D and 2D case in Fig. 2.4a and b, respectively.

Without energy transfer to the confined particles, for instance by thermal energy, the charge carriers cannot automatically overcome the potential barrier (unless via tunnelling effects). This is the essential feature of a quantum well, but also of quantum wires and quantum dots, which represent 1D or even 0D confinement potentials, respectively. Nevertheless, real systems do not provide ideal potential wells with infinite barrier height, but limited finite barriers, giving particles the chance to either diffuse into the barrier material by thermal activation or to tunnel through the barrier to other available low-lying energy states, as can be offered by defect states, band-structure valleys at different symmetry points or another tailored potential well.

In 3D crystals, i.e. bulk matter, excitons exhibit the wave vector **k** with a quasi-continuous distribution of states for the particles' momentum, respectively. In contrast, propagation of electron–hole pairs in quantum wells is restricted to the plane.

Thus, such confined particles cannot move in growth direction (here z-axis). Therefore, the lower dimensionality of the system results in a quantization of energy states in this direction. As a consequence, the 3D expression for the particles' momentum $\mathbf{k} = k_x\mathbf{e}_x + k_y\mathbf{e}_y + k_z\mathbf{e}_z$ (3D system with unit vectors $\mathbf{e}_{x/y/z}$) with quasi-continuous states no longer holds true in such structure. Instead, the expression $\mathbf{k} = \mathbf{k}_{||} + \mathbf{k}_{\perp}$ can be used to describe propagation of 2D particles, which takes into consideration an in-plane and out-of-plane component of the wave vector, respectively. The former component represents the transverse momentum $\mathbf{k}_{||}$ (which can be oriented in the x–y plane and be composed as $\mathbf{k}_{||} = k_x\mathbf{e}_x + k_y\mathbf{e}_y$) with quasi-continuous wave numbers for free in-plane propagation, while the latter one represents the quantized momentum in longitudinal direction $\mathbf{k}_{\perp}$ (in z direction e.g. $k_{z,i}\mathbf{e}_z$, with electronic quantum number i).

For the sake of simplicity, and supported by the fact that most polaritonic systems rely on ground-state excitons in the heavy-hole band of their host lattices, this book will not go into detail regarding heavy-hole and light-hole species and excited states, here. For more information on these peculiarities of excitonic systems, a vast number of excellent text books is available. Due to the reduced symmetry in growth direction and discontinuities in the energy band at the interfaces of quantum wells, the degeneracy of heavy-hole and light-hole excitons in the valence band is lifted at the Γ-symmetry point [19], which is the point of interest for cavity–polaritons in all common systems.[1] In the following, only heavy-hole excitons will be considered, owing to the fact that at low temperatures the light-hole band is typically not occupied as a consequence of different effective masses between the respective bands and those bands' energy separation due to the 2D confinement [20, 21]. Moreover, for the given structures, further considerations will only deal with the lowest exciton state with $i = 1$ and energy $E_{\mathrm{bind}} = \frac{\hbar^2}{2m_r a_{\mathrm{B,X}}^2}$ formed by electrons of the lowest conduction band together with holes of the highest valence band. Furthermore, both valence and conduction bands at the Γ point can be approximated by parabolas for small $\mathbf{k}$. Intra-band mixing, higher valence bands and e or h sub-bands can be neglected owing to the strong confinement of excitons [21].

At this point, one important parameter for the design of quantum wells should be highlighted: the thickness d of the well. This figure can have a significant impact on other parameters in quantum-well systems which will be briefly summarized. On the one hand, most noticeably, the eigen-energies of confined particles and the wave-function of charge carriers are altered as a function of d. This can be explained as follows: The thinner the well, the more the wave-function will be compressed. Accordingly, the wave-function's amplitude and energy levels in the potential well increase, and the standing wave's ends penetrate the surrounding material in compliance with the physics of finite potential wells. Herewith, the excitons energy will be increased in accordance to the energy difference $\Delta E_{\mathrm{e,h}}$ of the confined charge carriers. Is the thickness of the well to small, thermal emission of electrons and holes out of the film will be facilitated and the effective trapping will be reduced. Thus, the

[1] In 2D semiconductors such as monolayer transition-metal dichalcogenides, the relevant valleys for excitonic transitions are at the Brillouin-zone's corners, and not at the centre, i.e. the Γ point.

thickness should be carefully chosen with respect to the energy levels, so that a sufficiently low ground-state energy can be guaranteed. On the other hand, less obvious but of intriguing impact, the thickness affects the oscillator strength of excitons in quantum wells. The oscillator strength is an important figure concerning light–matter interaction and reaches significantly higher values than in bulk material due to the confinement effect [21]. With increasing confinement of charge carriers, i.e. with decreasing well thickness, the oscillator strength of the excitonic system, which will be briefly discussed later in the context of light–matter interaction, increases [22]. Thereby, the interaction of the dipole-like exciton with the light field is favoured. This increase, according to numerical calculations, is found to be proportional to $1/d$ [23]. This explains clearly why the thickness of the quantum well is a sensitive matter and that—in practice—an optimum needs to be found for the given material system.

In the case of an infinitely high potential well, the exciton Bohr radius of 2D confined excitons is halved compared to the 3D value given by (2.6), which leads to an increase of the binding energy by a factor of four compared to the 3D value given by (2.5) (cf. [24]). However, this represents an ideal case, while for real quantum well structures, this value is not reached and rather a reduced enhancement factor around 2 holds true for instance in GaAs (see [25]: $\approx$8 meV in 10-nm-thick well, 3D value $\approx$4 meV).

Nevertheless, the total energy of 2D excitons can be well approximated by

$$E_{\mathrm{X}}(\mathbf{k}, i) = \Delta E_{\mathrm{e,h}}(d) - 4E^{(3\mathrm{D})}_{\mathrm{bind},i} + \frac{\hbar^2 \mathbf{k}^2_{||}}{2m_{\mathrm{X,eff}}}, \tag{2.8}$$

whereas $\Delta E_{\mathrm{e,h}}(d)$ represents the energy difference of the involved 2D electron and hole levels, $E^{(3\mathrm{D})}_{\mathrm{bind},i}$ the 3D binding energy and $m_{\mathrm{X,eff}}$ the previously introduced effective mass of excitons. More precise values for the 2D Bohr radius $a^{(2\mathrm{D})}_{\mathrm{B}}$ and the binding energy $E^{(2\mathrm{D})}_{\mathrm{bind},i}$, respectively, can be determined when taking into account the finite quantum-well thickness [19, 25, 26].

Excitation Schemes

In solids, optical as well as electrical excitation can be applied to achieve exciton formation. In the simplest form of light–matter interaction, the absorption of a photon by the semiconductor crystal leads to the elevation of an electron from its valance-band level to an unoccupied conduction-band state, following momentum and energy conservation. If the photon energy $E_{\mathrm{photon}} \geq E_{\mathrm{X}}$ equals the optical band gap, i.e. the photon energy = the lowest lying excited state transition, one refers to this excitation scheme as resonant optical excitation, otherwise if excess energy is introduced to the system (>), non-resonant optical pumping is applied. In the latter case, non-resonantly excited electrons and holes have to dissipate excess energy by phonon emission and scattering processes before they can relax to become *hot* excitons

($k \gg 0$). With time, also hot excitons cool down to a Boltzmann distribution of occupied states.

In contrast to optical pumping, electrons and holes can also be injected into heterostructures of semiconductor material by electrical pumping. Therefore, a diode structure is employed which is operated in forward bias in order to flood the junction from both sides separately with electrons (n-doped side) and holes (p-doped side), respectively, in order to achieve exciton formation in the desired region, here quantum well. This principle will be further described in Chap. 6. For this purpose, well known principles of pn-junction and diode operation are exploited. However, electrical injection, if not provided otherwise by considerable efforts, is a non-resonant pumping scheme which injects a great amount of hot charge carriers into the structure. Under reverse bias, the junction is depleted and no current flows, but thereby the bands can be tilted, which allows one to exploit the quantum-confined Stark effect (QCSE) for research and applications (see Chap. 9).

When excitons are present in the system, they can either decay non-radiatively or emit a photon spontaneously by radiative recombination of their electron and hole on the time scale of their lifetime, setting free the total energy E_X in the form of a photon with the same energy. The mean spontaneous recombination time—the lifetime of the exciton—in quantum-well systems strongly depends on the material system as well as the structural quality and typically amounts to 0.1–1 ns [27, 28].

At high excitation densities, increased electronic screening leads to a reduction of the binding energy until a (Mott) transition from an insulator state (excitonic regime) to metallic state (free-charge-carrier plasma) occurs above the so-called Mott density. It shall be noted that excitons can only be considered as "good" bosons in the low-density regime when the fermionic character of their constituents is negligible. Moreover, in the presence of excess free carriers, negatively or positively charged excitons can be obtained (named trions due to their three-particle nature; two electrons and one hole bound together, or vice versa, respectively), whereas for elevated exciton densities (referred to as intermediate density regime) at low temperatures also biexcitons (XX, or X_2; exciton molecules similar to hydrogen or positronium molecules) are formed. Molecule formation leads to an additional reduction of the four-particle system's energy by the corresponding biexciton binding energy. After its radiative lifetime, a photon is emitted from the collapsed biexciton state and a single exciton is left behind. At elevated temperatures, they get typically ionized or dissociated via thermal excitation due to the comparably weak biexciton binding energy. Also relevant at higher densities are Auger effects and exciton–exciton annihilation that can cause an unintended non-radiative loss channel for the induced exciton population and shorten the exciton lifetime. However, these fundamental density-dependent effects in semiconductors and their quantum wells are not further discussed here after this brief summary (for further details on semiconductor excitations and optical properties see for instance [1]).

2.2.2 Confined Photons

In the following, photon confinement is introduced and the fundamentals of optical resonators for the generation, investigation and utilization of cavity–polaritons are summarized. As the expression *light–matter interaction* states it, matter excitation has to be brought together with light (or more generally, with electro-magnetic field states, i.e. optical resonances). In fact, this naturally occurs in vacuum when an emitter is exposed to the (3D) vacuum field. To achieve this kind of interaction in an effective and promising fashion, photon modes not only need to be spectrally resonant to the light-emitting and absorbing matter excitation, but also spatially resonant to it. Thus, localization of the quanta of light, photons, in cavities is indispensable. As an essential component of light–matter interaction, such strongly confined photons enable the exchange of energy between an electro-magnetic field and an emitter via dipole interaction locally. Note that strong light–matter interactions in the sense of Rabi oscillations can also occur when irradiating a two-level system with an unconfined but very-strong resonant and coherent light field.

The simplest picture of a resonator can be drawn using two highly reflective mirrors which oppose each other (see Fig. 2.5a). In-between, light will be reflected back and forth and the spatial expansion of light is strongly impeded. It is only due to finite reflectivities that photon leakage out of the cavity occurs and remains inevitable. If the resonator size is large (with length $L \gg \lambda$ the wavelength of light), the standing wave pattern at given wavelength will exhibit many nodes and anti-nodes inside the resonator, for which no particular localization of the cavity photons is achieved. This affects the mode volume of the optical field. In fact, the fundamental mode of the elongated resonator will lie at longer wavelength than for the shorter case. Moreover, the free-spectral range (FSR, here $\delta\nu$) gives rise to a comb of longitudinal modes (see Fig. 2.5b), with the spacing of modes decreasing with increasing resonator length. However, when the cavity is designed to have a resonator length on the order of the radiation wavelength, strong confinement is achieved.

In vacuum, the ground mode of the optical resonance (C, cavity mode) will be established in a lossless fashion between two perfectly reflecting mirrors of $L = \lambda_C/2$ separation, achieving the strongest localization in the direction of confinement. At this point, mode penetration into finitely reflecting real mirrors shall be neglected, which alters the cavity resonance conditions and should be taken into account when designing the actual cavity.[2]

The fundamental resonator mode (integer mode number $q = 1$) with its standing wave inside the resonator (with a single lobe/anti-node) corresponding to wavelength λ_C is characterized by the resonance frequency $\nu_{q=1,C} = c/\lambda_C$, with c the speed of light (see Fig. 2.5a). To account for the medium in which light propagates, the resonance wavelength is expressed as $\lambda_C = \lambda_0/n_C$, in dependence of the effective

[2]Such case can be well compared with the confinement of quantum-mechanical wave-functions in a quantum potential box with finite potential height of the barriers; in optics, for instance in waveguides, one refers to evanescent fields (modes) present across the highly reflective interfaces which decay exponentially with increasing distance to the interface.

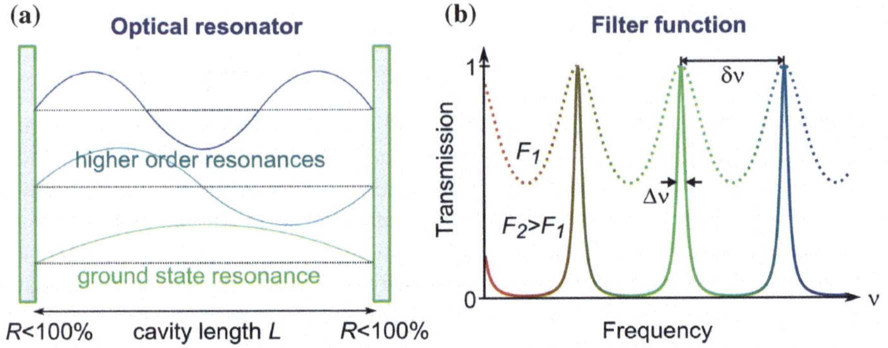

Fig. 2.5 **a** Schematic drawing of resonator modes between two plane mirrors forming a Fabry–Pérot cavity. The confined optical modes are standing light waves between the end mirrors of the resonator, with lowest resonance frequency ν_C and energies $E = qh\nu_C$, whereas q is an integer number, similar to vibrating guitar strings between two fixed ends (when ignoring polarization). **b** Sketch of the filter function (transmission spectrum) for a Fabry–Pérot interferometer/resonator with different cavity finesse F. Here, the examples show a reduction of the modes' linewidth for the case with high finesse compared to the case with low finesse. The mode spacing is determined by the cavity length and is referred to as free spectral range (FSR) $\delta\nu = c/(2n_C L)$, with n_C the refractive index of the cavity medium. The linewidth is defined as the full width at half maximum (FWHM) $\Delta\nu = 1/\tau$ (also given as decay rate times Planck's constant in units of energy, γ). From these figures, the finesse $F = \delta\nu/\Delta\nu$ can be derived. With the knowledge of the energy, also the quality factor is given, i.e. $Q = \nu/\Delta\nu$. Both, F and Q increase with increasing reflectivity of the mirrors

refractive index n_C of the cavity material, with λ_0 the wavelength in vacuum. In the following, the speed of light in medium is given by $c = c_0/n_C$, with c_0 referring to the speed of light in vacuum. For the sake of clarity, figures corresponding to the value in vacuum are indicated by a subscripted 0 in the following, whereas for energies, subscript 0 refers to the ground state, i.e. at $k_\parallel = 0$. Thus the cavity photon in its ground state exhibits the energy $E_{C,0} = hc/\lambda_C = h\nu_C = \hbar\omega_C$. To quickly translate between vacuum wavelength in nm and energy in eV, the following approximation is very useful: $E \approx 1239/\lambda_0$ ($hc_0 \approx 1239\,\mathrm{nm\,eV}$).

The most common ways to confine light are given by high-reflectivity (dielectric/Bragg) mirrors, which act as a one-dimensional photonic crystal with well defined stop band, and by total internal reflection at interfaces between two media, which feature a pronounced change of the refractive index for the propagating light. In the particle picture, an ideal optical confinement can also be described as an endless ping-pong-kind reflection of photons between the resonator mirrors. The quality of a real resonator determines the time during which photons prevail in the cavity. The lower the quality, the higher the photon leakage rate. In the wave picture, one can easily derive at which location the dipole interaction of matter with the electric field of the electro-magnetic standing wave is the strongest, namely at field maxima, i.e. at anti-node positions. It is worth noting that the effective standing-wave pattern itself can strongly depend on the cavity structure and design.

Many books and publications deal with the highly-interesting and technology-relevant topic of optical resonators, their fabrication, their properties as well as design, and their applications. The interested reader may consider for instance the books of K. Vahala [29], Y. Yamamoto et al. [18, 30], B. Deveaud-Plédran [31] and A. Kavokin et al. [24, 32, 33] for further information and for detailed insight into this matter, to name but a few. Among the most popular and widely-spread cavity types are Fabry–Pérot-type microcavities, the family of whispering-gallery-mode resonators such as microdisks, microspheres and micro-toroidal structures, and photonic-crystal microcavities (see [34]). With the transition from rather simple 2D resonators such as planar microcavities to more complex structural entities such as micropillars, buried photonic quantum boxes and photonic crystals for 0D confinement, the mode spectra can be strongly modified (also see Sect. 5.3). The peculiarities of such structures will be highlighted in the following chapters wherever necessary, but below shall be a focus on 2D structures for longitudinal confinement of light which give rise to cavity–polaritons.

Photonic Dispersion in Planar Microcavities

In a planar cavity, transverse propagation of photons can take place unrestricted, while photons are strongly confined in the longitudinal direction (z-axis $\perp$ mirror surface), the direction of growth in such microcavities. This gives rise to the dependence of the cavity photon's energy on the angle of emission, since the confinement energy $E_\mathrm{q}(k_\| = 0) = \hbar\omega_{\mathrm{q},0}$ (q is the mode number) is complemented by a transverse component of the photons momentum in the cavity, which has to be taken into account (see for instance [21]). In contrast, photons in (3D) vacuum follow a linear dispersion relation $E = \hbar\omega(k) = \hbar c|\mathbf{k}|$, which is referred to as light cone. With the expression $\mathbf{k} = \mathbf{k}_\| + \mathbf{k}_\perp$ for the separation into in-plane and out-of-plane component of the wave vector, respectively, the light cone can be easily analysed over the 2D phase-space plane with regard to 1D confinement effects. For discrete resonator eigen modes, $\mathbf{k}_\perp$ marks the energy states on the light cone which are accessible for planar-cavity photons. By taking a cross section in the direction of the $k_\|$ plane at the confinement energy $E_\mathrm{q}(\mathbf{k} = \mathbf{k}_\perp) = \hbar c|\mathbf{k}_\perp|$, the planar cavity-mode dispersion can be obtained, which is a hyperbola (see Fig. 2.6a).

The relation between the incidence angle $\theta_\mathrm{C} = \arcsin\left(\frac{\sin(\theta_0)}{n_\mathrm{C}}\right)$ (θ_0: emission angle in vacuum) on the mirrors inside the resonator and the cavity mode with transverse wave-number $k_\|$ can be given as follows:

$$k_\| = \frac{2\pi n_\mathrm{C}}{\lambda_0}\tan\left(\arcsin\left(\frac{\sin(\theta_0)}{n_\mathrm{C}}\right)\right), \tag{2.9}$$

which can be approximated for small in-plane vectors

$$k_\| \approx k_\perp\theta_0 \quad (\text{if } k_\| \ll k_\perp), \tag{2.10}$$

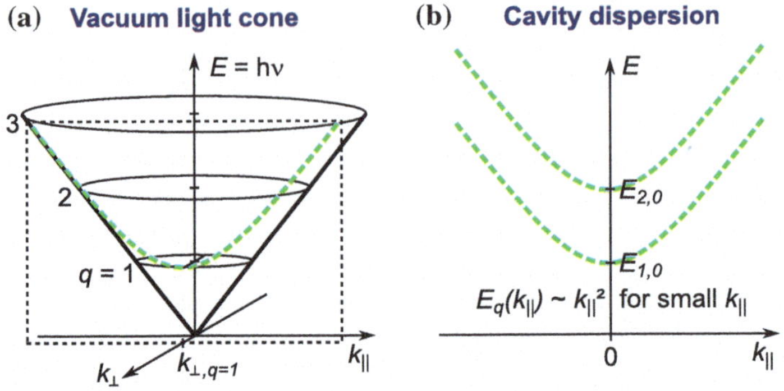

Fig. 2.6 **a** Schematic drawing of the vacuum light cone with energy plotted over a two-dimensional phase-space plane. Here, $\mathbf{k} = \mathbf{k}_{\|} + \mathbf{k}_{\perp}$. Thereby, the E–$k_{\|}$ dispersion of a confined cavity photon in a planar Fabry–Pérot resonator can be obtained from a cross section (hyperbola, green dashed) in the momentum space parallel to $k_{\|}$ at the out-of-plane vector ($k_{\perp}$) position corresponding to the cavity mode with respective mode number q. **b** Sketch of the hyperbola which represents the cavity-mode dispersion in the $k_{\|}$ plane with fixed ground-state energy $E_{q,0}$ corresponding to the confined mode's $k_{\perp}$. For small momenta, this dispersion can be approximated by a parabola. After [1]

with the longitudinal wave-number component $k_{\perp} = 2\pi/\lambda_{\mathrm{C}}$. This directly implies an energy dispersion of the cavity mode (C) with respect to the in-plane wave vector $k_{\|}$, which is described by the following equation:

$$E_{\mathrm{C}}(\mathbf{k}) = \left(\frac{hc_0}{\lambda_0(\theta_0)} = \frac{\hbar c_0}{n_{\mathrm{C}}}|\mathbf{k}| = \right)\frac{\hbar c_0}{n_{\mathrm{C}}}\sqrt{k_{\perp}^2 + k_{\|}^2}. \tag{2.11}$$

This energy dispersion of cavity photons can now be approximated for $k_{\|} \ll k_{\perp}$:

$$E_{\mathrm{C}}(\mathbf{k}) \approx \frac{\hbar c_0}{n_{\mathrm{C}}}k_{\perp}\left(1 + \frac{k_{\|}^2}{2k_{\perp}^2}\right) = E_{\mathrm{C}}(k_{\|} = 0) + \frac{\hbar^2 k_{\|}^2}{2m_{\mathrm{C,eff}}} = E_{\mathrm{C}}(k_{\|}). \tag{2.12}$$

For the sake of clarity, the equation is split into two components, one of which representing the confinement energy, i.e. the ground state energy of the cavity photon $E_{\mathrm{C},0} = E_{\mathrm{C}}(k_{\|} = 0) = (\hbar c_0/n_{\mathrm{C}})k_{\perp}$ for $|\mathbf{k}| = k_{\perp}$ (that is $k_{\|} = 0$), and the other one representing the in-plane component with parabolic dispersion $\hbar^2 k_{\|}^2/2m_{\mathrm{C,eff}}$, i.e. an in-plane kinetic energy term. Thus, the dispersion of photons in a planar microcavity clearly deviates from the linear dispersion of free photons (see Fig. 2.6b).

Starting from a ground-state energy level, with increasing angle, or k-vector ($k_{\|}$), the energy of the photon increases quadratically (for the small-angle regime). It is this fact which allows one to assign an effective mass to cavity photons in planar resonators in analogy to massive free particles, given the quadratic dispersion. The effective mass concept leads to a quantity, which resembles sort of a "rest mass" of photons in confined modes of a Fabry–Pérot resonator. For cavity photons, the

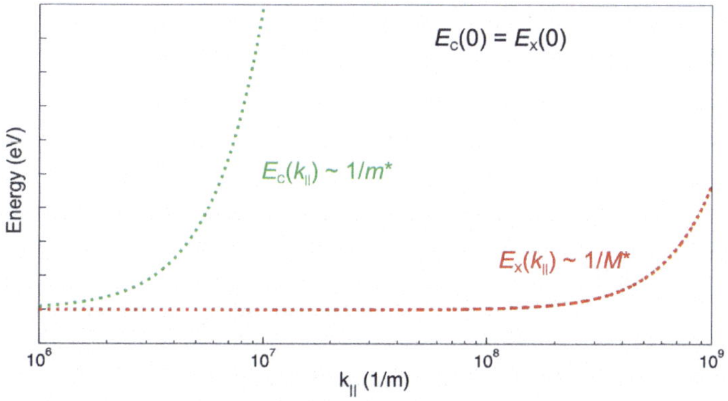

Fig. 2.7 Theoretical energy dispersion of a photon mode in a planar GaAs-based optical cavity (green dotted curve) in comparison to a QW exciton's dispersion at 1.42 eV, which features a much smaller curvature owing to its higher effective mass (red dashed curve). This becomes easily visible in such a semi-logarithmic plot, whereas in linear scaling, the exciton appears flat on the energy–momentum scales where the photon dispersion matters. Here, cavity and exciton effective masses are given as m^* and M^*, respectively. Adapted from [35]. Courtesy of the author

effective mass $m_{\mathrm{C,eff}}$ typically amounts to $10^{-5} \times m_e$ (m_e: free electron's mass), rendering them extremely light bosonic particles in solids:

$$m_{\mathrm{C,eff}} = \frac{\hbar k_\perp n_\mathrm{C}}{c_0} = \frac{2\pi \hbar n_\mathrm{C}}{\lambda_\mathrm{C} c_0} = E_\mathrm{C}(k_\| = 0)\frac{n_\mathrm{C}^2}{c_0^2} = \frac{E_{\mathrm{C},0}}{c^2}. \tag{2.13}$$

Figure 2.7 displays such photon mode dispersion with ground-state energy $E_{\mathrm{C},0} = 1.42\,\mathrm{eV}$ in a microcavity (green solid curve), revealing the photon energy's dependence on the angle and on the in-plane wave number $k_\|$, respectively. In comparison, the exciton dispersion shows a much smaller curvature owing to its significantly bigger effective mass (red dashed curve). Obviously, in the relevant momentum space region for photons, the exciton dispersion is nearly constant.

Cavity Losses

Since there is no chance to realize an ideal cavity with hundred percent confinement, optical microcavities will always face photon leakage. Nevertheless, an asymptotic approach to perfect cavities has been pursued as subject of a long-time endeavour in the optical-resonator community (cf. [34]). In real microresonators, a variety of loss channels exist, partly due to the finite reflectivity of the mirrors, the absorption in the resonator or in optically active regions of the microcavity structure (see Sect. 5.3), which limit the duration of time that cavity photons reside in the system. This finite lifetime of confined photons τ_C is one key figure, from which the actual figure of merit for resonators is derived, namely the Q-factor. This quality factor is a direct measure of the time which the photon will last inside the resonator, and is characterized by the ratio between cavity-mode energy E_C and its linewidth γ_C.

$$Q = \frac{E_C}{\gamma_C}. \tag{2.14}$$

Alternatively, the Q-factor can be expressed as $\lambda/\Delta\lambda$ using wavelengths to determine its magnitude ($\Delta\lambda$ denotes the linewidth). It is this important value which substantially determines the time of interaction for energy exchange between an emitter and the cavity-light field. In a $\lambda/2$ cavity with Q of 10,000, a photon theoretically experiences 10,000 round-trips inside the resonator before leaking out of it [21], which corresponds to a photon lifetime in the cavity of $\tau_C \approx 7$ ps at around $E = 1.55$ eV. Typical high-quality semiconductor planar microcavities feature photon lifetimes $\tau_C = 1$ to 10 ps, which set an upper limit to achievable polariton lifetimes in the hybrid regime, as will be elaborated on later.

2.3 Light–Matter Coupling

Without light–matter interaction, our world would of course be very boring, very unimaginable, definitely not as it is perceived by us—actually it would be not vivid at all lacking one fundamental means of energy exchange. Basically, the principles of vision, thermal radiation, photo-induced reactions or information processing in our world would have been obsolete, to name but a few simple examples, if there was not *absorption* and *emission* of light. In addition to these two basic processes, induced emission—the opposite process to (induced) absorption, well known as stimulated emission—gave rise to light amplification by stimulated emission of radiation, or in brief, laser operation. However, light–matter interaction is richer than these often recalled effects let one presume. In the following, light is shed on the weak and strong coupling that can occur between light and matter states, as an introduction to this topic.

As an important step, the strong increase of the spontaneous emission rate of an emitter in the presence of a spatially and spectrally resonant light field was already proposed back in 1946, by Purcell [36]. To harness this effect, the emitter should be placed into a cavity with resonant optical mode (cf. [37]). If the resonance is not given between an emitter and a strongly-confined optical light field of the cavity resonance, even inhibition of its spontaneous emission rate can take place. In other words, if there is no photonic state to emit into, emission will be suppressed. The effect of the enhancement and inhibition of spontaneous emission is a widely used feature of cavity quantum electrodynamics when it comes to, for instance, the design of nonclassical light sources based on microcavities, which allow one to tailor the photonic density of states (c-DOS) [38]. Its figure of merit is the Purcell factor (see for instance [39]). The mere presence of a vacuum field state/mode, i.e. an optical cavity with no photon inside, but a defined resonance, is enough for the resonant emitter to couple to it and to release its energy radiatively. This irreversible scenario described by the Purcell effect represents the case of a weakly-coupled system of emitter and light field, for which the loss rates are not small enough to facilitate a reversible process of light emission and re-absorption in the same cavity–emitter system. In such case, the coupling strength is not good enough, a figure, which usually needs to be maximized for pronounced cavity quantum electrodynamics effects [40]. Either, a very high Q-factor together with a very small mode volume V_{m} is necessary in the limit of a single emitter to result in a high vacuum field $\mathbf{E}_{\mathrm{vac}} \propto \sqrt{1/V_{\mathrm{m}}}$ with little enough photon decay rates, or the number of (resonant) emitters has to be raised in the case of moderate Q values. The increase in numbers affects the coupling strength directly, but leads to ensemble physics far away from vacuum-field coupling.

As soon as the coupling rate in the cavity–emitter system exceeds the given loss rates, the aforementioned energy exchange between emitter and field can become reversible. Such a regime, which is referred to as strong light–matter coupling, was observed for the first time in 1992 by Weisbuch et al., yet, for a QW-microresonator system [41]. Light–matter interaction in systems with discrete energy levels for photon and exciton mode is a well considered effect of cavity quantum electrodynamics [18] and will not be subject of this chapter. A schematic representation of such a system of cavity and emitter is shown in Fig. 2.8, in which the electromagnetic vacuum field and a resonant single exciton couple to each other giving rise to an energy exchange between optical mode and emitter. Strong coupling is achieved in such a system, if the rate of energy exchange exceeds the decay as well as decoherence rates of photon and emitter. In this regime, the system can be described as a superposition of both of its components and a new quasi-particle is formed, the *polariton*, which is partly light, partly matter. In contrast to the weak-coupling regime, the radiative decay of an exciton becomes reversible and periodic, given the periodic re-absorption of the emitted photon during its cavity lifetime (in a simplified picture). The frequency of energy exchange is referred to as Rabi frequency. This coherent energy exchange will last as long as the photon remains in the cavity or the exciton is preserved. As a damped oscillation, the probability to find the excitation in the cavity decays exponentially with time, rendering an envelope function onto the Rabi oscillations with given coupling strength (see Fig. 1.2c in Chap. 1).

Strong-coupling between oscillators/modes is easily identified spectrally by the formation of new eigen states. In the cavity–emitter system, these states are hybrid light–matter eigen states, the polariton modes, which are spectrally separated by the Rabi splitting (vacuum Rabi splitting in the quantum limit of strong coupling). This splitting is the signature of periodic energy exchange between light and matter. The stronger the coupling, the larger the splitting. The attributes of a coupled system will be discussed in the following for the case of quantum-well excitons coupled to microresonator photons as the relevant cavity–emitter system for cavity–polariton formation. Owing to the large number of involved emitters, this system represents a semi-classical oscillator system [20]. In the quantum limit of strong coupling, light–matter interaction with vacuum Rabi splitting is obtained between a single emitter, e.g. from a quantum-dot exciton, and an unoccupied discrete 0D photonic mode [42–44] (see Fig. 2.9). Far from the single emitter limit are structures based on quantum

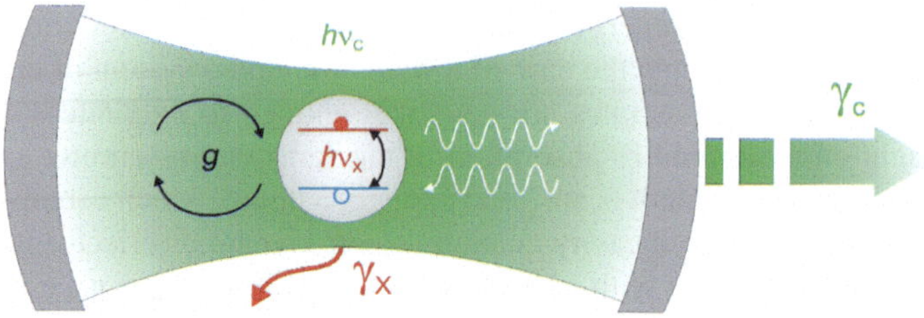

Fig. 2.8 Simplified picture of a coupled cavity–emitter system with coupling strength g, in which the excited state of matter interacts with the (vacuum) electro-magnetic field of the cavity (vacuum Rabi splitting is observed in the single-emitter limit with empty cavity mode). $\gamma_{X/C}$ and $h\nu_{X,C}$ represent decay rate and emission energy, respectively, for the exciton/photon mode

wells in planar microresonator structures, which employ large densities of quantum--well excitons, located for best possible interaction at anti-node positions of the resonator's electric standing-wave field. Thereby, usually the heavy-hole (hh) exciton with total angular momentum $J = 1$ ($J_{z,e} = \pm 1/2$ and $J_{z,hh} = \pm 3/2$) couples strongly with confined cavity photons, whereas angular-momentum conservation for optical transitions is observed (photon spin ± 1).

2.3.1 Exciton–Polaritons

In the regime of strong coupling between quantum-well exciton and microcavity photon, birth is given to the *exciton–polariton* (XP) (representing the system's new eigen states), which can be well described in a quasi-particle picture. More interestingly, out of *strong light–matter coupling* not only a new hybrid particle emerges, but—to be more precise—a composite boson, which is formed out of the bosonic photon and the bosonic exciton. It is worth highlighting that the exciton itself is also a composite boson, however, made of two coupled fermions. Yet, the concept of bosonic excitons only holds true in a diluted gas, in which screening and particle–particle interaction between its fermionic constituents is negligible.

The bosonic character of polaritons is particularly of interest for the investigation of many-particle effects and macroscopic quantum phenomena in solids, such as BEC (see Sect. **??**), to which effects like superfluidity, superconductivity and collective coherence are related. Moreover, the composite boson inherits attributes of its building blocks in correspondence to the weightings of its photonic and excitonic fraction. This feature can be well exploited for polaritons via the detuning between the photon and the exciton resonance, which can be set according to one's demand between the two available limits of having either predominantly a photon or an exciton in the system. By every intermediate setting, the photonic and excitonic character can be tuned and a strong impact on the polariton's characteristics achieved. This peculiarity provides polaritons a great advantage over other composite bosons, as will become clear in the course of the remaining chapters.

In the following part, a brief mathematical description of the polariton will be presented at first, before its characteristics and spectral properties are explained to provide a basis for the dealing with polariton systems. Indeed, a much more comprehensive description of light–matter interaction is given in excellent literature available on microcavities (see [18, 24, 31, 32], to name but a few).

Expressions for the New Eigen States

The linear Hamiltonian (Hamilton operator) of a polaritonic system, in the picture of second quantization, can be written as the sum of its constituent's Hamiltonians and an interaction term (cf. [31, 32, 45] for a cavity–polariton related discussion, and [1, 46, 47] for a general picture):

$$\hat{H}_{\mathrm{xp}} = \hat{H}_{\mathrm{p}} + \hat{H}_{\mathrm{x}} + \hat{H}_{\mathrm{i}}, \tag{2.15}$$

whereas the Hamiltonians $\hat{H}_{\mathrm{p}}$, $\hat{H}_{\mathrm{x}}$ and $\hat{H}_{\mathrm{i}}$ represent the resonator field, the emitter and the interaction between light and matter excitation in the system, respectively:

$$\hat{H}_{\mathrm{p}} = E_{\mathrm{C}}\,\hat{f}^{\dagger}\hat{f}, \tag{2.16}$$

$$\hat{H}_{\mathrm{x}} = E_{\mathrm{X}}\,\hat{e}^{\dagger}\hat{e}, \tag{2.17}$$

$$\hat{H}_{\mathrm{i}} = \hbar\Omega\,(\hat{f}^{\dagger}\hat{e} + \hat{f}\hat{e}^{\dagger}), \tag{2.18}$$

which contain the cavity-mode energy E_{C}, the excitation energy of the emitter E_{X}, the interaction energy $\hbar\Omega$, and the creation as well as annihilation operators $\hat{f}^{\dagger}/\hat{e}^{\dagger}$ and $\hat{f}/\hat{e}$ (field/emitter), respectively. In planar many-particle systems, one has to sum over all particles and k-vectors (transverse wave-vectors $k_{\parallel}$) in each Hamiltonian in order to describe the complete system. However, for the sake of clarity, the detailed notation has been skipped here, but can be found in the literature (see [18, 31, 32]).

Here, $\hbar\Omega$ represents the coupling strength, which is proportional to the dipole interaction frequency Ω of the cavity–emitter system. Ω depends on the dipole transition-matrix element

$$M = |\langle\varphi_X|\,\mathbf{d}\cdot\mathbf{E}\,|\varphi_C\rangle|\,, \tag{2.19}$$

as summarized briefly, and is referred to as *Rabi frequency*, in accordance to historical investigations on atom–field systems by I. I. Rabi and the respective studies of Rabi oscillations in atom cavities [2, 48]. This frequency is equivalent to the standard Rabi frequency known for a two-level oscillator with $\omega_{\mathrm{R}} = e|\mathbf{E}_0|M/(m\hbar\omega_0)$, where M is the dipole matrix element of (2.19), and $\mathbf{E}_0$ the electric-field amplitude of the photon with energy $\hbar\omega_0$ resonant to the exciton, respectively [13]. Hereby, M depends on the mean electric dipole moment of the emitter $\mathbf{d}$ and the electric-field strength $\mathbf{E}$ at the location of interaction. Furthermore, M depends on the state of the system (also see "Fermi's golden rule", e.g. in [24]). This can be explained in a simple manner when considering the system being prepared in either of the two possible states $|\varphi_C\rangle$ (electronic system in its ground state) and $|\varphi_X\rangle$ (excited system), representing the system in a state where it contains a photon or an exciton, respectively [32]: An optical transition (absorption/emission) is favoured if the system contains a photon (i.e. is in state $|\varphi_C\rangle$) for absorption or an exciton (i.e. is in state $|\varphi_X\rangle$) for emission. Particularly, a measure of coupling of light to matter in a specific system is given by the oscillator strength f of the electric dipole. This material parameter is given for a certain resonance energy E_{res} and the electric quantisation volume V in the 3D case by

$$f^{(3D)} = \frac{2m_r E_{res}}{\hbar^2} M^2 \frac{V}{\pi a_B^3}, \tag{2.20}$$

whereas m_r and the cube of the exciton Bohr radius a_B^3 represent the reduced effective mass of the dipole (here exciton) and the dipole's volume, respectively [21, 30]. Compared to an unbound pair of electron and hole, the wave-function overlap in an exciton is enhanced, which leads to an enhancement of the interaction as represented by $V/\pi a_B^3$.

However, since quantum-well excitons exhibit a confinement in one direction, the cavity mode experiences these 2D-localized excitations with an altered oscillator strength, which can be given in dependence of the QW number N_{QW} (cf. [49, 50]):

$$f^{(2D)} = N_{QW} \frac{8 f^{(3D)}}{\pi a_{B(3D)}^2}. \tag{2.21}$$

Here, the 2D oscillator strength is proportional to the 3D oscillator strength per unit area $f^{(3D)}/\pi a_{B(3D)}^2$ [41].

For strong coupling, the parameter $\hbar\Omega$ given in the Hamiltonian for the interaction between photon and emitter is of importance, which is also named coupling constant $g = \hbar\Omega$ (in the literature, the Rabi frequency Ω is sometimes denoted as $\Omega_0 = 2g/\hbar = 2\Omega$). This figure directly depends on the oscillator strength f and thereby on the number of quantum wells for 2D systems (see [20, 32, 51]):

$$g \propto \sqrt{\frac{f^{(2D)}}{m_e n_{eff}^2 L_{eff}}} \propto \sqrt{N_{QW}}, \tag{2.22}$$

whereas n_{eff} and L_{eff} denote the effective refractive index of the cavity material and the effective cavity length, respectively. Since the dipole interaction between photon and exciton only takes place for equal $k_\parallel$ due to selection rules for optical absorption and emission processes [20], $M \neq 0$ for modes with the same $k_\parallel$ and with it the coupling strength g.

The eigen states of such aforementioned system in strong coupling can be derived as solution of the linear polariton system [52, 53]: A diagonalization of the previously described Hamiltonians can be achieved with the following transformation (cf. [21, 24, 31, 46])

$$\hat{l}_{k_\parallel} = +X_{k_\parallel}\hat{e}_{k_\parallel} + C_{k_\parallel}\hat{f}_{k_\parallel}, \tag{2.23}$$

$$\hat{u}_{k_\parallel} = -C_{k_\parallel}\hat{e}_{k_\parallel} + X_{k_\parallel}.\hat{f}_{k_\parallel}, \tag{2.24}$$

with coefficients C and X quantifying the contributions from the field and emitter to the hybrid states, respectively. From these terms, a linear Hamiltonian for the polariton results:

$$\hat{H}_{\mathrm{xp}} = \sum_{N,k_\parallel} E_{\mathrm{LP}}(k_\parallel)\hat{l}^{\dagger}_{k_\parallel}\hat{l}_{k_\parallel} + \sum_{N,k_\parallel} E_{\mathrm{UP}}(k_\parallel)\hat{u}^{\dagger}_{k_\parallel}\hat{u}_{k_\parallel}. \tag{2.25}$$

Here, the notation with a sum over all possible states describing the full system includes the $k_\parallel$ dependence of the parameters and the number of particles N. This equation reflects the formation of two new eigen states of the single-particle Hamiltonian in the coupled system, with creation and annihilation operators $\hat{l}^\dagger/\hat{u}^\dagger$ and $\hat{l}/\hat{u}$, respectively, which describe the quasi-particles referred to as lower (LP) and upper (UP) polariton with energies E_{LP} and E_{UP}, respectively.

In matrix representation, the stationary coupled system's eigen-value equation with diagonalized Hamiltonian $\hat{H}_{\mathrm{xp}}$ and eigen-energy E_{xp} can be expressed as follows:

$$\hat{H}_{\mathrm{xp}}\left|\Psi\right\rangle = \begin{pmatrix} E_{\mathrm{C}} & \hbar\Omega \\ \hbar\Omega & E_{\mathrm{X}} \end{pmatrix}\begin{pmatrix} C \\ X \end{pmatrix} = E_{\mathrm{xp}}\left|\Psi\right\rangle. \tag{2.26}$$

According to (2.23) and (2.24), the polaritons are characterized by their photonic ($|C|^2$) and excitonic ($|X|^2$) fractions, which give the amplitudes of the components in the diagonalized system. These squared amplitudes are referred to as Hopfield coefficients [52] and they sum up to unity:

$$\left|C_{k_\parallel}\right|^2 + \left|X_{k_\parallel}\right|^2 = 1. \tag{2.27}$$

Thereby, a polariton can be considered a linear superposition of an exciton with a photon of the same transverse wave vector $k_\parallel$. Thus, colloquially spoken, the exciton is dressed with a photon. As will be seen later, this has severe implications for the exciton-based system's dynamics and properties. In the following, the polaritonic states shall be elaborated on in the context of their spectral properties and relevant features.

2.3.2 Detuning Dependencies of Polariton Modes

The spectral detuning is a key parameter in systems designed for strong light–matter coupling, as it determines the spectral overlap of the two components' resonances

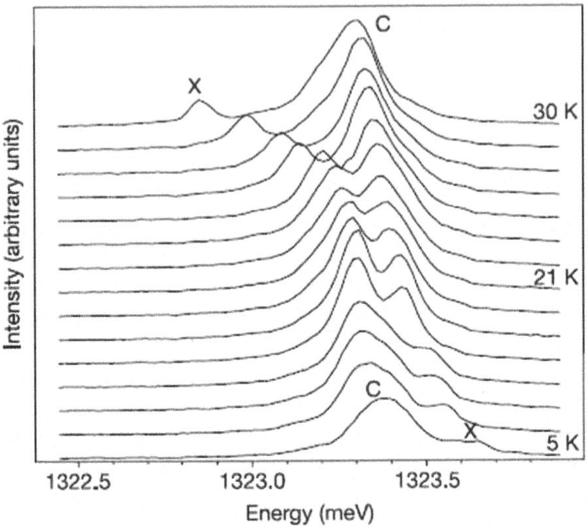

Fig. 2.9 Example of an anti-crossing behaviour for an exciton–photon system in a QD-micropillar in the single-particle limit. In this demonstration of a strong-coupling regime, the detuning of the modes with respect to each other was controlled by the temperature, which affects the QD line stronger than the cavity mode. Hence, the exciton from a single quantum dot is brought into spectral resonance with the cavity field and a vacuum Rabi splitting on the deep-sub-meV scale is resolved. It is worth noting that the strength of the coupling not only depends on the exciton's properties but also on the spatial overlap of the emitter with the light field. For micropillars, quantum dots in the centre of the pillar naturally experience a stronger coupling to the light field than those located at the edges due to the lateral profile of the resonator fundamental mode. One major difference in quantum dot systems compared to those with quantum wells is the loss of dispersion. Thus, when referring to cavity–polaritons, one usually thinks of quantum-well polaritons. Typically, their coupling is stronger owing to a higher number of resonant emitters, i.e. an ensemble of equivalent excitons. Tuning mechanisms for the achievement of resonance conditions in polariton systems can be the temperature, external electric or magnetic fields, or in the case of planar (i.e. two-dimensional) and bulk systems the momentum/wave-vector $k_{\parallel}$ and $\mathbf{k}$, respectively (see Fig. 2.1). Reused with permission. Reference [42] Copyright 2004, Springer Nature

(see Fig. 2.9). It can be generally determined by the following expression using cavity and exciton resonances E_{C} and E_{X}, respectively, at given in-plane wave-vector $k_{\parallel}$:

$$\Delta(k_{\parallel}) = E_{\mathrm{C}}(k_{\parallel}) - E_{\mathrm{X}}(k_{\parallel}) \qquad (2.28)$$

In a polariton system, the detuning determines the quasi-particles' attributes, since the Hopfield coefficients strongly depend on this magnitude [21, 32]:

Table 2.1 Overview on detuning cases in a resonator–emitter system

$\Delta = E_C - E_X$	$\lvert C\rvert^2$	$\lvert X\rvert^2$	Reference	Regime
<0	>**0.5**	<0.5	*Negative* (red) detuning	Photonic
=0	=0.5	=0.5	*Zero* (resonance) detuning	Hybrid
>0	<0.5	>**0.5**	*Positive* (blue) detuning	Excitonic

$$\left|C_{k_\parallel}\right|^2 = \frac{1}{2}\left(1 - \frac{\Delta(k_\parallel)}{\sqrt{\Delta(k_\parallel)^2 + (2\hbar\Omega)^2}}\right), \tag{2.29}$$

$$\left|X_{k_\parallel}\right|^2 = \frac{1}{2}\left(1 + \frac{\Delta(k_\parallel)}{\sqrt{\Delta(k_\parallel)^2 + (2\hbar\Omega)^2}}\right). \tag{2.30}$$

This set of equations clearly shows the equality of both Hopfield coefficients if the detuning is zero, i.e. when both eigen states are half photonic and half excitonic,

$$\Delta = 0 \Rightarrow \left|C_{k_\parallel}\right|^2 = \left|X_{k_\parallel}\right|^2 = 0.5. \tag{2.31}$$

Three important cases can be easily identified which reflect the system's most prominent detuning situations, all of which unravel a strong impact on the respective polaritons' features. They are complemented by the two cases when the system is either almost purely photonic or purely excitonic. An overview over different representative detunings is given both in a schematic diagram and in a table (see Fig. 2.10 and Table 2.1). These characteristic detuning cases will be discussed in more detail in the following, yet, mainly with respect to mode energies, lifetimes and effective masses.

The $k_\parallel$-dependent eigen-energies of the coupled system can be given by the solution of the linear equation (2.25) of the diagonalized $\hat{H}_{\mathrm{xp}}$.

$$E_{\mathrm{LP,UP}}(k_\parallel) = \frac{1}{2}\left[E_{\mathrm{C}}(k_\parallel) + E_{\mathrm{X}}(k_\parallel) \mp \sqrt{(2\hbar\Omega)^2 + (\Delta(k_\parallel))^2}\right]. \tag{2.32}$$

These energy dispersions for polaritons (representing LP and UP) are of utmost importance for the characterization of polaritons in planar microresonator structures (cf. [21, 32]) and will be continuously in the focus of polariton-related work. This is why it is worth taking a closer look.

In this system, the aforementioned Hopfield coefficients ((2.29) and (2.30)) primarily describe the polariton's character in the lower energy branch (LP), while the respective photonic and excitonic fractions of the upper polariton are complementary. A visualization of the Hopfield coefficients for three different detunings Δ is

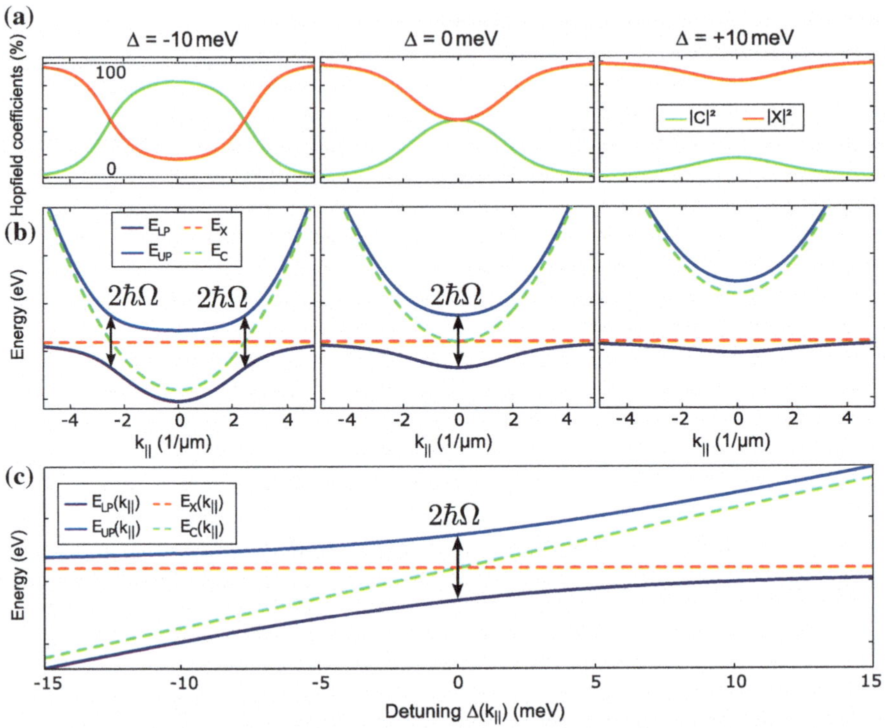

Fig. 2.10 **a** Hopfield coefficients at three different detuning situations together with **b** corresponding polariton dispersions in a QW-microresonator. A mode splitting of $2\hbar\Omega$ is obtained at resonance conditions for the cavity and exciton modes. **c** In a diagram of the system's resonances as a function of the detuning Δ, an anti-crossing behaviour of the modes is exhibited in the strong-coupling regime, which is represented by the establishment of new eigen states, the polariton modes. Adapted from [35]. Courtesy of the author

exemplarily presented in Fig. 2.10a with the corresponding model dispersions of the polaritonic system according to (2.32) in Fig. 2.10b. From left to right, the detuning is altered and reveals how the character of the polariton changes from a more photonic to a more excitonic configuration. While at negative detunings, the dispersion of the LP is strongly curved similar to the cavity dispersion in the given momentum-space range, the photonic content dominates the ground-state polariton's behaviour (see left column in Fig. 2.10); at positive detunings, the excitonic content determines the LP's properties, which exhibits a flattened and exciton-like dispersion (see right column). The opposite description holds true for the UP and its dispersion relation. In resonance, when the polariton is most hybrid, both components are equally represented, which is reflected by the properties of the LP and UP in their respective ground states of their dispersion relations at $k_{\|} = 0$ (see centre column in Fig. 2.10).

The dependence of the mode energies on the cavity–exciton detuning $\Delta_{k_{\|}}$ can be directly found in (2.32). Moreover, under resonance conditions, an energy spitting between the two eigen states (see (2.32)) is clearly featured in the energy spectrum and amounts to $E_{\mathrm{Rabi}} = 2\hbar\Omega$, the *Rabi splitting*. In contrast, only the crossing of both modes would have been observed when tuning them through one-another in the weak-coupling regime. Thus, this anti-crossing behaviour of the photonic and excitonic modes with minimum splitting in case of full resonance is a—if not *the* most—considerable signature of strong coupling (see Figs. 2.1 and 2.10). Spectral anti-crossing as an indicator for strong coupling is almost only rivalled by Rabi-flopping signatures (oscillations) in the time domain, the measurement of which is less trivial (two representative experiments from the literature are mentioned in Sects. 7.4 and 9.3 that used digital holography of photoluminescence and ultra-fast transient absorption, respectively). Naturally, anti-crossing behaviour can be also observed in the energy dispersions at non-zero k-vectors, explaining the following expression for zero-detuning splitting in QW-microcavities, with g the aforementioned coupling constant.

$$\left[(E_{\mathrm{UP}} - E_{\mathrm{LP}})_{k_{\parallel}}\right]_{\Delta_{k_{\parallel}}=0} = 2\hbar\Omega = 2g = E_{\mathrm{Rabi}}. \tag{2.33}$$

In the following, the description of the detuning situation will focus on the situation at $k_{\parallel} = 0$, which provides a characteristic potential minimum of the polaritonic system and is of strong relevance to polariton physics, thus $\Delta = \Delta(k_{\parallel} = 0)$, if not declared else.

As can be seen well for $\Delta < 0$, the predominantly photonic dispersion of the LP branch approaches the dispersion of the pure exciton mode at higher wave vectors $k_{\parallel}$ (see Fig. 2.10b). Thereby the curvature of the LP dispersion changes towards high $k_{\parallel}$, with the dispersion exhibiting an inflection point, correspondingly. In the regions where $|\Delta| \gg 2\hbar\Omega$, the dispersion curves of the polariton branches are very similar to the regular cavity and exciton dispersion relations, respectively.

Similar to cavity and exciton modes, effective masses can be assigned to polariton branches [21, 32]. In regions of small wave numbers $k_{\parallel}$, for which $\hbar^2 k_{\parallel}^2 / 2m_{\mathrm{C,eff}} \ll 2g$ holds true, the polariton dispersions of (2.32) can be approximated by a Taylor series up to the second-order term.

$$E_{\mathrm{LP/UP}}(k_{\parallel}) \cong E_{\mathrm{LP/UP}}(0) + \frac{\hbar^2 k_{\parallel}^2}{2m_{\mathrm{LP/UP,eff}}}, \tag{2.34}$$

whereas $m_{\mathrm{LP/UP}}$ approximate the effective masses for the LP and UP, respectively. These effective masses can be described by weighted harmonic mean values of the exciton and cavity-photon masses:

$$\frac{1}{m_{\mathrm{LP,eff}}} = \frac{|C|^2}{m_{\mathrm{C,eff}}} + \frac{|X|^2}{m_{\mathrm{X,eff}}}, \tag{2.35}$$

$$\frac{1}{m_{\mathrm{UP,eff}}} = \frac{|X|^2}{m_{\mathrm{C,eff}}} + \frac{|C|^2}{m_{\mathrm{X,eff}}}. \tag{2.36}$$

With the effective cavity mass $m_{\mathrm{C,eff}}$ being much smaller than the excitonic counterpart $m_{\mathrm{X,eff}} \approx 10^4 \times m_{\mathrm{C,eff}}$ [21], the polariton effective masses in the ground state can be well approximated by

$$m_{\mathrm{LP}}(k_{\parallel} \approx 0) \cong \frac{m_{\mathrm{C}}}{|C|^2}, \tag{2.37}$$

$$m_{\mathrm{UP}}(k_{\parallel} \approx 0) \cong \frac{m_{\mathrm{C}}}{|X|^2}. \tag{2.38}$$

For higher $k_{\parallel}$, the comparably small effective mass of the LP at $k_{\parallel} \approx 0$ undergoes a dramatic change to $m_{\mathrm{LP}}(k_{\parallel} \gg 0) \approx m_X$ by an increase of four orders of magnitude from small to large wave numbers. This characteristic of the polariton dispersion hence plays an important role in the relaxation dynamics of a polaritonic system, leading for instance to a bottleneck effect for scattering processes, which can become a major obstacle for thermalization and macroscopic ground-state population [54–57].

Also the lifetimes of the single components, of which a polariton is composed, can significantly determine the dynamics of the system, as they directly influence the quasi-particles' lifetime. Particularly, the overall decay time of the composite boson can have a pronounced effect on polariton condensation phenomena, which will be introduced in the following. Moreover, the decay rates of photons and excitons also affect the eigen-energies of the coupled system (2.32). This can be seen in an expression for the eigen states, which takes lifetimes into account:

$$\begin{aligned} E_{\mathrm{LP,UP}}(k_{\parallel}) &= \frac{1}{2} \times \\ &\left(E_{\mathrm{C}}(k_{\parallel}) + E_{\mathrm{X}}(k_{\parallel}) + \mathrm{i}(\gamma_{\mathrm{C}} + \gamma_{\mathrm{X}}) \pm \sqrt{(2\hbar\Omega)^2 + \left(\Delta(k_{\parallel}) + \mathrm{i}(\gamma_{\mathrm{C}} - \gamma_{\mathrm{X}})\right)^2} \right) \end{aligned} \tag{2.39}$$

whereas γ_{C} and γ_{X} denote the homogeneous linewidth of the cavity and exciton modes, respectively, which are directly linked to the corresponding lifetimes via Heisenberg's uncertainty principle $\tau \equiv \delta t \geq \hbar/\gamma$ ($\gamma = h\Delta\nu$ or $\gamma = \hbar\Delta\omega$, typical lifetimes are $\tau_{\mathrm{X}} \approx 1\,\mathrm{ns}$ and $\tau_{\mathrm{C}} \approx 1$–$10\,\mathrm{ps}$) [21, 32]. The decay rate of the photons $1/\tau_{\mathrm{C}} \leq \gamma_{\mathrm{C}}/\hbar$ is governed by the cavity losses (absorption and leakage), i.e. the Q-factor, while the decay rate of the excitons $1/\tau_{\mathrm{X}} \leq \gamma_{\mathrm{X}}/\hbar$ is determined by radiative and non-radiative decay channels. Thus, a prerequisite for the observation of a splitting and strong coupling is the magnitude of the coupling strength $g = \hbar\Omega > |(\gamma_{\mathrm{C}} - \gamma_{\mathrm{X}})/2|$, in order to yield a real solution for the square root term of (2.39). For $g \gg \gamma_{\mathrm{C}} \gg \gamma_{\mathrm{X}}$, (2.32) acts as a good approximation for the XP energies. In contrast, the system is in the weak-coupling regime, if the eigen modes remain the cavity and exciton modes and no coherent energy exchange between emitter and field takes place. In this case, the radiative decay of excitons is described by the transition-matrix element M [21]. Typically, photon losses dominate the decay channels and the strong-coupling regime can be achieved simply when $g > \gamma_{\mathrm{C}}/2$.

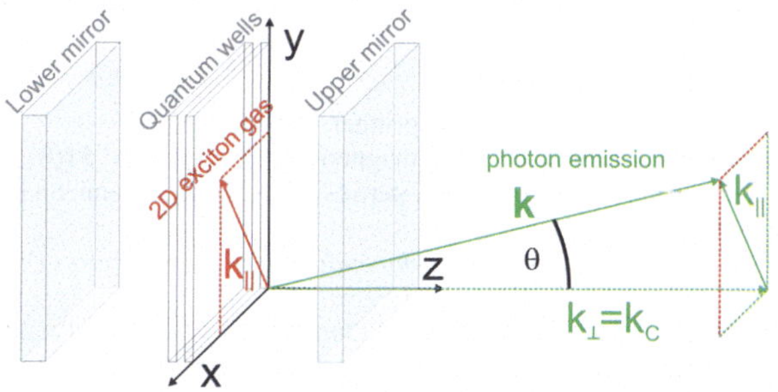

Fig. 2.11 Schematic drawing of a QW-microcavity system based on a planar Fabry–Pérot resonator design. Here, the cavity photon's momentum is set in relation to the emission angle. The projection onto a vertical axis and parallel plane with respect to the QW plane shows the momentum-components' strength in the respective directions corresponding to the angle θ. In the quantum wells, excitons with their in-plane momentum $k_{\parallel}$ can only couple to cavity photons of total photon momentum $\mathbf{k}$ when their in-plane momentum projection $k_{\parallel}$ matches

According to the superposition principle, one can also put the linewidth of the polariton modes for a given $k_{\parallel}$ directly in relation to the components' linewidths [21, 32]:

$$\gamma_{\mathrm{LP}} = |C|^2\gamma_{\mathrm{C}} + |X|^2\gamma_{\mathrm{X}}, \tag{2.40}$$

$$\gamma_{\mathrm{UP}} = |X|^2\gamma_{\mathrm{C}} + |C|^2\gamma_{\mathrm{X}}. \tag{2.41}$$

In case of the ground state, the significantly shorter lifetime of photons allows one to estimate the LP lifetime via $\tau_{\mathrm{LP}} = \hbar/\gamma_{\mathrm{LP}} \cong \hbar/(|C|^2\gamma_{\mathrm{C}}) = \tau_{\mathrm{C}}/|C|^2$. After the characteristic lifetimes $\tau_{\mathrm{LP,UP}}$ expires, being primarily governed by cavity losses, the polariton decays via photon emission with energy $E_{\mathrm{LP,UP}}(k_{\parallel})$, also obeying momentum conservation. It is this feature of the polariton system, i.e. that the internal polariton mode and the externally outcoupled light field are directly related to each other, which is exploited for its characterization. The light from a strongly-coupled QW-microcavity system is emitted exactly under that angle (relative to the growth direction z, i.e. perpendicular to the QW plane) that corresponds to the in-plane momentum component of the cavity mode according to (2.9). This is sketched in Fig. 2.11. Such an experimental access to its properties is quite unique for a quasi-particle system in solids. It is the consequence of conservation of energy and momentum, i.e. the direct relationship between the photon's in-plane momentum with that of the coupled exciton forming the polariton. Thereby, information about the energy dispersion of polaritons can be gathered in a comparably easy manner, allowing one to extract characteristics of the system by means of optical spectroscopy (see Chap. 7), which give insight into occupation numbers, effective masses and statistical distributions of particles in the system. This, and the advantage of providing very light bosonic particles based on matter excitation, i.e. the excitons dressed with photons, explain the strong interest in QW-microcavities for polariton studies.

References

1. C .F. Klingshirn, *Semiconductor Optics* (Springer, Berlin, 2012)
2. I.I. Rabi, Space quantization in a gyrating magnetic field. Phys. Rev. **51**(8), 652–654 (1937)
3. L.V. Keldysh, A.N. Kozlov, Collective properties of excitons in semiconductors. Sov. Phys. JETP **27**, 521–528 (1968)
4. G. Bastard, *Wave Mechanics Applied to Semiconductor Heterostructures* (Wiley-Interscience, New York, 1991)
5. H. Haug, S.W. Koch, *Quantum Theory of the Optical and Electronic Properties of Semiconductor*, 5th edn. (World Scientific Publishing Co, Singapore, 2009)
6. J.I. Frenkel, *Wave Mechanics: Elementary Theory* (Clarendon Press, Oxford, 1932)
7. E.L. Ivchenko, *Optical Spectroscopy of Semiconductor Nanostructures* (Springer, Berlin, 2004)
8. E.L. Ivchenko, *Optical Spectroscopy of Semiconductor Nanostructures* (Alpha Science, Chelmsford, 2005)
9. E. Hanamura, H. Haug, Condensation effects of excitons. Phys. Rep. **33**(4), 209 (1977)
10. G.H. Wannier, The structure of electronic excitation levels in insulating crystals. Phys. Rev. **52**, 191 (1937)
11. N.F. Mott, Conduction in polar crystals. II. The conduction band and ultra-violet absorption of alkali-halide crystals. Trans. Faraday Soc. **34**, 500 (1938)
12. H.M. Gibbs, G. Khitrova, S.W. Koch, Exciton-polariton light-semiconductor coupling effects. Nat. Phot. **5**(5), 273 (2011)
13. D. Sanvitto, V. Timofeev, *Exciton Polaritons in Microcavities: New Frontiers* (Springer, Berlin, 2012)
14. M. Kira, W. Hoyer, T. Stroucken, S.W. Koch, Exciton formation in semiconductors and the influence of a photonic environment. Phys. Rev. Lett. **87**(17), 176401 (2001)
15. R.A. Kaindl, M.A. Carnahan, D. Hägele, R. Lövenich, D.S. Chemla, Ultrafast terahertz probes of transient conducting and insulating phases in an electron-hole gas. Nature **423**(6941), 734–738 (2003)
16. S. Chatterjee, C. Ell, S. Mosor, G. Khitrova, H.M. Gibbs, W. Hoyer, M. Kira, S.W. Koch, J.P. Prineas, H. Stolz, Excitonic photoluminescence in semiconductor quantum wells: plasma versus excitons. Phys. Rev. Lett. **92**(6), 067402 (2004)
17. J.H. Davies, *The Physics of Low-dimensional Semiconductors: An Introduction* (Cambridge University Press, Cambridge, 1998)
18. Y. Yamamoto, F. Tassone, H. Cao, *Semiconductor Cavity Quantum Electrodynamics* (Springer, Berlin, 2000)
19. R.L. Greene, K.K. Bajaj, D.E. Phelps, Energy levels of Wannier excitons in $GaAs - Ga_{1-x}Al_xAs$ quantum-well structures. Phys. Rev. B **29**(4), 1807 (1984)
20. M.S. Skolnick, T.A. Fisher, D.M. Whittaker, Strong coupling phenomena in quantum microcavity structures. Semicond. Sci. Technol. **13**, 645–669 (1998)
21. H. Deng, H. Haug, Y. Yamamoto, Exciton-polariton Bose-Einstein condensation. Rev. Mod. Phys. **82**(2), 1489 (2010)
22. L.C. Andreani, A. Pasquarello, Accurate theory of excitons in GaAs-GaAlAs quantum wells. Phys. Rev. B **42**, 8928–8938 (1990)
23. Y. Masumoto, M. Matsuura, S. Tarucha, H. Okamoto, Direct experimental observation of two-dimensional shrinkage of the exciton wave function in quantum wells. Phys. Rev. B **32**(6), 4275–4278 (1985)
24. A.V. Kavokin, J.J. Baumberg, G. Malpuech, F.P. Laussy, *Microcavities* (Oxford University Press, Oxford, 2007)
25. R.P. Leavitt, J.W. Little, Simple method for calculating exciton binding energies in quantum-confined semiconductor structures. Phys. Rev. B **42**(18), 11774 (1990)
26. C. Priester, G. Allan, M. Lannoo, Wannier excitons in GaAs-$Ga_{1-x}Al_xAs$ quantum-well structures: influence of the effective-mass mismatch. Phys. Rev. B **30**(12):7302 (1984)
27. J. Dreybrodt, A. Forchel, J.P. Reithmaier, Excitonic transitions in GaInAs/GaAs surface quantum wells. Journal de Physique IV **3**(C5), 265–268 (1993)

28. L.S. Dang, D. Heger, R. André, F. Bœuf, R. Romestain, Stimulation of polariton photoluminescence in semiconductor microcavity. Phys. Rev. Lett. **81**(18), 3920 (1998)
29. K. Vahala, *Optical Microcavities* (World Scientific, Singapore, 2004)
30. Y. Yamamoto, A. Imamoglu, *Mesoscopic Quantum Optics* (Wiley, Hoboken, 1999)
31. B. Deveaud. *The Physics of Semiconductor Microcavities* (Wiley-VCH Verlag, Hoboken, 2007)
32. A. Kavokin, G. Malpuech, *Cavity Polaritons* (Academic Press, Cambridge, 2003)
33. A.V. Kavokin, J.J. Baumberg, G. Malpuech, F.P. Laussy, *Microcavities*, vol. 1 (Oxford University Press, Oxford, 2017)
34. K.J. Vahala, Optical microcavities. Nature **424**(6950), 839–846 (2003)
35. A. Rahimi-Iman. *Nichtlineare Eekte in III/V Quantenlm-Mikroresonatoren: Von dynamischer Bose–Einstein-Kondensation hin zum elektrisch betriebenen Polariton-Laser* (Cuvillier Verlag, Göttingen, 2013)
36. E.M. Purcell, Spontaneous emission probabilities at radio frequencies. Phys. Rev. **69**(11–12), 681 (1946)
37. J.M. Gérard, B. Sermage, B. Gayral, B. Legrand, E. Costard, V. Thierry-Mieg, Enhanced spontaneous emission by quantum boxes in a monolithic optical microcavity. Phys. Rev. Lett. **81**(5), 1110–1113 (1998)
38. S. Reitzenstein, A. Forchel, Quantum dot micropillars. J. Phys. D: Appl. Phys. **43**(3), 033001 (2010)
39. B. Gayral, J.M. Gérard, Comment on single-mode spontaneous emission from a single quantum dot in a three-dimensional microcavity. Phys. Rev. Lett. **90**(22), 229701 (2003)
40. G. Khitrova, H.M. Gibbs, M. Kira, S.W. Koch, A. Scherer, Vacuum Rabi splitting in semiconductors. Nat. Phys. **2**(2), 81–90 (2006)
41. C. Weisbuch, M. Nishioka, A. Ishikawa, Y. Arakawa, Observation of the coupled exciton-photon mode splitting in a semiconductor quantum microcavity. Phys. Rev. Lett. **69**(23), 3314 (1992)
42. J.P. Reithmaier, G. Sek, A. Löffler, C. Hofmann, S. Kuhn, S. Reitzenstein, L.V. Keldysh, V.D. Kulakovskii, T.L. Reinecke, A. Forchel, Strong coupling in a single quantum dot-semiconductor microcavity system. Nature **432**(7014), 197–200 (2004)
43. T. Yoshie, A. Scherer, J. Hendrickson, G. Khitrova, H.M. Gibbs, G. Rupper, C. Ell, O.B. Shchekin, D.G. Deppe, Vacuum Rabi splitting with a single quantum dot in a photonic crystal nanocavity. Nature **432**(7014), 200–203 (2004)
44. E. Peter, P. Senellart, D. Martrou, A. Lemaître, J. Hours, J.M. Gérard, J. Bloch, Exciton-photon strong-coupling regime for a single quantum dot embedded in a microcavity. Phys. Rev. Lett. **95**(6), 067401 (2005)
45. V. Savona, F. Tassone, C. Piermarocchi, A. Quattropani, P. Schwendimann, Theory of polariton photoluminescence in arbitrary semiconductor microcavity structures. Phys. Rev. B **53**(19), 13051 (1996)
46. H. Haug, S.W. Koch, *Quantum Theory of the Optical and Electronic Properties of Semiconductors* (World Scientific, Singapore, 1994)
47. N. Peyghambarian, S.W. Koch, A. Mysyrowicz, *Introduction to Semiconductor Optics* (Prentice Hall, Upper Saddle River, 1993)
48. M. Brune, F. Schmidt-Kaler, A. Maali, J. Dreyer, E. Hagley, J.M. Raimond, S. Haroche, Quantum rabi oscillation: a direct test of field quantization in a cavity. Phys. Rev. Lett. **76**(11), 1800–1803 (1996)
49. J. Bloch, T. Freixanet, J.Y. Marzin, V. Thierry-Mieg, R. Planel, Giant Rabi splitting in a microcavity containing distributed quantum wells. Appl. Phys. Lett. **73**(12), 1694–1696 (1998)
50. M. Saba, C. Ciuti, J. Bloch, V. Thierry-Mieg, R. Andre, L.S. Dang, S. Kundermann, A. Mura, G. Bongiovanni, J.L. Staehli, B. Deveaud, High-temperature ultrafast polariton parametric amplification in semiconductor microcavities. Nature **414**(6865), 731–735 (2001)
51. V. Savona, L.C. Andreani, P. Schwendimann, A. Quattropani, Quantum well excitons in semiconductor microcavities: unified treatment of weak and strong coupling regimes. Solid State Commun. **93**(9), 733–739 (1995)

52. J.J. Hopfield, Theory of the contribution of excitons to the complex dielectric constant of crystals. Phys. Rev. **112**(5), 1555 (1958)
53. G. Panzarini, L.C. Andreani, Quantum theory of exciton polaritons in cylindrical semiconductor microcavities. Phys. Rev. B **60**(24), 16799–16806 (1999)
54. F. Tassone, C. Piermarocchi, V. Savona, A. Quattropani, P. Schwendimann, Bottleneck effects in the relaxation and photoluminescence of microcavity polaritons. Phys. Rev. B **56**(12), 7554 (1997)
55. A.I. Tartakovskii, M. Emam-Ismail, R.M. Stevenson, M.S. Skolnick, V.N. Astratov, D.M. Whittaker, J.J. Baumberg, J.S. Roberts, Relaxation bottleneck and its suppression in semiconductor microcavities. Phys. Rev. B **62**(4), R2283 (2000)
56. H. Deng, D. Press, S. Götzinger, G.S. Solomon, R. Hey, K.H. Ploog, Y. Yamamoto, Quantum degenerate exciton-polaritons in thermal equilibrium. Phys. Rev. Lett. **97**(14), 146402 (2006)
57. G. Roumpos, C.-W. Lai, T.C.H. Liew, Y.G. Rubo, A.V. Kavokin, Y. Yamamoto, Signature of the microcavity exciton-polariton relaxation mechanism in the polarization of emitted light. Phys. Rev. B **79**(19), 195310 (2009)

Chapter 3
On the Condensation of Polaritons

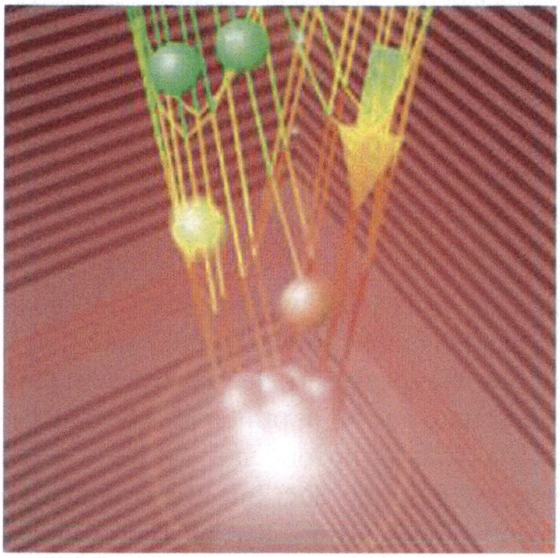

Abstract For polaritons, the phase transition to a macroscopically-occupied ground-state has attracted immense attention in the research community given the fact that exciton–polaritons in microcavities are ideal testbeds for investigations of many-particle physics in solids. As composite bosons formed out of the strong light–matter coupling in quantum-well microcavity systems, they enjoy prominent features such as a very light effective mass and thereby an elevated critical temperature for BEC when compared to other experimentally available particles for such studies. Being part light and part matter, their excitation, dynamics and condensation behaviour deserves special consideration. Firstly, bosonic many-particle features such as BEC and the effect of stimulated ground-state scattering will be explained. Consecutively, excitation and relaxation dynamics are discussed, which are fundamental to polariton condensation. In this context, the significance of overcoming the so-called relaxation bottleneck towards the ground state will be explained and stimulated scattering into a macroscopically occupied Bose state discussed.

A. Rahimi-Iman, *Polariton Physics*, Springer Series in Optical Sciences 229,
https://doi.org/10.1007/978-3-030-39333-5_3

3.1 Bosonic Many-Particle Features

Bosonic many-particle effects such as the Bose–Einstein condensation, superfluidity and superconductivity have fascinated scientists for about a century. It is natural that great efforts were devoted to the realization of experimental platforms for the demonstration as well as study of these effects. Polaritons serve this purpose well, since they are bosonic composite quasi-particles in solids. They are comparably light bosons and experimentally well accessible which explains the developments in the field of polariton physics. In this context, this section may serve as an introduction to the intriguing many-particle physics in polariton systems and provide a brief summary on the fundamentals of polariton BEC. A more general and detailed view on the subject of BEC can be well acquired for instance through [1, 2].

3.1.1 Condensation of a Bose Gase

The experimental demonstration of BEC was one of the key goals of the past decades. Naturally, its prediction and demonstration triggered intensive fundamental research activities in the field of bosonic many-particle systems. A macroscopic occupation of a system's energy ground state can be in principle achieved for any bosonic system owing to the quantum statistics of bosons and a process referred to as stimulated scattering. Due to the condensation effect into the common ground state, a many-particle system can become macroscopically coherent, which is explainable by the formation of a common wave-function for the condensed particles. In the ideal case, at zero temperature, the whole gas in its condensed state would be even described by a single giant matter wave, the Bose condensate. The bosonic particles' thermal distribution of such a system at thermodynamic equilibrium is then described by the Bose–Einstein distribution function (see [2–5])—since bosons obey Bose–Einstein-statistics.

$$f(E) = \frac{1}{\exp\left(\frac{E-\mu}{k_B T}\right) - 1}, \tag{3.1}$$

with k_B the Boltzmann constant and μ the chemical potential,[1] instead of being described by a classical thermal distribution of particles according to Boltzmann statistics:

$$f(E) = \frac{1}{\exp\left(\frac{E-\mu}{k_B T}\right)}. \tag{3.2}$$

[1]A negative number, if the lowest value of E is zero. This value resembles the energy needed to add a particle to the system, which becomes zero at the phase transition to a condensate.

The underlying concept was first proposed 1925 by Einstein for non-interacting, massive bosons on the basis of Bose's quantum statistical studies of massless particles [6, 7]. Nevertheless, it took 70 years until the first experimental demonstration of BEC in 1995, for which an ultracold diluted gas of Rubidium, Lithium or Sodium atoms was employed at $T \approx 0.2\,\mu K$ [8–10]. Almost simultaneously, cavity–polaritons entered the research stage after numerous efforts on exciton BEC in solids did not bear any fruits. Remarkably, in the following decade, a BEC-like phase transition was also achieved in the significantly lighter exciton–polariton system in QW-microcavities at (at that time) elevated temperatures of a few Kelvin [11, 12]. Moreover, in the same decade, the microscopic concept of an interacting Bose gas, which was proposed by Bogoliubov in 1947 in consideration of BEC, was experimentally confirmed for atoms [13] and polaritons [14], respectively. Additionally, researchers found that the phase transition toward a BEC-like polariton condensate is strongly connected to exotic properties of the condensed gas such as long-range order of the system [15–17], superfluidity [18–21] and the formation of quantized vortices and vortex–antivortex pairs [22–24].

Besides (quasi-)BEC ("quasi" accounts for the low-dimensionality in cavity–polariton systems, explained later), a number of other quantum phases have been discussed for polaritons such as superfluidity, but also the Berezinskii–Kosterlitz–Thouless (BKT) and Bardeen–Cooper–Schrieffer (BCS) phase, respectively [25] (see [26–29]). In this context, the polariton system has become a very promising testbed for experimental studies of these phase transitions. However, while no bosonic final state exists for bulk polaritons due to their linear energy-dispersion relation at low momenta which converges to zero, cavity–polaritons naturally provide a ground state for the formation of quantum degeneracy, conveniently located at $k_{\|} = 0$ (see energy–momentum dispersions in Chap. 2). In the following, the focus will be on condensation of a Bose gas regarding its fundamentals and prerequisites, so that later, the concept of the polariton laser can be described (see Chap. 4).

3.1.2 Criteria for Condensation

In principle, all bosonic systems have the ability to undergo Bose condensation, be it atoms, quasi-particles, or Higgs bosons. However, there are certain prerequisities for this to happen. The figures of merit for the underlying effect of spontaneous symmetry breaking—accumulating a macroscopic population in a single quantum state and forming a phase-coherent gas of indistinguishable bosons—are basically the temperature and the particle density. The critical temperature T_{crit} describes the temperature, below which a condensate for the given system can be formed, while the critical particle density υ_{crit} determines a condensation-threshold condition at a given temperature. For a 3D system, the following expression represents the critical condition as a function of the particle density υ and the particle mass m (see for instance [2]):

$$T_{\mathrm{crit}} = \left(\frac{\upsilon}{2.612}\right)^{2/3} \frac{h^2}{2\pi m k_{\mathrm{B}}}, \tag{3.3}$$

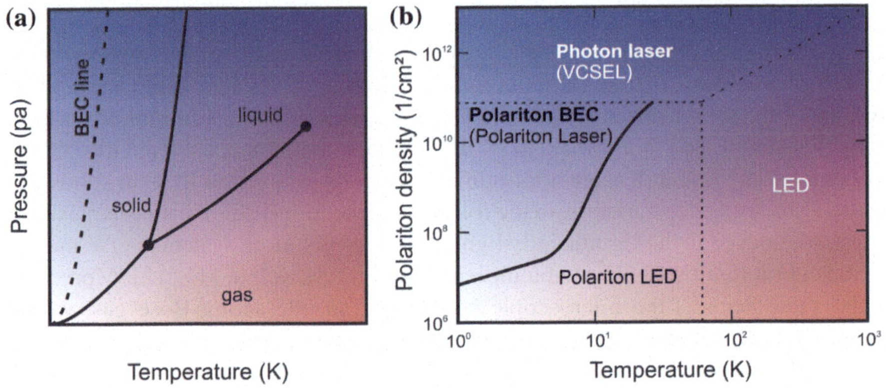

Fig. 3.1 **a** Schematic representation of a general phase diagram with gas, liquid and solid phases, together with the BEC line, after [2]. **b** A similar phase diagram is shown for a polaritonic system in GaAs after [18]. Indeed, for a Bose gas of atoms, the BEC phase can only be obtained in a very cold, diluted metastable state, provided that three-particle collisions are negligible and the transition to the solid state under equilibrium conditions thereby prevented. In contrast to atoms, polaritons can condense at orders of magnitude higher temperatures. The transition from a polariton LED to a polariton laser is indicated by a solid curve. In the corresponding phase diagram, the limits of the strong-coupling regime are indicated by vertically and horizontally drawn dashed lines, respectively. In real polariton systems, observation of condensation effects above certain particle densities and temperatures is impossible, because boundaries are imposed by Coulomb interaction between charge carriers, of which the excitons are composed, and thermal dephasing of excitons, respectively. Correspondingly, beyond these boundaries, the system enters the weak-coupling regime in which the operation mode of the system can be described as (conventional) photon laser (at high particle densities) or LED (below the lasing threshold), with the threshold condition indicated by a dotted line

with k_B the Boltzmann constant. This equation clearly illustrates the importance of the Bose-particle's mass with respect to the condensate threshold. To provide an exemplary overview over common phase transitions for a bosonic system, a schematic representation of a general phase diagram with gas, liquid and solid phases, together with the BEC line, respectively, is shown in Fig. 3.1a (see [2]), while a similar chart is given specifically addressing polaritons (b) (see [18]). Here, the phase transition's dependence on the pressure (a) and particle density (b) becomes obvious, respectively. At given temperature, a Bose gas only condenses when the particle density exceeds the critical density υ_{crit}. This corresponds to the situation, when the mean particle distance undercuts the thermal de-Broglie wavelength

$$\lambda_{\text{dB}}(T) = \sqrt{2\pi\hbar^2/(mk_{\text{B}}T)} \tag{3.4}$$

of that bosonic particle [30–32]. How crucial the role of the effective mass is, can be derived from Table 3.1. A schematic overview over the Bose gas' state as a function of the three relevant parameters, i.e. temperature, density and mass, has been given in Fig. 1.4 in Chap. 1.

Table 3.1 Overview on relevant parameters for atoms, excitons and polaritons with respect to Bose condensation, after [25]

Parameters	Atoms	Excitons	Polaritons
Effective mass m^*/m_e	10^3	$\approx 10^{-1}$	10^{-4} to 10^{-5}
Critical temperature T_{crit}	nK to μK	mK range	≈1 K up to 300 K
Bohr radius	10^{-2} nm	10^1 nm	≈ 10^2 to 10^3 nm
Thermalization time/Life time	1 ms/1 s ≈ 10^{-3}	10 ps/1 ns ≈ 10^{-2}	(1–10) ps/(1–10) ps ≈ 0.1–10

Although the assumptions concerning BEC hold true in a 3D system in which long-range order of the quantum gas can be effectively established, the ideal 2D system of bosons is actually deprived of this "privilege" to undergo a phase transition to a BEC at finite temperatures in the thermodynamic limit [33, 34]. The theoretical impossibility of BEC formation was shown in 1967 by Hohenberg, since he demonstrated that the common equations at $T > 0$ K diverge. Thus, a 2D system is not expected to macroscopically occupy the ground-state and to evolve off-diagonal long-range order (ODLRO), i.e. spatial coherence, as discussed in general for a system with dimensionality lower than two by Mermin and Wagner [33], after which the Mermin–Wagner theorem is named. This is explained by thermal fluctuations of the quantum state, which would prevail at any temperature and suppress condensation (see discussion of BEC for polaritons in [3, 35]). Nevertheless, a phase transition of this kind is still possible, provided that the constant density of states (DOS) in a 2D system is modified by a local potential variation [1, 36].

Local confinement potentials were found to suppress the aforementioned thermal fluctuations by introducing a finite energy splitting between single-particle ground state and excited states as the modes become discrete in the potential [37], thereby giving rise to condensate formation. Such potential landscapes are naturally given in real 2D structures, which are prone to inhomogeneities in size and composition, thus automatically introducing a break of symmetry and local confinement effects. In addition to natural confinement potentials, artificial traps can be tailored. The fact that condensation in 2D is facilitated by modulation of the potential landscape strongly explains the pronounced interest in the experimental realization of transverse confinement potentials for polariton systems. With such driving force provided, various types of lateral confinement structures have been demonstrated and tested for polaritons, as is evidenced throughout this book.

On the other hand, the establishment of long-range order is possible in 2D systems, given the special transition to a phase referred to as BKT phase (named after Berezinskii [38], Kosterlitz and Thouless [39]). In such a scenario, the remarkable effect of vortex–antivortex pair formation screens local phase defects, as was demonstrated experimentally for polariton systems [23, 24]. Furthermore, single quantized vortices [40] (as observed for other BEC systems and superfluid Helium), vortex arrays [41] and even half-quantum vortices [42] have been studied. Additionally, the existence of obstacles in the propagation path of a superfluid polariton gas, referred to as solitons, were experimentally evidences [41].

The BKT phase transition was discussed in the context of polariton condensation in [18], and represents a phase transition which typically occurs after condensate droplets formation or local BEC within the laser spot. Above the critical BKT temperature, quasi-condensate droplets form prior to the establishment of a superfluid phase, which requires a statistical connection of two points in space via a phase-coherent path between them. However, at the critical BKT temperature, a sudden percolation of condensate droplets takes place accompanied by vortex pairing and clustering (i.e. vanishing of free vortices, which prevent the establishment of long-range order), thereby enabling the formation of a superfluid. Below the critical BKT temperature, both normal and superfluid phases coexist [18]. Nevertheless, the assumption of the BKT phase transition only seems meaningful as topological transition in perfectly homogeneous systems and might be rethought as interpretation for polariton condensation, as discussed in the literature [3], since disorder and polariton localization are always present.

As microresonators are an ideal model system for the investigation of condensation phenomena and the properties of condensates as well as dynamical many-particle systems based on polaritons, the rich physics can be accessed experimentally in order to gain insight into the principles of BEC and to offer unique application possibilities. Owing to the fact that an experimental system only exhibits a finite number of single-particle states, the critical conditions for a quasi BEC in finite 2D systems can be defined by the following equations [25]: The critical 2D particle density is given by

$$\upsilon_{\mathrm{crit}}^{(2D)} = \frac{1}{S}\sum_{i\geq 2}\frac{1}{e^{\epsilon_i/k_{\mathrm{B}}T}-1}, \tag{3.5}$$

with discrete energy levels ϵ_i (with $i = 1, 2, \ldots$), chemical potential ϵ_1 and the area $S = L^2$ with length L of a 2D square. For such a 2D square, the critical conditions for $T_{\mathrm{crit}} > 0$ can be fulfilled at the critical density of

$$\upsilon_{\mathrm{crit}}^{(2D)} = \frac{2}{\lambda_T^2}\ln\frac{L}{\lambda_T}. \tag{3.6}$$

In such a 2D system, the phase transition can show similar properties like a BEC in the thermodynamic limit as soon as the particle number N is sufficiently high and the temperature T sufficiently low [31].

As mentioned above, the formation of vortices offers another unique phase transition, i.e. to the BKT phase between the normal state and superfluid state, in addition to quasi BEC. Local order is supported at low T by thermally excited vortices, while at temperatures undercutting a critical BKT temperature, pairwise occurrence of counter-circulating vortices stabilizes the phase and leads to a local finite superfluid [25], with the critical temperature at superfluid density υ_{s} given by

$$k_{\mathrm{B}}T_{\mathrm{crit}}^{(KBT)} = \upsilon_{\mathrm{s}}\frac{\pi\hbar^2}{2m^2}. \tag{3.7}$$

This equation can be also written in terms of the thermal de-Broglie wavelength $\upsilon_s \lambda_T (T_{\text{crit}})^2 = 4$ [43].

3.1.3 Dynamical Bose–Einstein Condensation of Polaritons

For many years, there has been a strive for the demonstration and characterization of BEC in semiconductor structures using excitons as well as polaritons as bosonic particles in solids [11, 44–47]. This hunt for condensation of excitonic particles was triggered by promising studies concerning collective properties of excitons in solids by Keldysh and Kozlov in 1968 [26]. However, for a very long time, the phase transition could not be obtained owing to various limitations existing for excitons, and difficulties were faced towards the successful investigation of excitonic condensates. Particularly, sufficiently high densities at experimentally accessible temperatures can hardly be reached without exciton dissociation into electron and hole plasma, or recombination prior to condensation [48]. On the one hand, the effective exciton mass sets the critical temperature to the mK range which first of all needs to be experimentally reached, while at the same time optical access to the Bose gas is necessary to probe quantum degeneracy. On the other hand, structural qualities and potential landscapes can have a strong influence on the localized particles' properties such as mobility and lifetime, resulting in a limited outreach to the next neighbours via propagation and an insufficient interaction time for excitons. Moreover, direct excitons are naturally prone to fast radiative recombination, and the potential landscape easily uncouples excitons from each other owing to trapping and energy shifts. Nevertheless, after strong efforts and a long stamina, first indicators of exciton condensation were recently demonstrated by High et al. employing indirect excitons in an exciton trap [49, 50].

In contrast to excitons, polaritons in QW microcavities offer an excellent testbed for condensation studies and were proposed as candidates for the demonstration of BEC and the realization of a new kind of coherent light source, namely the polariton laser [51, 52]. Its little effective mass, its dispersion with natural ground state at zero momentum and its short lifetime are key attributes which distinguishes the polariton gas from its atomic or excitonic counterpart. Moreover, as the polariton gas is not isolated from its environment and is affected by the host lattice, it differs significantly from its atomic pendant, which can be isolated in vacuum, in terms of equilibrium conditions. This leads to the formation of dynamical BEC, which can be practically probed at relatively well accessible temperatures using optical spectroscopy, while the condensate is typically not close to thermal equilibrium with the host lattice, although thermal equilibrium within the polariton gas is possible. Owing to energy and momentum conservation, insight into the dynamics and properties of such Bose gas in the solid-state system under continuous radiative decay is provided. The prerequisite for polariton studies is that at given critical pump rate and operation temperature the bosonic character is still preserved, which corresponds directly to the existence of excitons in a strongly-coupled regime with the light field of an optical microcavity system.

Precursory studies showed, that polaritons indeed represent "good" composite bosons up to densities high enough for the observation of quantum degeneracy [53] and even up to relatively high temperatures [54]. However, it is important to note that the accessible temperature range strongly depends on the semiconductor system and materials. Also, common obstacles towards excitonic BEC can be circumvented using polaritons, such as the impact of crystal defects and structural inhomogeneities on the exciton gas [25]. The matter excitation dressed with the cavity photon as a consequence of strong coupling between light and matter becomes delocalized in the optical medium and thereby the implications related to inhomogeneous broadening of excitonic transitions can be strongly softened [45].

Furthermore, the amount of available quantum-well excitons for polariton condensation can be increased by the employment of multiple coupled quantum wells, which allows to ramp the polariton density above the required critical density while the actual exciton density per quantum well remains below the exciton saturation density and diluted. Thereby, one can preserve strong coupling and facilitate the desired phase transition for these composite bosons [55]. This saturation density marks the point above which polaritons break up due to exciton ionisation as a consequence of Coulomb screening and particle–particle interaction [56, 57]. As the number of free states in the electronic bands is limited, the so-called phase-space filling effect [58, 59] reduces the exciton oscillator strength f [56], and thereby the coupling strength g. This takes place in a system at higher excitation rates which is not anymore representing a low particle density regime (i.e. with many unoccupied states for electrons and holes). The corresponding transition from an insulator state in the bosonic regime, where the neutral excitons exist, to a metallic plasma in the fermionic regime above a critical particle density, where a hot uncoupled gas of electrons and holes prevails, is referred to as Mott transition [60, 61]. Above the Mott density ($\approx 10^{-11}\,\text{cm}^{-2}$ in GaAs QWs), the system naturally enters a weak-couping regime (approximately, when mean exciton distance comparable to Bohr radius) and instead of polaritons, photonic modes govern the emission spectra of the QW-microcavity system.

Figure 3.1b shows a schematic representation of a phase diagram for polaritons in the GaAs material system after [18], which depicts the relevant regimes of operation for a QW microresonator in the strong and in the weak coupling regime. Dashed vertical and horizontal lines mark a box that includes densities and temperatures for which strong coupling is preserved. Correspondingly, an uncondensed gas of polaritons can undergo a phase transition to quasi BEC (threshold represented by a line) in the strong-coupling range. However, beyond the density or temperature margin set for polaritons in this diagram, a dotted line indicates the threshold for photonic lasing from a conventional QW-laser diode, since the system is governed by weak coupling at too high temperatures and densities.

Based on intensive efforts in the early years of the last decade, dynamical BEC or Bose condensation away from thermal equilibrium was eventually obtained experimentally for polaritons at cryogenic temperatures around 5 K both in the GaAs

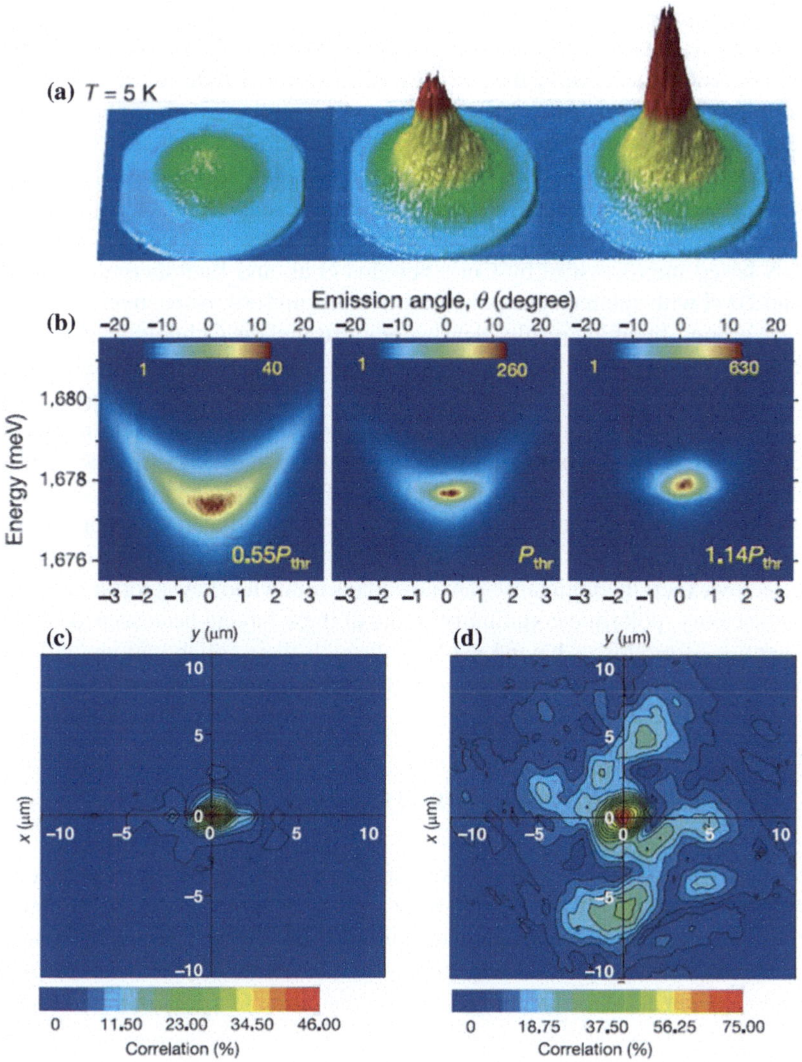

Fig. 3.2 Experimental demonstration of a dynamical BEC of exciton–polaritons in an optical CdTe-based microcavity system. A clear k-space narrowing occurs in a density-dependent emission characterization in Fourier-space projection (**a**). **b** Corresponding energy–k-resolved false-colour contour diagrams indicate also spectral narrowing above the condensation threshold. The nonlinearity in the input–output characteristics of a polariton system is typically accompanied by a linewidth narrowing around the threshold and a gradual increase of the ground-state energy, see Fig. 8.6(**a**) in Sect. 8.2. Higher densities typically spoil the linewidth and energy position of the condensate owing to interaction effects in the system. Actual proof for a BEC-like state of the polariton gas is given by the demonstration of Bose–Einstein-like particle distribution in the energy-dependent occupancy of states, see Fig. 8.6(**b**) in Sect. 8.2, and, most importantly, spatial correlation measurements. Using optical interferometry as shown in (**c**) and (**d**) for excitation densities below and above condensation threshold, respectively, spontaneous coherence build-up has been revealed. Reused with permission. Reference [12] Copyright 2006, Springer Nature

(III/V) [11, 62] and the CdTe (II/VI) material systems [12, 63], respectively. After first condensate demonstrations, the key to the claim of a BEC-like state was primarily given by studies concerning the particles' occupation distribution and the system's coherence, most importantly spatial coherence representing long-range order, as presented for the first time by Kasprzak, Richard, Kundermann, Baas et al. in 2006 [12] for their CdTe structure (see Fig. 3.2). Shortly after, similar evidence was given for condensates in GaAs structures by both Deng et al. and Balili et al. [62, 64]. In addition, dynamic condensation of polaritons was also reported at room temperature for GaN based microcavities by Christopoulos et al. and Baumberg et al. in 2007 [65] and 2008 with spontaneous polarization build-up [66], respectively.

These ground-breaking results clearly demonstrated the technological advance in the field of microcavity fabrication and characterization, at that time driven by several key groups around the world, such as the groups of Yamamoto (Stanford University), Deveaud-Plédran, Morier-Genoud and Grandjean (EPF de Lausanne), André and Dang (University of Grenoble), Lemaître and Bloch (CNRS/LPN Marcoussis), Skolnick and Roberts (University of Sheffield), Forchel (University of Würzburg), Snoke (University of Pittsburgh), and Pfeiffer and West (Princeton University, before at Bell Labs), to name but a few. Indeed, many more experimental groups—typically supported by expert theoreticians—are nowadays renowned key actors and contributors in the cavity–polariton community, many of these having been seeded out of the aforementioned groups or having evolved through their collaborations with them, and a selection of their work will be subject of the following chapters.

3.2 Excitation and Relaxation Dynamics

To achieve polariton lasing by providing critical particle densities or by undercutting critical temperatures, some obstacles have to be overcome owing to the nature of polaritons. Indeed, that is partly the reason why the observation of condensation effects was impeded so much although first polariton structures were available already in the 1990s—put aside the unambiguous demonstration of a polaritonic condensate.

3.2.1 Excitation of Polaritons

One main issue for all polariton systems can be addressed more or less easily in two ways, which, however, exhibit some drawbacks of their own. The successful implementation sometimes strongly relies on seemingly hidden system parameters, structural design aspects and experimental limitations. It is about the steady excitation of an open, thus, leaky system, in order to compensate for losses and to maintain the polariton gas. As mentioned before in Sect. 2.2.1, optical and electrical pumping schemes are available. In short, polaritons can be conveniently injected into the system by a pump laser or current. However, while optical pumping can be applied to excite polaritons locally, and to generate charge carriers at a certain excitation energy level using non-resonant pumping or to directly inject polaritons in one of the polariton branches by resonant techniques, electrical pumping is a totally different scenario. Even if in principle optically-pumped condensates could be achieved in a given structure, this was not readily possible in the same structure by means of electrical pumping, even though technical preparations for current injection were made.

Electrical pumping requires strong efforts in order to design effective injection schemes, which flood the sample with hot charge carriers in a typically non-resonant diode-style pumping scheme while at the same time have to efficiently fill multiple quantum wells in those microresonators homogeneously. Additionally, a well determined current flow through the structure requires further technical considerations such as patterning. Nevertheless, electroluminescence from polariton LEDs had been demonstrated multiple times in the last decade [67–69], while non of the structures could enter a regime of polariton condensation in the early stage of device development. Eventually, electrically driven condensates—condensates on demand—came into reach after the successful demonstration of a polariton laser diode in 2013 [70, 71], using current injection into optimized structures with p-i-n-type heterojunction.

3.2.2 Relaxation Towards the Energy Minimum

The other main issue arises when relaxation processes shall lead to effective relaxation of the (weakly-interacting and low-density) bosonic quasi-particles into the common final state. This can be addressed by setting a proper detuning in order to provide certain attributes to polaritons: such as a higher scattering cross section and interaction propability for efficient thermalization; such as longer lifetimes in order to overcome all necessary relaxation steps within the particles lifetime; such as a suitable energy dispersion to avoid or promote interaction with lattice phonons and to set the depth of the ground-state dip in the lower polariton branch. It is worth noting that the *proper* parameters and a *suitable* dispersion to a great extent depend on the type of structure that is given, the quality of the structure and the desired output from the device, as there is no standard operation mode for a microcavity in order to achieve condensation. Instead, a few rules of thumb allow one to optimize the device for certain features. This has a lot to do with the hybrid nature of polaritons, which is not only detrimental with regard to the relaxation process, but also to the type of quantum degeneracy in the final state achieved. This is strongly affected by the experimental settings.

Even when effective relaxation is possible, one obtains a dynamical system in which macroscopic ground-state population not only has to be accomplished, but even continuously maintained: In microcavities, polaritons decay radiatively through the cavity mirrors with finite reflectivity and depopulate the polariton gas in the excited system. As mentioned before, this is very helpful when it comes to the characterization of polaritons. In contrast, it is the major obstacle with respect to ground-state

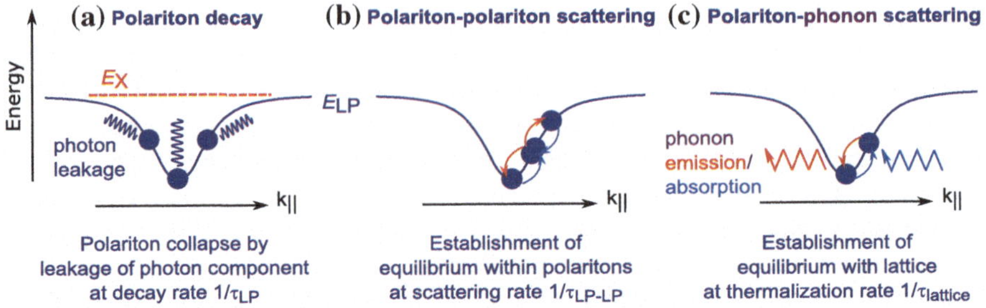

Fig. 3.3 Sketch of the polaritonic decay and relaxation by two prominent scattering processes. From left to right, polariton decay, polariton–polariton scattering and polariton–phonon scattering. Both scattering processes play an important role in the thermalization process of the polariton gas. While interactions between polaritons efficiently establish equilibrium conditions within the gas of particles, interactions with the phonon bath further thermalize the total system, so that equilibrium of the condensate with the lattice can get established. After [72]

relaxation and condensation, and more importantly towards the establishment of an equilibrium state in any polaritonic system (see Fig. 3.3). Thermalization with the lattice is hardly possible when polaritons decay on time scales (typically $< 10\,\mathrm{ps}$) faster than the relaxation time (in case of linear LP-phonon scattering 10–50 ps). At best, in typical high-quality microcavity systems the lifetime approaches or approximates the thermalization time ($\approx 10\,\mathrm{ps}$), whereas excitonic recombination in an uncoupled regime would take place on a much longer time scale (up to ns) (characteristic times according to [25]).

Indeed, according to (2.40), the polariton lifetime in the LP branch $\tau_{\mathrm{LP}} = \hbar/\gamma_{\mathrm{LP}}$ can be tuned to reach longer lifetimes in a more excitonic regime, yet at the price of stronger localization and heavier effective mass. In addition, the lower-polariton lifetime is always shorter in the proximity of $k_{\parallel} = 0$ than in higher LP energy states, since the photonic component dominates the polariton's properties around there. This has strong implications for the relaxation into the ground state of the LP dispersion, for which several scattering processes are necessary. While the photonic fraction of the polariton increases further and further the more the polariton approaches the ground states, its scattering cross section and thereby the interaction rate with particles or the lattice reduces more and more. Thus, polaritons experience a relaxation bottleneck on the way to the dispersion ground state. Moreover, the more photonic the polariton, the faster its radiative decay rate, which makes the bottleneck feature well visible in angle-dependent spectra. These lifetime and relaxation aspects are schematically illustrated in Fig. 3.3 and summarized in Table 3.2.

Table 3.2 Relationship between time scales for relaxation τ_{LP-LP}, thermalization $\tau_{lattice}$ and spontaneous polariton decay τ_0 with regard to the observation of condensation. Illustrations of the relevant processes are provided in Fig. 3.3. After [72]

Regime	Non-equilibrium	Quasi-equilibrium among polaritons	Thermal equilibrium with host lattice
Time scales	$\tau_0 < \tau_{LP-LP} < \tau_{lattice}$	$\tau_{LP-LP} < \tau_0 < \tau_{lattice}$	$\tau_{LP-LP} < \tau_{lattice} < \tau_0$
Temperatures	T_{LP} not defined	$T_{LP} > T_{lattice}$	$T_{LP} \cong T_{lattice}$
Gas state	(Bottleneck)	Quasi-condensate	(Quasi-)thermal equilibrium condensation
Lifetime optimization towards...	Matter laser	Dynamical condensate	BEC-like state

3.2.3 *The Bottleneck Effect*

Depending on the excitation scheme, several relaxation steps precede the scattering processes where the bottleneck comes into play (Fig. 3.4). These are summarized below. Usually, non-resonant excitation creates hot charge carriers which relax via

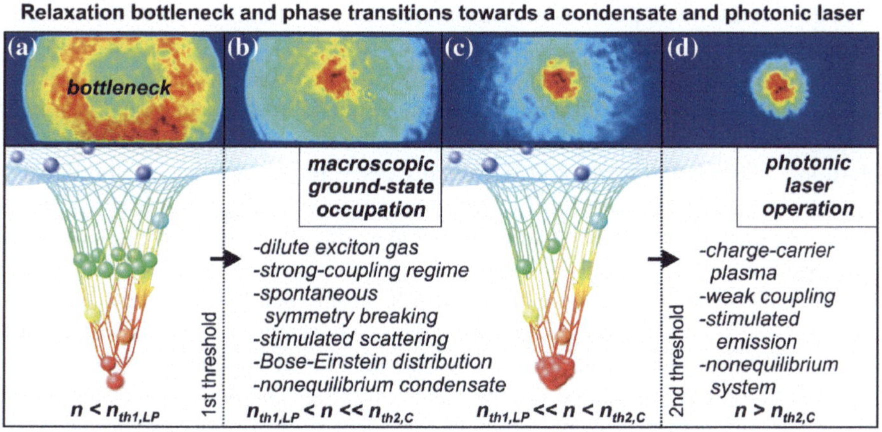

Fig. 3.4 Fourier-space-resolved EL-emission profile (within the detectable k-space circle) of a polariton system with relaxation bottleneck occurrence (**a**) prior to a phase transition to a condensate (**b**). False-colour plots with intensity increase from blue to red. Strong energy and momentum narrowing is experienced well above the first threshold, as indicated by the momentum distribution of the polaritonic emission in (**c**). At higher excitation densities, a second phase transition to lasing in the weak-coupling regime (**d**) usually takes place due to the Mott transition from an insulator to a metallic phase of the electronic system. Characteristics of these transitions are briefly summarized, whereas corresponding schematic lower polariton branch occupations for the below and above threshold density regimes are displayed next to it. Coloured balls represent polaritons of different energy on their dispersion branch (coloured mesh). Adapted from [73]. Courtesy of the author

optical-phonon scattering to hot excitons, incoherent electron–hole pairs, at high k-vectors. After further scattering processes, these excitons cool down to reach k-space regions at which interactions with cavity photons become possible [74, 75]. There, an exciton–polariton reservoir is formed, from which the quasi-particles make their way to the energy ground state in the LP dispersion. Alternatively, optical side-pumping can directly inject polaritons into this reservoir to prevent a strong background of hot charge carriers (further examples of excitation schemes shall be given in the following chapters and discussed later wherever appropriate).

Towards the ground state, a more and more pronounced photonic character, which corresponds to a decreasing matter component, leads to a reduced scattering efficiency with acoustic phonons. At low pump rates, acoustic phonons can be seen as the main relaxation process. At the same time, optical phonon states can be neglected as an alternative scattering mechanism at cryogenic temperatures [76, 77]. Since the process of phonon emission is elastic, energy and momentum conservation needs to be fulfilled [75]:

$$E_{\mathrm{LP}}(k_{\parallel}^{(\mathrm{ph})}) = E_{\mathrm{LP}}(k_{\parallel}^{(\mathrm{i})}) - E_{\mathrm{LP}}(k_{\parallel}^{(\mathrm{f})}), \tag{3.8}$$

$$k_{\parallel}^{(\mathrm{ph})} = k_{\parallel}^{(\mathrm{i})} - k_{\parallel}^{(\mathrm{f})}. \tag{3.9}$$

with i and f denoting the initial and final scattering state, respectively. With the dispersion of phonons almost flat on the energy scale compared to the polariton branch, and some limitations in phonon emission steps, the maximum energy dissipated per single phonon emission is less than 2 meV in GaAs [75]. The transition from the reservoir to the ground state is thus governed by a bottleneck effect, which occurs in the k-space region around the inflection point of the LP dispersion owing to a typically one order of magnitude higher ground-state relaxation time compared to the polaritons' lifetime at this point [78]. In many cases, the occurrence of a bottleneck can be prevented by the choice of a detuning between small negative and positive values. The deeper the LP dispersion ground-state dip, the more likely a bottleneck effect becomes.

Finally, if polaritons overcome the bottleneck within their lifetime, they can accumulate in the ground state (Fig. 3.4). However, the occurrence of a bottleneck makes this an inefficient process with little chance to reach threshold conditions for a nonlinear effect, namely stimulated ground-state scattering. The processes experienced by the externally-driven system, which can be excited optically or electrically, are summarized in a sketch of relaxation to the ground state from higher energy states in Fig. 3.5. To keep it as general as possible, non-resonant excitation of hot charge-carriers is assumed here, and the polariton system is depicted for zero detuning.

3.2.4 Stimulated Ground-State Scattering

Up to this point, polariton relaxation was described purely in the linear regime without any nonlinear effects. Indeed, the dynamics of relaxation can be summarized

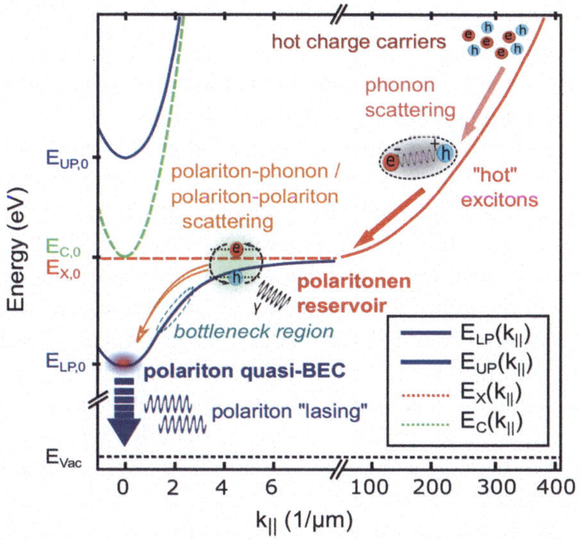

Fig. 3.5 Chain of processes experienced by excited charge carriers towards exciton–polariton formation and condensation. Via phonon scattering, electrons and holes cool down to form hot Coulomb-bound pairs on the exciton dispersion which further cool down until they enter the light-cone region. There, light–matter interaction gives rise to polaritons at finite $k_{\|}$, where a reservoir is formed. The light–matter coupled species, indicated by a green cloud behind an excited two-level matter system, can further dissipate momentum and energy by polariton–phonon or polariton–polariton scattering. Typically, a bottleneck occurs at the inflection point of the lower-polariton dispersion owing to slow relaxation times in comparison to increasingly shortened polariton lifetimes. Stimulated final-state scattering, as a mechanism to overcome the bottleneck and efficiently populate the ground state, is indicated by an orange arrow into the dispersion dip towards zero in-plane momentum. When a condensate is formed, decay of polaritons from a coherent bath results in polariton lasing, i.e. leakage of correlated photons. Adapted from [73]. Courtesy of the author

by the following rate equation even neglecting nonlinearities, since those are only obtained above aforementioned threshold conditions. In principle, these simplified equations do not change for different pump densities, while under stimulated scattering conditions the underlying relaxation rates change drastically corresponding to an efficient mechanism to accumulate bosons into their common ground state. The model can be described as follows:

$$\frac{\partial N_R}{\partial t} = P(t) - \frac{N_R}{\tau_R} - \frac{N_R}{\tau_{relax}}, \tag{3.10}$$

$$\frac{\partial N_0}{\partial t} = -\frac{N_0}{\tau_0} + \frac{N_R}{\tau_{relax}}. \tag{3.11}$$

with N_R and N_0 the number of reservoir particles and final-state particles, respectively, τ_R and τ_0 the lifetimes in the corresponding state, and τ_{relax} the relaxation time, respectively. Concerning the former equation, $P(t)$ denotes the pump rate, while the second term summarizes losses from the reservoir and the third term represents the transfer of particles from reservoir states to the final state $|0\rangle$. In the latter equation, the first term represents the losses in the final state and the second term the gain obtained from refilling by relaxation from reservoir states.

In semi-classical Boltzmann equations presented in [25], the rate equation term concerning relaxation N_R/τ_{relax} is further split into two components, one representing LP–LP scattering and the other one LP–phonon scattering, which also take into account the occupancy of the states, i.e. stimulated scattering. LP–LP scattering is explained to be the driving force behind thermalization of the LP gas within a few ps, while it cannot lower the total energy of the LP gas further. Cooling of the polariton gas is mainly accomplished by LP–phonon scattering, which has a distinct role in the gas' thermalization with the lattice. In pulsed experiments such as those of Deng et al. in 2002 and 2006, the obtained polariton signal well represents such build-up and decay of the ground state emission and such modelled results from Boltzmann kinetics turned out to be in good agreement with the experimental data [79]. In the simplified rate equation model above, τ_{relax} summarizes these characteristic ground-state scattering times in a mean value for a given experimental pump rate. For more details about theoretical considerations, the interested reader may refer to [25] and the references within.

Stimulated final-state scattering of bosonic polaritons is the final step to polariton condensation, provided that, at high excitation densities around the critical density for condensation, the strong-coupling regime prevails. This nonlinear effect is detrimental when it comes to the efficient accumulation of higher-energy LPs in their dispersion ground state. At high particle densities, LP–LP interaction via the particles' excitonic component scatter a macroscopic amount of polaritons into their lowest energy state. The interaction between excitons at higher densities is mainly governed by exchange Coulomb interactions, which lead to repulsive interaction between polaritons of same spin and provide a fast scattering mechanism.

This nonlinear stimulation effect, which strongly depends on the number of particles in the same quantum state, drastically reduces LP–LP scattering times and strongly accelerates the ground-state relaxation process (reducing τ_{relax} in (3.11) significantly). To utilize this effect, polaritons have to be able to make their way into the final state in order to form a seed population. If the final (f) state is occupied by N_f bosonic particles, scattering into this quantum state is increased by a factor of $1 + N_f$, in which the particles are described by a macroscopic wave-function.

In principle, having one particle in the final state attracts further particles to pile up in this state, as bosons underly Bose–Einstein statistics. With a seed population provided, a quantum statistically degenerate state is obtained as soon as the number of polaritons N_f in the final state is much larger than 1. Under such condition, the final state population attracts polaritons massively, with stimulated scattering dominating spontaneous relaxation processes by a factor of $1 + N_f$ [25]. This can be represented in the semi-classical Boltzmann kinetics by the replacement of υ_R/τ_{relax} (υ_R denoting the exciton density in the reservoir) by a modified term which includes final-state density-dependent (stimulated) scattering (factor $1 + \upsilon_0$, with ground-state density υ_0) and different scattering terms such as for exciton–phonon (a), exciton–exciton

(b), and exciton–electron (free charge carriers) (c) scattering, respectively:

$$\frac{\partial \upsilon_{\mathrm{R}}}{\partial t} = P_{\mathrm{R}} - \frac{\upsilon_{\mathrm{R}}}{\tau_{\mathrm{R}}}$$

$$-a\upsilon_{\mathrm{R}}(1+\upsilon_0) + a\mathrm{e}^{\frac{\Delta_{\mathrm{esc}}}{-k_{\mathrm{B}}T}}\upsilon_{\mathrm{R}}\upsilon_0 - b\upsilon_{\mathrm{R}}^2(1+\upsilon_0) - c\upsilon_{\mathrm{e}}'\upsilon_{\mathrm{R}}(1+\upsilon_0), \tag{3.12}$$

$$\frac{\partial \upsilon_0}{\partial t} = -\frac{\upsilon_0}{\tau_0}$$

$$+a\upsilon_{\mathrm{R}}(1+\upsilon_0) - a\mathrm{e}^{\frac{\Delta_{\mathrm{esc}}}{-k_{\mathrm{B}}T}}\upsilon_{\mathrm{R}}\upsilon_0 + b\upsilon_{\mathrm{R}}^2(1+\upsilon_0) + c\upsilon_{\mathrm{e}}'\upsilon_{\mathrm{R}}(1+\upsilon_0). \tag{3.13}$$

Δ_{esc} represents the energy difference between final-state and reservoir, from which reservoir refilling can take place (by escaping the ground state), and P_{R} denotes the pumping rate into the reservoir. Constants a, b, and c are prefactors for the processes (a) to (c), respectively.

Quantum degeneracy is achieved below the critical temperature for condensation, and the condensate fraction determines the purity of the condensed state, given by the ratio of condensed particles to the total amount of those particles in the system $\upsilon_0/\upsilon_{\mathrm{tot}}$. On the one hand, owing to quantum depletion of the final state or strong interactions between particles depending on the given configuration, polariton condensates not necessarily obtain a good condensate fraction. On the other hand, since the effective mass is small and with it the energy-density of states (DOS) of polaritons, a quantum degenerate seed of low-energy polaritons in the LP branch is achieved relatively easily.

One should note that in the leaky photon-emitting QW-microcavity system, in which the particle number is not preserved but needs to be constantly uphold through a refilling process from incoherent reservoir states, the polariton ensemble also suffers noise related to these losses as a direct consequence of the fluctuation–dissipation theorem [80]: Accordingly, polariton condensation cannot occur at arbitrarily low densities. It is also stressed in [80] that inevitable polariton losses from the condensed state can affect its low-frequency collective behavior, rendering it the most important difference between superfluid quantum liquids (such as for Helium) and Bose condensates of exciton–polaritons.

References

1. A. Griffin, D.W. Snoke, S. Stringari, *Bose-Einstein Condensation* (Cambridge University Press, Cambridge, 1995)
2. L. Pitaevskii, S. Stringari, *Bose-Einstein Condensation* (Oxford University Press, Oxford, 2003)
3. B. Deveaud, *The Physics of Semiconductor Microcavities* (Wiley, Hoboken, 2007)
4. A.V. Kavokin, J.J. Baumberg, G. Malpuech, F.P. Laussy, *Microcavities* (Oxford University Press, Oxford, 2007)
5. H. Haug, S.W. Koch, *Quantum Theory of the Optical and Electronic Properties of Semiconductors*, 5th edn. (World Scientific Publishing Co., Singapore, 2009)

6. S.N. Bose, Plancks Gesetz und Lichtquantenhypothese. Z. Phys. **26**, 178 (1924)
7. A. Einstein, Quantentheorie des einatomigen idealen Gases. 2. Abhandlung. Sitz. Ber. Preuss. Akad. Wiss. **22**, 261 (1925)
8. M.H. Anderson, J.R. Ensher, M.R. Matthews, C.E. Wieman, E.A. Cornell, Observation of Bose-Einstein condensation in a dilute atomic vapor. Science **269**(5221), 198–201 (1995)
9. C.C. Bradley, C.A. Sackett, J.J. Tollett, R.G. Hulet, Evidence of Bose-Einstein condensation in an atomic gas with attractive interactions. Phys. Rev. Lett. **75**(9), 1687 (1995)
10. K.B. Davis, M.O. Mewes, M.R. Andrews, N.J. van Druten, D.S. Durfee, D.M. Kurn, W. Ketterle, Bose-Einstein condensation in a gas of sodium atoms. Phys. Rev. Lett. **75**(22), 3969 (1995)
11. H. Deng, G. Weihs, C. Santori, J. Bloch, Y. Yamamoto, Condensation of semiconductor microcavity exciton polaritons. Science **298**(5591), 199–202 (2002)
12. J. Kasprzak, M. Richard, S. Kundermann, A. Baas, P. Jeambrun, J.M.J. Keeling, F.M. Marchetti, M.H. Szymańska, R. André, J.L. Staehli, V. Savona, P.B. Littlewood, B. Deveaud, L.S. Dang, Bose-Einstein condensation of exciton polaritons. Nature **443**(7110), 409–414 (2006)
13. D.L. Feder, A.A. Svidzinsky, A.L. Fetter, C.W. Clark, Anomalous modes drive vortex dynamics in confined Bose-Einstein condensates. Phys. Rev. Lett. **86**(4), 564–567 (2001)
14. S. Utsunomiya, L. Tian, G. Roumpos, C.W. Lai, N. Kumada, T. Fujisawa, M. Kuwata-Gonokami, A. Löffler, S. Höfling, A. Forchel, Y. Yamamoto, Observation of Bogoliubov excitations in exciton-polariton condensates. Nat. Phys. **4**(9), 700–705 (2008)
15. L. Landau, E. Lifshitz, *Statisticheskai Fizika* (Fizatgiz, Moskau, 1951)
16. O. Penrose, On the quantum mechanics of Helium II. Phil. Mag. **42**, 1373–1377 (1951)
17. O. Penrose, L. Onsager, Bose-Einstein condensation and liquid helium. Phys. Rev. **104**(3), 576 (1956)
18. A. Kavokin, G. Malpuech, F.P. Laussy, Polariton laser and polariton superfluidity in microcavities. Phys. Lett. A **306**(4), 187–199 (2003)
19. I. Carusotto, C. Ciuti, Probing microcavity polariton superfluidity through resonant rayleigh scattering. Phys. Rev. Lett. **93**(16), 166401 (2004)
20. Y.G. Rubo, A.V. Kavokin, I.A. Shelykh, Suppression of superfluidity of exciton-polaritons by magnetic field. Phys. Lett. A **358**(3), 227–230 (2006)
21. A. Amo, J. Lefrere, S. Pigeon, C. Adrados, C. Ciuti, I. Carusotto, R. Houdré, E. Giacobino, A. Bramati, Superfluidity of polaritons in semiconductor microcavities. Nat. Phys. **5**(11), 805–810 (2009)
22. K.G. Lagoudakis, M. Wouters, M. Richard, A. Baas, I. Carusotto, R. Andre, L.S. Dang, B. Deveaud-Plédran. Quantized vortices in an exciton-polariton condensate. Nat. Phys. **4**(9), 706–710 (2008)
23. M.D. Fraser, G. Roumpos, Y. Yamamoto, Vortex-antivortex pair dynamics in an exciton-polariton condensate. New J. Phys. **11**, 113048 (2009)
24. G. Roumpos, M.D. Fraser, A. Löffler, S. Höfling, A. Forchel, Y. Yamamoto, Single vortex-antivortex pair in an exciton-polariton condensate. Nat. Phys. **7**(2), 129–133 (2011)
25. H. Deng, H. Haug, Y. Yamamoto, Exciton-polariton Bose-Einstein condensation. Rev. Mod. Phys. **82**(2), 1489 (2010)
26. L.V. Keldysh, A.N. Kozlov, Collective properties of excitons in semiconductors. Sov. Phys. JETP **27**, 521–528 (1968)
27. G. Malpuech, Y.G. Rubo, F.P. Laussy, P. Bigenwald, A.V. Kavokin, Polariton laser: thermodynamics and quantum kinetic theory. Semicond. Sci. Technol. **18**, 395 (2003)
28. D. Sarchi, V. Savona, Collective excitation kinetics in the condensation of polaritons. Phys. Stat. Sol. B **243**(10), 2317–2321 (2006)
29. D. Sarchi, V. Savona, Long-range order in the Bose-Einstein condensation of polaritons. Phys. Rev. B **75**(11), 115326 (2007)
30. D.S. Fisher, P.C. Hohenberg, Dilute Bose gas in two dimensions. Phys. Rev. B **37**(10), 4936 (1988)
31. W. Ketterle, N.J. van Druten, Bose-Einstein condensation of a finite number of particles trapped in one or three dimensions. Phys. Rev. A **54**(1), 656–660 (1996)

32. F. Tassone, Y. Yamamoto, Exciton-exciton scattering dynamics in a semiconductor microcavity and stimulated scattering into polaritons. Phys. Rev. B **59**(16), 10830 (1999)
33. N.D. Mermin, H. Wagner, Absence of ferromagnetism or antiferromagnetism in one- or two-dimensional isotropic heisenberg models. Phys. Rev. Lett. **17**(22), 1133 (1966)
34. P.C. Hohenberg, Existence of long-range order in one and two dimensions. Phys. Rev. **158**(2), 383 (1967)
35. A.V. Kavokin, J.J. Baumberg, G. Malpuech, F.P. Laussy, *Microcavities*, vol. 1 (Oxford University Press, Oxford, 2017)
36. V. Bagnato, D. Kleppner, Bose-Einstein condensation in low-dimensional traps. Phys. Rev. A **44**, 7439–7441 (1991)
37. J. Lauwers, A. Verbeure, V.A. Zagrebnov, Bose-Einstein condensation for homogeneous interacting systems with a one-particle spectral gap. J. Stat. Phys. **112**(1–2), 397–420 (2003)
38. V.L. Berezinskii. Destruction of long-range order in one-dimensional and two-dimensional systems possessing a continuous symmetry group. II. Quantum systems. J. Exp. Theor. Phys. **34**, 610 (1972)
39. J.M. Kosterlitz, D.J. Thouless, Ordering, metastability and phase transitions in two-dimensional systems. J. Phys. C: Solid State Phys. **6**(7), 1181 (1973)
40. G. Tosi, D. Sanvitto, M. Baudisch, E. Karimi, B. Piccirillo, L. Marrucci, A. Lemaître, J. Bloch, L. Viña, Vortex stability and permanent flow in nonequilibrium polariton condensates. J. Appl. Phys. **109**, 102406 (2011)
41. A. Amo, S. Pigeon, D. Sanvitto, V.G. Sala, R. Hivet, I. Carusotto, F. Pisanello, G. Leménager, R. Houdré, E. Giacobino, C. Ciuti, A. Bramati, Polariton superfluids reveal quantum hydrodynamic solitons. Science **332**(6034), 1167–1170 (2011)
42. K.G. Lagoudakis, T. Ostatnický, A.V. Kavokin, Y.G. Rubo, R. André, B. Deveaud-Plédran, Observation of half-quantum vortices in an exciton-polariton condensate. Science **326**(5955), 974–976 (2009)
43. I. Khalatnikov, *An introduction to the theory of superfluidity* (W.A. Benjamin, New York, 1965)
44. L.S. Dang, D. Heger, R. André, F. Bœuf, R. Romestain, Stimulation of polariton photoluminescence in semiconductor microcavity. Phys. Rev. Lett. **81**(18), 3920 (1998)
45. Y. Yamamoto, A. Imamoglu, *Mesoscopic Quantum Optics* (Wiley, Hoboken, 1999)
46. S.A. Moskalenko, D.W. Snoke, *Bose-Einstein Condensation of Excitons and Biexcitons: And Coherent Nonlinear Optics with Excitons* (Cambridge University Press, Cambridge, 2000)
47. L.V. Butov, A.L. Ivanov, A. Imamoglu, P.B. Littlewood, A.A. Shashkin, V.T. Dolgopolov, K.L. Campman, A.C. Gossard, Stimulated scattering of indirect excitons in coupled quantum wells: signature of a degenerate Bose-Gas of excitons. Phys. Rev. Lett. **86**, 5608 (2001)
48. P.B. Littlewood, X. Zhu, Possibilities for exciton condensation in semiconductor quantum-well structures. Phys. Scr. **1996**(T68), 56 (1996)
49. A.A. High, J.R. Leonard, A.T. Hammack, M.M. Fogler, L.V. Butov, A.V. Kavokin, K.L. Campman, A.C. Gossard, Spontaneous coherence in a cold exciton gas. Nature **483**(7391), 584–588 (2012)
50. A.A. High, J.R. Leonard, M. Remeika, L.V. Butov, M. Hanson, A.C. Gossard, Condensation of excitons in a trap. Nano Lett. **12**(5), 2605–2609 (2012)
51. A. Imamoglu, R.J. Ram, S. Pau, Y. Yamamoto, Nonequilibrium condensates and lasers without inversion: Exciton-polariton lasers. Phys. Rev. A **53**(6), 4250 (1996)
52. S. Pau, H. Cao, J. Jacobson, G. Björk, Y. Yamamoto, A. Imamoglu, Observation of a laserlike transition in a microcavity exciton polariton system. Phys. Rev. A **54**(3), R1789 (1996)
53. P.G. Savvidis, J.J. Baumberg, R.M. Stevenson, M.S. Skolnick, D.M. Whittaker, J.S. Roberts, Angle-Resonant stimulated polariton amplifier. Phys. Rev. Lett. **84**(7), 1547 (2000)
54. M. Saba, C. Ciuti, J. Bloch, V. Thierry-Mieg, R. Andre, L.S. Dang, S. Kundermann, A. Mura, G. Bongiovanni, J.L. Staehli, B. Deveaud. High-temperature ultrafast polariton parametric amplification in semiconductor microcavities. Nature **414**(6865), 731–735 (2001)
55. J. Bloch, T. Freixanet, J.Y. Marzin, V. Thierry-Mieg, R. Planel, Giant Rabi splitting in a microcavity containing distributed quantum wells. Appl. Phys. Lett. **73**(12), 1694–1696 (1998)

56. R. Houdré, J.L. Gibernon, P. Pellandini, R.P. Stanley, U. Oesterle, C. Weisbuch, J. O'Gorman, B. Roycroft, M. Ilegems, Saturation of the strong-coupling regime in a semiconductor microcavity: Free-carrier bleaching of cavity polaritons. Phys. Rev. B **52**(11), 7810 (1995)
57. J. Bloch, B. Sermage, C. Jacquot, P. Senellart, V. Thierry-Mieg, Time-Resolved measurement of stimulated polariton relaxation. Phys. Stat. Sol. A **190**, 827–831 (2002)
58. D. Huang, J.I. Chyi, H. Morkoç, Carrier effects on the excitonic absorption in GaAs quantum-well structures: Phase-space filling. Phys. Rev. B **42**(8), 5147–5153 (1990)
59. M. Choi, K.C. Je, S.Y. Yim, S.H. Park, Relative strength of the screened Coulomb interaction and phase-space filling on exciton bleaching in multiple quantum well structures. Phys. Rev. B **70**(8), 085309 (2004)
60. N.F. Mott, *Metal Insulator Transitions*, 2nd edn. (Taylor & Francis, Milton Park, 1990)
61. L. Kappei, J. Szczytko, F. Morier-Genoud, B. Deveaud, Direct observation of the mott transition in an optically excited semiconductor quantum well. Phys. Rev. Lett. **94**(14), 147403 (2005)
62. R. Balili, V. Hartwell, D. Snoke, L. Pfeiffer, K. West, Bose-Einstein condensation of microcavity polaritons in a trap. Science **316**(5827), 1007–1010 (2007)
63. M. Richard, J. Kasprzak, R. André, R. Romestain, L.S. Dang, G. Malpuech, A. Kavokin, Experimental evidence for nonequilibrium Bose condensation of exciton polaritons. Phys. Rev. B **72**(20), 201301 (2005)
64. H. Deng, G.S. Solomon, R. Hey, K.H. Ploog, Y. Yamamoto, Spatial coherence of a polariton condensate. Phys. Rev. Lett. **99**(12), 126403 (2007)
65. S. Christopoulos, G.B.H. von Högersthal, A.J.D. Grundy, P.G. Lagoudakis, A.V. Kavokin, J.J. Baumberg, G. Christmann, R. Butté, E. Feltin, J.-F. Carlin, N. Grandjean, Room-temperature polariton lasing in semiconductor microcavities. Phys. Rev. Lett. **98**(12), 126405 (2007)
66. J.J. Baumberg, A.V. Kavokin, S. Christopoulos, A.J.D. Grundy, R. Butté, G. Christmann, D.D. Solnyshkov, G. Malpuech, G.B.H. von Högersthal, E. Feltin, J.-F. Carlin, N. Grandjean, Spontaneous polarization buildup in a room-temperature polariton laser. Phys. Rev. Lett. **101**(13), 136409 (2008)
67. S.I. Tsintzos, N.T. Pelekanos, G. Konstantinidis, Z. Hatzopoulos, P.G. Savvidis, A GaAs polariton light-emitting diode operating near room temperature. Nature **453**(7193), 372–375 (2008)
68. D. Bajoni, E. Semenova, A. Lemaître, S. Bouchoule, E. Wertz, P. Senellart, J. Bloch, Polariton light-emitting diode in a GaAs-based microcavity. Phys. Rev. B **77**(11), 113303 (2008)
69. A.A. Khalifa, A.P.D. Love, D.N. Krizhanovskii, M.S. Skolnick, J.S. Roberts, Electroluminescence emission from polariton states in GaAs-based semiconductor microcavities. Appl. Phys. Lett. **92**(6), 061107 (2008)
70. C. Schneider, A. Rahimi-Iman, N.Y. Kim, J. Fischer, I.G. Savenko, M. Amthor, M. Lermer, A. Wolf, L. Worschech, V.D. Kulakovskii, I.A. Shelykh, M. Kamp, S. Reitzenstein, A. Forchel, Y. Yamamoto, S. Hoefling, An electrically pumped polariton laser. Nature **497**, 348 (2013)
71. P. Bhattacharya, B. Xiao, A. Das, S. Bhowmick, J. Heo, Solid state electrically injected exciton-polariton laser. Phys. Rev. Lett. **110**(20), 206403 (2013)
72. H. Deng, Dynamic condensation of semiconductor microcavity polaritons. Ph.D. thesis, Department of Applied Physics, Stanford University, 2006
73. A. Rahimi-Iman, *Nichtlineare Eekte in III/V Quantenlm-Mikroresonatoren: Von dynamischer Bose–Einstein-Kondensation hin zum elektrisch betriebenen Polariton-Laser* (Cuvillier Verlag Göttingen, Göttingen, 2013)
74. R. Butté, G. Delalleau, A.I. Tartakovskii, M.S. Skolnick, V.N. Astratov, J.J. Baumberg, G. Malpuech, A. Di Carlo, A.V. Kavokin, J.S. Roberts, Transition from strong to weak coupling and the onset of lasing in semiconductor microcavities. Phys. Rev. B **65**(20), 205310 (2002)
75. D. Bajoni, Polariton lasers. Hybrid light-matter lasers without inversion. J. Phys. D: Appl. Phys. **45**(31), 313001 (2012)
76. F. Tassone, C. Piermarocchi, V. Savona, A. Quattropani, P. Schwendimann, Bottleneck effects in the relaxation and photoluminescence of microcavity polaritons. Phys. Rev. B **56**(12), 7554 (1997)
77. M. Richard, J. Kasprzak, R. Romestain, R. André, L.S. Dang, Spontaneous coherent phase transition of polaritons in CdTe microcavities. Phys. Rev. Lett. **94**(18), 187401 (2005)

78. A. Kavokin, G. Malpuech, *Cavity Polaritons* (Academic Press, Cambridge, 2003)
79. H. Deng, D. Press, S. Götzinger, G.S. Solomon, R. Hey, K.H. Ploog, Y. Yamamoto, Quantum degenerate exciton-polaritons in thermal equilibrium. Phys. Rev. Lett. **97**(14), 146402 (2006)
80. V.B. Timoffeev. Bose Condensation of Exciton Polaritons in Microcavities. Fizika i Tekhnika Poluprovodnikov, **46**(7), 865–883 (2012). (engl. translation in Semiconductors **46**(7), S. 843, Springer, Berlin, 2012)

Chapter 4
The Concept of Polariton Lasing

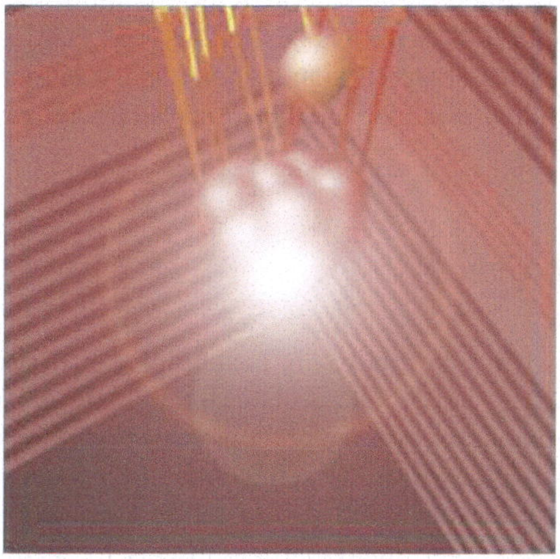

Abstract Polariton-laser research has attracted a considerable amount of attention in the last decade and has led to remarkable insight into the physics of QW microresonators and exciton–polaritons. Technological advances allowed the design and fabrication of more precise and more suitable structures until the demonstration of the long desired effect of polariton condensation was observed. Since then, a plethora of condensate studies has been performed utilizing polaritons. As a driven-dissipative system, polaritons constantly decay, delivering a stream of coherent photons out of a quantum-degenerate state, and need to be refilled to uphold the quasi-condensate regime. Different examples of a polariton laser have been presented, both optically and electrically pumped, and even operated up to room temperature. However, the idea of a laser-kind device which utilizes bosonic polaritons and their phase transition into a macroscopically-occupied quantum-degenerate gas was proposed years before, back in the 1990s. This led to the birth of a new class of coherent light sources, named "polariton laser". In the following, the principles of such a device are summarized, and its differences and similarities to conventional lasers and the means of identification for an unambiguous demonstration of polaritonic lasing explained.

A. Rahimi-Iman, *Polariton Physics*, Springer Series in Optical Sciences 229,
https://doi.org/10.1007/978-3-030-39333-5_4

4.1 Polariton Lasers—Electrically-Driven, Please!

After having used the term *polariton laser* several times in previous chapters and having motivated a considerable fraction of polariton research with the goal of demonstrating such novel coherent light source, this section will explain what it is actually about and what makes a polariton laser a *polariton laser*. Thereafter, the next section will show similarities and differences to a conventional microcavity laser. Thereby, we can understand why the achievement of such a device is favorable given the vast amount of existing semiconductor laser types. In this context, a brief overview is acquired by the reader concerning the strengths, but also weaknesses of such a device; in other words, with this background one can judge in which regard it is superior to its photon counterpart, and when it is yet outperformed. For further experimental insight into this matter and the optical characterization of condensates, see Chap. 8.

4.1.1 What Is It About?

If we ask ourselves, what polariton lasing is about, the list of answers will consist the following statements: First of all it is about light–matter interaction. The system employs strongly-coupled excitons and photons which are the building-blocks for a new quasi-particle that is formed in the semiconductor device, called polariton.

Moreover, it is about dynamic Bose–Einstein condensation in solids, since a polariton laser utilizes bosonic many-particle effects to accumulate a high number of the involved quasi-particles into the system's ground state to achieve a strong output of monochromatic, highly-directional light from this state.

In this system, a stimulated scattering process is involved to achieve efficient particle relaxation towards a phase-coherent polariton gas for the generation of coherent radiation, which originates from the bosonic quantum degeneracy of polaritons due to the macroscopic occupation of the polaritonic energy ground-state.

Out of such a system, a beam of coherent photons is emitted simply due to polaritonic decay from a coherent state owing to the finite mirror reflectance in the microcavity structure. Thus, the dynamical condensate state is continuously repopulated to compensate for optical losses via optical pumping or electrical current injection.

For practical applications, electrically-driven polariton lasers are more desired than their optically-pumped pendants. Electrical pumping simply promises condensates on demand on a device level, enabling the implementation of polaritonic devices into electro-optical systems and integration into an on-chip architecture, paving the way for the field of *polaritonics*.

It is worth noting that any practical realization of a polariton laser delivers one device with three operation modes (see Fig. 1.7 of Chap. 1): incoherent polariton light emission, polariton lasing above a critical particle density and below the break-up density of strong coupling (Mott transition), and conventional (photonic) lasing above the particle densities for "population inversion" or "transparency", respectively. Thus,

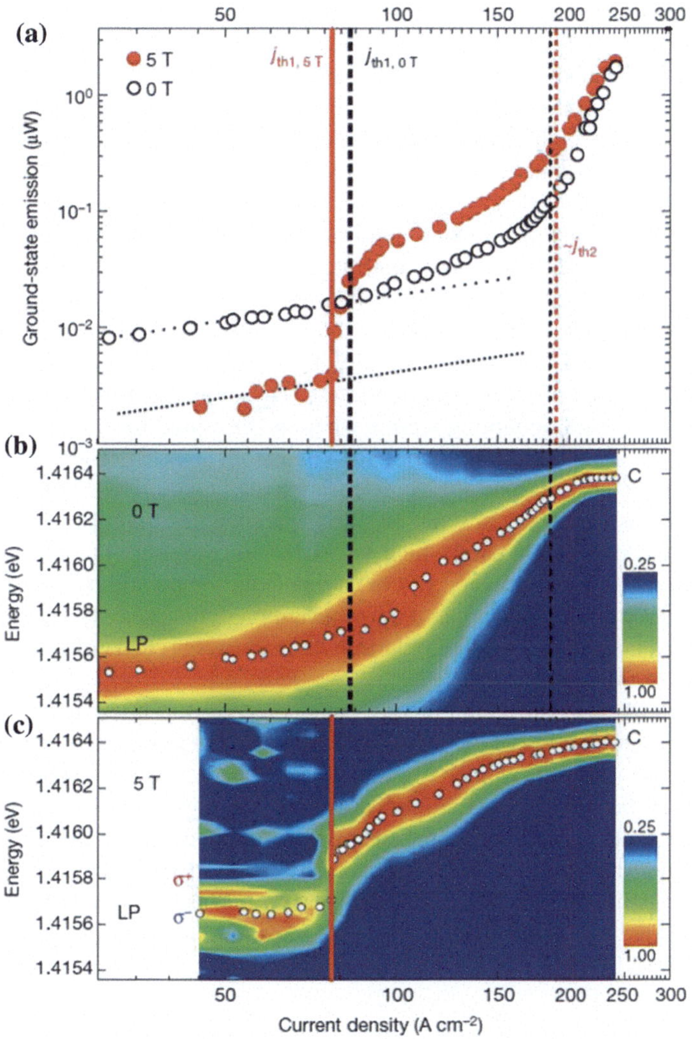

Fig. 4.1 Input–output characteristics with three emission regimes shown for an electrically-driven polariton-laser device. **a** Below condensation threshold, the device operates as a polariton LED in the so-called linear regime underlying Boltzmann distribution. This is indicated by the linear slope. Above the first nonlinearity threshold, for which the matter component in the emitting resonance is still preserved, the device acts as a polariton laser. Above the second threshold, conventional microcavity lasing operation takes place. This two-threshold behaviour is even more pronounced when an external out-of-plane magnetic field is applied. **b** and **c** show spectral signatures for the current-density-dependent emission from a polariton-laser diode at 0 and 5 T, respectively. Reused with permission. Reference [1] Copyright 2013, Springer Nature

a double-threshold feature is very characteristic for a typical polariton-laser device (see Fig. 4.1).

With both equivalent terms often used to mark the onset of lasing in semiconductor lasers representing a non-equilibrium state of the free electron and hole gas in the active medium, the actual laser threshold is determined by the relation between gain and losses in the structure. In high-quality resonators, this threshold can indeed coincide with the Mott density. Moreover, using high-quality microcavities with low mode-volumes and quantum emitters such as quantum dots even enables lasing based on excitons, and at ultra-low pump-threshold values. A definition of the laser threshold can be found in [2].

With conventional lasing in microcavities referred to as lasing in the weak-coupling regime, polariton lasing naturally has to occur at charge-carrier densities much lower than that of weak-coupling lasers. Thus, threshold-less lasing has been proposed for the strong-coupling regime where no threshold for transparency needs to be overcome, as for instance reported for a one-atom laser [3], whereas for a gas of polaritons in QW-microcavities the condensation effect would resemble a transition to coherent emission in the absence of a transparency threshold [4]. Nevertheless, practically, the particle density which is needed to overcome the critical density for stimulated scattering renders a form of threshold for polariton lasing, although expectedly much smaller than that of the (conventional) microcavity-laser threshold. This explains the attractiveness of polariton lasers as a source of coherent light with ultra-low operation threshold.

Finally, the key to an unambiguous claim of polariton-laser demonstration is given by the identification of the strong-coupling regime above the threshold for the system's nonlinearity attributed to stimulated scattering. Later in this chapter, one will learn which difficulties arise when it comes to delivering proof of polariton lasing, and how this problem is solved using the matter component in the polariton system. As will be seen, the existence of excitons can be evidenced above the threshold for the first of two distinct nonlinearities in the system by means of a simple but intriguing technique. Thus, this first nonlinearity (phase transition) accordingly corresponds to polariton lasing, before a transition to photon lasing takes place at a second threshold.

4.1.2 The Stimulated Scattering Process

In a polariton device, the stimulated final-state scattering of bosonic particles leads to a macroscopic number of polaritons in the polariton ground state, which is given by the LP dispersion minimum, as polariton–polariton interactions become sufficiently strong when the matter-wave overlap between the excitonic fractions of the polariton gas is good enough. The more particles (particle number N_0) are in the ground state, the more likely (proportional to $1 + N_0$) other particles scatter into this quantum state, wherein polaritons are described by a macroscopic wave-function. This is the process behind the realization of a novel coherent light source, the "polariton laser", the concept of which was proposed by Imamoglu et al. in 1996 [4], more than 20 years ago. Instead, conventional lasing requires population inversion in the gain medium, which amplifies the light field by stimulated emission of radiation, and thereby produces coherent, monochromatic and directional light output.

This concept of a polariton laser achieved from QW microresonators is based on Bose condensation of polaritons, while the actual differentiation to a BEC is made with regard to the dynamics, the degree of thermalization, and the range of spatial phase order. In the following, the concept of coherent-light generation from dynamic polariton condensates is further summarized and characteristic differences to BEC highlighted. The attributes of polaritons and their relaxation dynamics are described in Chaps. 2 and 3, respectively.

Thermalization with the environment is the key to achieving a phase in thermodynamic equilibrium, which is a prerequisite for the observation of real Bose–Einstein condensation. However, polaritons with their short lifetime τ_0 at best thermalize within the Bose gas, but not with the lattice. Typical thermalization times of $\tau_{th} \approx 1$–10 ps for cavity–polaritons are yet longer than τ_0 near the ground state which lies on the order of a few ps [5, 6]. For a more excitonic regime of the polariton system, the lifetime around $k_\parallel = 0$ may be larger than even 10 ps for good cavities. However, such regime comes at the cost of higher effective masses, stronger ground-state–reservoir interactions and reduced delocalization. While for atoms and excitons according to Deng et al. [7], the ratio of thermalization time to lifetime τ_{th}/τ_0 amounts to approximately 10^{-3} and 10^{-2}, respectively, the corresponding value for polaritons is specified to lie somewhere between 0.1 and 10. Thus, a typical polariton gas is usually not in thermal equilibrium with its host lattice (gas temperature $\neq$ lattice temperature), while macroscopic ground-state occupation via stimulated scattering is obtained. Moreover, atom gases in magnetic traps at ultra-low temperature for condensation studies are uncoupled from other matter and can therefore naturally achieve a purer condensate with high condensate fraction [8–10] very close to a thermodynamic equilibrium, as the particle numbers in the system are virtually preserved.

With simple considerations, three scenarios can be outlined by the ratio of the gas thermalization time to the ground-state lifetime τ_{th}/τ_0 [4, 11].

In the primary case, when $\tau_0 \gg \tau_{th}$, the Bose gas can thermalize completely with its surroundings. For a condensate in thermodynamic equilibrium, a chemical potential μ can be determined. Also, the condensed phase establishes long-range order over the whole size of the system. This thermodynamic equilibrium marks the regime of a *Bose–Einstein condensate* (BEC) (corresponding to quasi-BEC for polaritons in a driven-dissipative symmetry-broken 2D system, see Sect. 3.1).

Provided that $\tau_0 \geq \tau_{th}$, quasi thermal equilibrium is obtained for polaritons, whereas the thermalized Bose gas approximately reaches the lattice temperature. This is considered a (metastable) transition regime and the temporary effect of *dynamic condensation* is obtained towards the transition to a BEC. Accordingly, dynamic condensates feature many signatures of a BEC and are ideal testbeds for the study of condensate physics as well as BEC signatures.

However, is the lifetime of polaritons too short, i.e. $\tau_0 \ll \tau_{th}$, the condensed Bose gas cannot thermalize with the host lattice (phonon reservoir). The system is far from thermodynamic equilibrium. Nevertheless, some signatures of condensates are preserved. For instance, characteristics such as the macroscopic occupation of the ground state and an increased phase correlation can be observed. Such a system is labelled *matter laser*.

The polariton laser with its excitonic content can be considered as such case, as it relies on coherence of the matter system for coherent output of light, which occurs via exciton (exciton–polariton) decay from a dynamic out-of-equilibrium condensate state. Imagoglu et al. proposed both exciton lasers and polariton lasers in their original work, which has promised inversion-less lasing, i.e. coherent radiation generated from an excitonic system strongly based on stimulation of scattering processes. Simultaneously, they also draw attention to the fact that *amplification* (of the output of light) is only possible in an exciton–polariton laser, when the exciton–phonon system is inverted [4]. Moreover, as soon as the density of particles goes up and screening takes place, so that excitons cannot anymore be considered as bosons, the concept of inversion-less lasing from an excitonic (polaritonic) condensate becomes invalid.

Imamoglu et al. concluded, that when the coherence in the matter system vanishes, as is the case in the weak-coupling regime above the Mott density and for polaritons with predominantly photonic character, an electronic population inversion becomes necessary for the generation of coherent light. This transition is well observed for a polariton laser structure in the input–output characteristics, when a two-threshold behaviour occurs which naturally reveals the photon-laser threshold at high densities (see Fig. 4.1). Several studies have discussed the orders of magnitude in difference between conventional lasing (weak-coupling lasing) and polariton lasing (strong-coupling lasing) in microcavities regarding the excitation densities required to overcome the respective threshold [12], which is often used as a means of discrimination between the two regimes and to emphasize low-threshold operation. For instance in GaN-based systems, threshold densities for the two observed nonlinearities in the investigated, different polariton structures were separated by up to 3 orders of magnitude [13–15], while in GaAs structures differences between the threshold densities of about 1 to 3 orders of magnitudes were reported [1, 12, 16–18]. Both simultaneously demonstrated first electrically-driven polariton lasers showed this common two-threshold behaviour [1, 18], which was analogous to the behaviour of their previously reported optically-pumped counterparts in terms of density differences.

It is worth noting that weak-coupling lasing in microcavities can be thought of as a form of non-equilibrium condensation of the bosonic photons in which interaction between photons is mediated by the optically active medium, while this is still no undisputed idea. Stimulated emission is in fact related to the bosonic nature of photons. However, as photons do not interact, they cannot self-thermalize. Nevertheless, work on photon BEC has been conducted and the demonstration of such can be found with detailed descriptions in the literature [19].

The aforementioned three regimes, that are BEC, dynamical condensate and polariton laser, respectively, are characterized all together by a macroscopic occupation of the ground state, an increased range of phase order and spatial coherence [6, 20, 21], although they differ significantly in terms of thermalization degree. The transition from one regime to the other is not very sharp in experimental systems, and a differentiation between them in experiments is only feasible via the determination of the gas temperature. Although currently no universal statements concerning the temporal coherence of (polariton) BEC exists, which can be investigated by the second-order temporal autocorrelation function (cf. [5, 22]), it is expected that the degree of temporal coherence decreases for decreasing lifetimes. Moreover, intensity noise and the coherence degree are understood to be directly related to particle–particle interactions (e.g. ground-state–reservoir interactions) in the condensate system (see Sect. 8.2). Thus, the temporal coherence from a polariton laser should be lower than that of a dynamical condensate or BEC-like state. Many basic properties of such systems are well summarized in the literature [7, 23–26] and diverse analysis of the different regimes has been pursued over the years.

In this context, it is understandable that a polariton-laser state is typically considered a condensate state far from thermodynamic equilibrium, with finite spatial expansion and limited range of order, while having properties similar to that of a photon laser. Owing to the phase coherence of polaritons in the macroscopically occupied quantum state, emission from a condensate exhibits a very narrow momentum distribution and a well pronounced spatial and temporal coherence. Furthermore, the critical particle density for condensation can be obtained at significantly lower pump rates than the threshold power for laser operation, which requires conditions in semiconductor lasers defined by Duraffourg and Bernard [27], i.e. transparency of the gain medium. Thus, this has promised the realization of an energy-efficient, inversion-less source of coherent light, which yields its coherent radiation from the decay of coherent particles in a macroscopically occupied quantum state, as proposed in 1996 [4]. At that time, a hunt for the demonstration of polariton lasing began [28].

Of technological interest has indeed been the achievement of an electrically-driven polariton laser, which is more practical from the applications point of view, but also for applied sciences which go beyond the fundamental investigations on polariton BEC. Additionally, also condensate studies could experience a fresh breeze with electrically-pumped structures being able to deliver well-defined condensates on demand. Non-resonant electrical excitation via electron–hole injection from two opposite sides leads to a sequence of various relaxation processes in the underlying diode structure, before macroscopic ground-state occupation can be exploited for the generation of coherent light, yet similar to experiments under optical non-resonant excitation [6, 22].

The relaxation processes involved can be summarized as follows: Firstly, electrons and holes have to get homogeneously distributed in the structure in order to fill the available quantum wells. Hot charge carriers relax via scattering processes with optical phonons to form hot, incoherent excitons with large momenta, which can cool down further via acoustic phonon scattering into k-space regions where interaction with cavity photons becomes feasible [29, 30]. It is there, where the polariton reservoir is formed at moderate k-vectors and energies, where the polariton is still highly excitonic. As the polariton's photonic character begins to raise on the way to the ground state, scattering efficiencies, which are linked to the matter component of the polariton, start to decrease.

If thermalization is successful without the appearance of a bottleneck effect, yet the particle density in the ground state has to exceed a critical value in order to boost stimulated scattering effects and to kick off condensation. However, a bottleneck effect (cf. [31, 32]) can be commonly observed in most of the structures instead of efficient ground-state relaxation, impeding condensation in the ground state. Nevertheless, in well-designed structures with high resonator quality and thus polariton lifetimes and suitable operation conditions (temperature and detuning), at high enough polariton densities a new relaxation channel can be opened up by the stimulated polariton–polariton scattering process. This stimulated process facilitates overcoming the bottleneck and enables efficient ground-state filling [30, 33, 34], yet below the Mott density. Thereby, above the critical density υ_{crit} for condensation, the lifetime of polaritons is undercut by the relaxation time and a condensate is formed [5, 6, 32, 35] (see Chap. 3). A schematic drawing of the processes involved from injection of charge carriers to ground-state occupation of polaritons is shown in Fig. 3.5 of Chap. 3, in which also rate equations for the kinetics of the system are summarized.

Whether an obtained nonlinearity in the ground-state occupation of polaritons is attributed to a regime of polariton condensation, or polariton lasing, has not been undisputed. Assuming that the matter component is preserved, and that the nonlinearity occurs below the Mott density, it is nowadays generally accepted to label the observed feature a dynamical polariton condensate. However, the validity of the claim to have obtained BEC[1] or dynamic quasi-BEC of polaritons with respect to polariton condensation has been strongly discussed in the literature owing to the low degree of thermalization and the absence of thermodynamic equilibrium conditions, even though pronounced coherence has been evidenced for condensed polaritons [36, 37]. A thermal equilibrium between condensed and uncondensed fractions of the system is considered to be another important criterion for quasi-BEC, besides the spontaneous evolution of off-diagonal long-range order (ODLRO) out of an incoherent (precursor) state—such as the incoherent gas of polaritons originating from reservoir states. Thus, claims of BEC obtained from resonant pumping schemes, which can imprint the pump source's coherence onto the polariton gas, are even more disputed.

Nevertheless, the terms condensation and polariton lasing both generally apply well to systems, for which stimulated final-state scattering resulted in ground-state accumulated coherent polaritons, as both can be basically considered as equivalent [30, 34]. In fact, both terms have been used very often in the literature in order to describe, discuss and refer to the same phenomenon, either in theoretical or experimental work. Typically, the term dynamical BEC becomes acceptable when spontaneous symmetry breaking with established ODLRO and Bose–Einstein distribution of polaritons is demonstrated, and with polariton lifetimes sufficiently high to allow the assertion of gas temperatures being similar to lattice temperatures. Yet, the requirements for the successful demonstration of polariton lasing remain less high than for (quasi-)BEC, with the necessary proof being discussed in the final section of this chapter.

[1] Irrespective of the fact that particle numbers are not conserved and theoretically no true BEC can evolve in 2D systems (discussed in Sect. 3.1), one often finds Bose condensation of polaritons at (near-to-)thermal equilibrium with the host lattice being labelled as BEC, or quasi-BEC.

Having summarized the key features of a polariton laser and introduced its working principle, it comes to ones mind to give it a well-matching tangible acronym, which represents the nature of the device and its output. While some attempts were made in the literature to introduce an acronym such as POLLAS (*POL*ariton *LAS*er) or PLASER (*P*olariton *LASER*), the key characteristics are better represented by POCCCER, POC^{3}ER, or POC3R, the acronym for *POlariton Condensation-Caused Coherent Emission of Radiation* (the author's suggestion and preference). This gives an individual and catchy acronym for a novel and unique source of coherent light, or short POCER in the sense of *PO*lariton *C*ondens*ER*. Still, *polariton laser* will remain a useful label for this device, even though not based on stimulated emission, as will the comparison between photonic and polaritonic lasing in the following section show.

4.2 Comparison with Photon Lasing (Lasing in the Weak-Coupling Regime)

In this section, a comparison of polariton lasing with its photonic counterpart is in the focus. In order to understand similarities and differences, a brief summary will be given on *the laser*. An excellent overview on lasers can be found in the vast literature available on this topic (the interested reader can find detailed information in books such as [38]).

4.2.1 What Is a Laser?

Laser is the acronym for *Light Amplification by Stimulated Emission of Radiation*. It is one of the most famous resources of the late 20th century and has become an asset to the industrialized world. The invention of the laser is based on early quantum mechanics and the development of the theory of light–matter interaction, to which A. Einstein and his contemporary colleagues contributed significantly. Decades later, stimulated emission of radiation, which was postulated in 1917 bei Einstein, was experimentally demonstrated, first for a *microwave* emitter—predecessor of the laser named *maser* (envisioned and demonstrated by C. Townes and his team in the 1950s) [39, 40]—and later in 1960 for visible light attributed to the work by T. H. Maiman based on synthetic ruby (Al_2O_3:Cr) [41]. The invention of the laser (at the beginning labelled "a solution *seeking* a problem" by Maiman) has led to the steady emergence of a plethora of applications, which particularly have required exclusive properties of this special light source. The demand for the exclusive benefits of lasers and their specific features increased drastically in the last decades, with the laser having been employed and further developed for many novel techniques in science, medicine, telecommunication and other domains of our lives. The boost of industrial and scientific capabilities has strongly relied on the invention of various types of lasers, ranging from gas lasers (for further details and information, the interested reader is referred to corresponding literature, e.g. [42]), solid-state lasers (e.g. [43]) and semiconductor lasers (e.g. [44, 45]) to optical frequency combs [46, 47] and single-atom lasers [3].

This trend has been pushed further forward by achievements in the field of semiconductor technologies which introduced reliable, cheap and mass-producible laser diodes to the portfolio of the laser world, most of which can operate directly at room temperature since the late 20th century. In addition, miniaturization and diversification of employable gain materials have lead to more compact devices, higher efficiencies and wavelength versatility. Among lasers, semiconductor lasers are the most compact and robust systems which can be well integrated in electronic devices. On this path, many milestones have been reached, such as the demonstration of the first helium–neon continuous-wave (cw) gas laser at Bell Telephone Laboratories by A. Javan, W. R. Bennett, and D. R. Herriott in 1960 [48], the first semiconductor diode laser by R. N. Hall et al. in 1962 [49], the development of the first heterostructure laser enabled by the work of H. Krömer [50] and Z. I. Alferov [51], room-temperature operation of a diode laser and many more, such as the single-atom laser [3]. Together with the contributions by T. W. Hänsch and J. L. Hall on frequency combs [46, 47], the contribution to the quantum theory of optical coherence by R. Glauber [52] and the development of ultra-short pulsed lasers, numerous Nobel prices have been awarded to (laser) physicists for their ground-breaking achievements. In fact, the list of contributions to the achievements in the field of lasers outlined here for overview purposes cannot be complete.

The class of semiconductor lasers can be further separated into two main subclasses. These two main types of laser diodes have indeed revolutionized our everyday life in many regards. They can be specified as edge emitters and surface emitters, according to the optical output direction obtained from the respective device with regard to the semiconductor chip.

Edge emitters are particularly easy to fabricate and benefit from the high gain cross section owing to the lateral propagation of light through the in-plane active region, which allows amplification via feedback from low-reflectivity mirrors. In such a case, the mirrors are the chip facets provided by cleaving, i.e. the refractive index contrast between crystal and air acts as the mirror of a low quality Fabry–Pérot resonator. Unfortunately, edge emitters generally lack a good beam profile since the output aperture of edge emitters, which is simply given by the cleaved edge of the chip, gives rise to strong divergence of the beam in the direction perpendicular to the plane of the laser mode. Yet, the realization of a diode-pumping scheme is achieved in a comparably easy fashion, typically using a p-i-n doping scheme to form a diode junction across the wave-guide structure. Thereby, the intrinsic (i) active region exhibits the minimum amount of dopants in order to keep the losses and heating in the gain structure at a minimum.

Surface emitters are typically realized as vertical-cavity surface-emitting laser (VCSEL) diodes, which typically also employ Fabry–Pérot resonators. However, as the name suggests, optical feedback is provided in the vertical direction, perpendicular to the plane of the active medium. The corresponding resonators are achieved in a stacked arrangement with the active medium sandwiched between planar mirrors consisting of sophisticated multilayer heterostructures for interference effects, known as distributed Bragg reflectors (DBRs). With optical modes being confined vertically, optical output through the top mirror is also vertical, i.e. in the growth direction of the semiconductor structure. Naturally, in order to obtain unidirectional emission and a high output efficiency, the top mirror is typically less reflective than the bottom mirror. As vertical emitters commonly employ spherical output facets or top apertures, they benefit strongly from their excellent beam quality with Gaussian beam profiles and low divergence. Similar to edge emitters, a p-i-n doping scheme usually provides a diode junction, with the mirrors consisting of oppositely doped semiconductors, while the active region remains intrinsically doped for better gain performance.

For the implementation in optical devices and the achievement of compact systems for various applications, electrical pumping has become the most desired excitation scheme. This excitation scheme has become particularly important in the sphere of semiconductor lasers, while optical pumping naturally delivers an alternative to electrical pumping. It is worth noting that many laser types are actually optically pumped, such as atom lasers and solid-state lasers, and optical excitation of active media in some cases still poses a reasonable alternative pumping scheme even for semiconductor lasers, such as in the case of semiconductor disk lasers [53–58]. However, the attractiveness of electrical pumping is understandable in the context of applications, where compactness and integrability is desired and electrically-driven lasers are commonly implemented in a superior system, which is particularly the case in telecommunications.

In the field of semiconductor lasers, one major trend can be identified towards ultra-low-threshold or even threshold-less lasers [2, 59–61], and "green" (ecologically friendly) diode lasers [62]. The fabrication of high-quality microresonators [63], low-mode-volume microcavities [64] and the incorporation of low-dimensional gain structures (see for instance [65]) such as QWs and more importantly quantum dots (QDs) allowed for more design flexibility, wavelength versatility as well as quantum efficiency and very low laser thresholds, such as in the case of QD-micropillar [66] or QD-microdisk [67] lasers, seeking to achieve the ultimate nanolaser [68]. Ideally, gain from a threshold-less laser is to be achieved by a single quantum dot inside a microcavity, a concept driving the developments in the field of microresonator technology [69]. In fact, (single) quantum dots are not only of relevance for lasing [61, 70].

In contrast to the ultra-efficiency schemes, at the high-power end for example, ensemble-QD systems in VCSEL-like structures with external cavity and appropriate thermal management can even show record high output powers from a single QD-laser chip [71], whereas their QW counterparts were demonstrated to even exceed the 100-W level due to their naturally higher density of emitters [55]. Additionally, QD-laser structures can be attractive for mode-locked laser operation [72]. While mode-locking can be addressed in the conventional laser regime, this is not imaginable for polariton lasers (based on the current understandings), unless a means of phase-locking between energetically-different (coexisting) polariton condensates (e.g. occurring in tailored potential landscapes) of a device becomes accessible.

4.2.2 *Stimulated Emission, Laser Conditions and Coherence Properties*

Fundamental interactions between light and matter are the basis for laser physics and shall be considered briefly in an ideal two level system with discrete electronic energy states $|1\rangle$ and $|2\rangle$, respectively. In such, three mechanisms describe the possible interaction between the two-level system and an electro-magnetic radiation field with spectral energy density $u \equiv u(\nu)$: (a) (induced) absorption, (b) induced/stimulated emission, (c) spontaneous emission. (b) is the reverse process to (a) involving a spatially and spectrally resonant light field, where an incident photon induces an optical transition, while (c) is determined by the mean particle lifetime in the excited state and describes spontaneous radiative decay which returns the system to its ground state. A schematic representation of these processes is provided in Fig. 4.2.

Stimulated Emission

In such a two-level system, the occupation densities n_1 and n_2 of the electronic states are subject to these fundamental processes (a), (b) and (c), respectively, and can be described by the following rate equations:

$$\frac{\partial n_2}{\partial t} = +n_1 A_{12} u - n_2 I_{21} u - n_2 S_{21}, \tag{4.1}$$

$$\frac{\partial n_1}{\partial t} = -n_1 A_{12} u + n_2 I_{21} u + n_2 S_{21}. \tag{4.2}$$

In this rate equation, the Einstein coefficients A_{12}, I_{12} and S_{12} describe the transition probabilities between $|1\rangle$ and $|2\rangle$ at the presence of the light field u. As becomes obvious, the occupancies are changed by three processes, which represent (induced) absorption $n_1 A_{12} u$, induced emission $n_2 I_{21} u$ and spontaneous emission $n_2 S_{21}$. The process with coefficient A_{12} elevates an electron from the ground state to its excited state by a photon, which provides the transition energy to overcome the gap $h\nu = E_2 - E_1$. Similarly, but opposite to this process, by stimulation one electron relaxes from the excited state to the ground state under emission of a photon with $h\nu = E_2 - E_1$ with probability I_{21}, provided that a resonant photon is present in the system. Thereby, a clone of the existing photon is created within the stimulated emission process, which has equal energy, phase, polarization and propagation direction, which means that the resulting light field is coherent.

At this stage, it becomes obvious that spontaneous emission with coefficient S_{21} is a parasitical effect, which depopulates $|2\rangle$ whereas the energy is lost by emission of an incoherent photon when the excitation lifetime is expired. Radiation from this process is isotropic (in 3D vacuum) and its contribution to amplification in a laser medium is negligible. It has rather negative implications regarding lasing efficiency or heat generation in laser structures. It is worth noting that the spontaneous emission

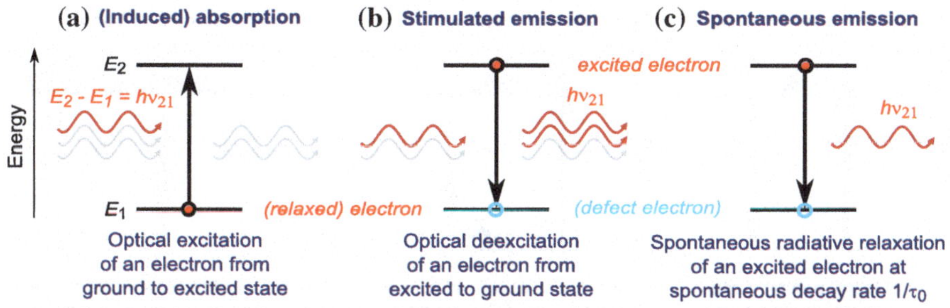

Fig. 4.2 Sketch of fundamental light–matter interaction processes. Three mechanisms describe the possible interaction between an ideal two-level system and a resonant electro-magnetic radiation field: (induced) absorption, induced/stimulated emission, spontaneous emission. While the first two processes depend on the presence of an occupied photonic mode, i.e. number of photons in the mode > 0, the last process occurs spontaneously. The spontaneous emission rate is defined by the inverse excitation lifetime. Quantum-mechanically speaking, this process equals stimulated emission triggered by zero-point fluctuations of the electro-magnetic (vacuum) field. The presence of modes that are not occuppied by actual photons effectively causes the relaxation of the excited emitter by offering a possible radiative decay channel. If such decay channel is removed, e.g. by tailoring low-dimensional optical cavities (photonic quantum boxes) to yield an off-resonant atom-like photon-mode density of states, this process of spontaneous emission can be accordingly suppressed (one aspect of the Purcell effect). While a resonant empty light field can reduce the spontaneous lifetime considerably, an occupied resonant field induces/stimulates an optical transition as known for laser operation. In fact, the spontaneous emission rate in vacuum directly depends on the 3D vacuum density of states for light. This influence of vacuum fluctuations is an interesting aspect which can be even linked to the so-called Casimir effect, where the pressure from virtual photons is understood to press two plane parallel metallic plates together due to the lower photonic density of states in the 2D space between the plates compared to the 3D space surrounding the plates

rate can be modified by the Purcell effect, which can enhance or suppress a quantum emitter's radiative recombination [73]. What is special about these coefficients which represent interaction cross sections (e.g. $\sigma_{12} = A_{12} h\nu_{12}/c$), is, that they are related to each other as follows:

$$A_{12} = I_{21} = \frac{c^3}{8h\pi\nu^3} S_{21}. \tag{4.3}$$

Indeed, σ_{12} only equals σ_{21} if states $|1\rangle$ and $|2\rangle$ exhibit the same degree of degeneracy.

To achieve amplification of light by stimulated emission of radiation in a given optical medium—thus, referred to as active medium or gain medium—one fundamental condition has to be fulfilled: The stimulated emission rate has to be higher than the absorption rate, i.e. propagation of light through the optically active medium needs to result into a net gain. However, this is only possible if the occupation density n_2 in the excited state is larger than the density n_1 for ground-state occupation, since the Einstein coefficients for absorption and stimulated emission are equal (see (4.3)). Such a situation is called inversion of the system and is not possible in a two-level system in thermal equilibrium. At best, a two-level system will be obtained equally

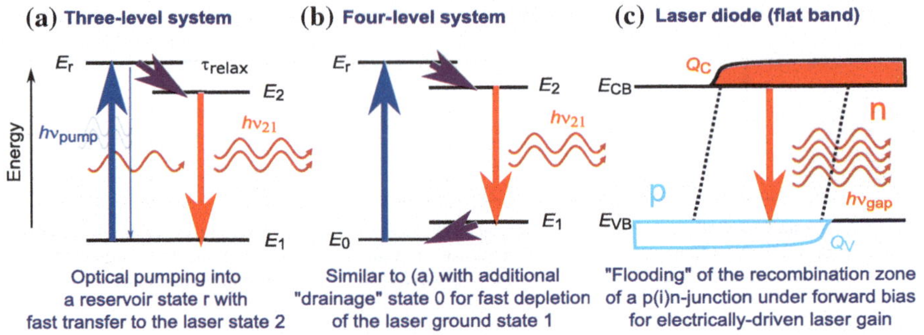

Fig. 4.3 Schematic representation of the energy levels in a three-level (**a**), four-level system (**b**) and a laser-diode pn-junction under a forward driving bias (**c**). In multilevel systems, population inversion of the laser-relevant states is achieved by fast injection of carriers from higher-energy reservoir states, which are populated by a pumping process. By depletion of the laser ground state by a fast transfer to a drainage-like lower state, the degree of inversion can be increased. In p(i)n-doped diode junctions (with intrinsic active region), conditions for lasing are achieved by driving the system into a highly non-equilibrium state. Thereby, hot electrons (holes) are injected from the n-doped (p-doped) side through external electrical contacts for inversion in the charge-recombination zone, in which quasi-Fermi levels are separated further than the band gap. Similarly, non-resonant optical pumping can excite hot charge carriers into the gain region

populated in its ground and excited state (under irradiation of resonant light), as can be understood from the rate equations (4.1) and (4.2). Thus, lasing requires a system with at least three levels (e.g. ruby laser), better even four (e.g. helium–neon laser) (see Fig. 4.3a and b, respectively). Under such condition, it becomes feasible to invert the density distribution and bring the system out of thermodynamic equilibrium by a pump process, which lifts electrons from the ground state into a reservoir state $|r\rangle$. From that reservoir state, fast non-radiative decay or energy transfer into the laser state creates a situation in which $n_2 > n_1$. Moreover, if the laser ground-state $|1\rangle$ is depleted quickly by non-radiative drainage into a lower-lying energy state $|0\rangle$, the extent of inversion of the laser states can be increased. Indeed, inversion of the gain medium has to be maintained via pumping in order to continuously supply energy to the system. For instance, this can be achieved via optical or electrical pumping.

Prerequisites for Laser Operation

Besides population inversion and stimulated emission, lasing requires another condition that has to be established, namely optical feedback. This can be achieved in an optical resonator configuration, which in its simplest form consists of two parallel opposing mirrors that surround the gain medium. Usually, one mirror is highly reflective, while the other one with lower reflectivity acts as an out-coupling mirror. In an optical resonator, feedback is only provided for the respective resonator modes which have discrete frequencies (energies/wavelengths) according to the free spectral range (FSR). To obtain lasing, the resonator mode and the gain spectrum (laser transitions of the active medium) have to be spectrally resonant. Then, for each round-trip of photons through the resonator, the active medium amplifies the light-field by adding photons through stimulated emission to the system. The quality of a resonator is thereby an important figure when it comes to laser optimization, as it determines the period in which light propagates back and forth between the mirrors through the gain structure, or in other terms, which defines the decay time of the standing-wave resonator light field. Finally, a critical prerequisite for laser operation is defined by the ratio of gain and loss in the laser resonator. As soon as gain exceeds losses, the threshold for lasing can be overcome and emission from the resonator results in an intense, coherent, monochromatic and directional output of light.

Different media can act as gain region in lasers, such as atoms, molecules, quantum wells and others. The choice of the gain medium strongly depends on the desired output wavelength, intensity as well as other laser properties and are usually defined by the application which demands laser light. Similarly, a plethora of resonator configurations are known nowadays, ranging from those with conventional dielectric mirrors, Bragg reflectors to sophisticated microresonator structures. The pump source is chosen with respect to the laser medium and application. Mostly, lasers or other powerful light sources are used for optical pumping, while electrical pumping is usually achieved by injection schemes such as for diodes. Indeed, one can also use other pumping methods which fulfil the purpose of energy transfer to the active medium in order to achieve a high population density in excited states. To make a laser work well, many different design aspects come into play, with the thermal management, the excitation scheme and other factors related to the resonator and gain structure having their role in the optimization processes.

Coherence Properties

One remarkable feature of lasers is their coherent output of light, which shall be briefly summarized in this section for the sake of completeness of the presentation of laser properties.

Coherence is evidenced spatially by the occurrence of speckles, or generally said, interference patterns, which mark constructive and destructive interference of phase-coherent wave-packages when they overlap spatially according to the path difference they experienced. Inside the resonator, the laser mode is macroscopically occupied with coherent photons, which are the consequence of amplification by stimulated emission of radiation. Light out-coupled from this coherent bath of photons usually exhibits a strong phase relation with the standing wave electromagnetic field inside the resonator. This can lead to a photon stream with spatial coherence length in the range of metres, up to kilometres and more for certain laser types that are run on a single frequency (frequency-stabilized lasers). In general, the narrower the laser line and the better its energy stability is, i.e. the sharper the laser transition, the stronger the coherence. For systems which lase on many modes simultaneously or the light of which originates from energetically-broadened laser transitions, a reduced coherence length is expected. Emission lines can be homogeneously (limited by energy–time uncertainty, thus dependent on a quantum emitter's lifetime) or inhomogeneously broadened (affected by disorder, structural quality, distribution of quantum emitter ensemble).

To give an example, a long coherence length of a laser is for instance useful for holography, where the phase information of the light field reflected from an object is saved in a special photo plate and retrieved with the same laser with which the information was recorded. A very narrow linewidth is also very favourable in experiments with atoms, where a single electronic transition is targeted in order to have a very high absorption cross section, for instance for laser cooling, atomic clocking and other experiments.

Spatial coherence can be well investigated using interferometry techniques such as the Michelson interferometer (originally applied in astronomy [74]), which puts the amplitude of the light wave at different locations (path lengths) in relation to each other. This is strongly related to the first-order temporal coherence, which puts the amplitude of time-lapsed laser fields in correlation with each other and reflects the linewidth of the laser mode.

Additionally, laser radiation is known for its random output of photons with no temporal relation to each other, while other light sources such as thermal emitters exhibit a temporally bunched output. This is characterized by photon statistics, which has its origin in the quantum description of light, pioneered by Nobel laureate R. Glauber [52]. Temporal coherence refers to the correlations in the temporal sequence of emitted photons, with a laser being in a *Glauber state* (coherent state) [75, 76]. For a stable laser, the statistical distribution of photons is described by a Poissonian distribution. In other words, the probability to find a certain number of photons per light pulse is totally random (that corresponds to a normal distribution centred around the expectation value). This is in stark contrast to a light bulb (black-body radiation with super-Poissonian distribution, i.e. thermal source), or a non-classical light source (single-photon emitter, sub-Poissonian distribution), which emits down to only one single photon per light pulse.

The temporal photon statistics is accessible in a second-order temporal autocorrelation experiment, which puts intensities of the light wave at different times in relation to each other. The technique commonly used is referred to as Hanbury–Brown-and-Twiss (HBT) technique [77], which also originated from the domain of astronomy as an improved means of size determination for stars (cf. Michelson stellar interferometer). In fact, the HBT technique has now become a wide-spread tool in quantum optics, where the non-classical state (*Fock state*) has to be distinguished from a thermal or coherent state for quantum emitters. Nevertheless, photon statistics measurements can be also useful for the characterization of ultra-efficient nanolasers and nearly-threshold-less laser devices. Above the threshold of such a laser (when stimulated emission becomes the dominating factor of light generation), a distinct transition from the thermal state to the coherent state is obtained in the photon statistical measurements (see [78, 79]). Indeed, an unstable laser's photon trace (with intensity noise) may deviate from that of a pure coherent state.

4.2.3 Bernard–Duraffourg Condition in Semiconductors

While most laser types involve stimulated emission from discrete energy states in the amplification process, the situation is quite different for semiconductor lasers. Their gain structures are made of semiconductor material which exhibit a band structure with an energy–momentum distribution of electronic states. Optical transitions take place between the valence band and the conduction band, which are separated by the energy band gap. In this gap region in the band structure no electronic states are present. Thus, the minimum energy for excitation (or de-excitation) from one band to the other is determined by the band gap, which separates the upper edge of the valence band from the lower edge of the conduction band. In bulk, both bands consist of a quasi-continuum of a large number of densely packed energy states (in low-dimensional systems, the density of states is altered based on the degree of confinement), which can only be occupied by two electrons of opposite spin, since fermions underly the Pauli principle (identical particles repel each other, opposite to the bunching effect for bosons). In such a system, the occupation distribution of these states determines the rate of optical transitions. If the valence band (VB) was completely filled, then the conduction band (CB) in a charge neutral system would be completely empty (the case at 0 K). At higher temperatures, the probability to find electrons excited in the conduction band, and defect electrons (holes) available in the valence band, is determined by the Fermi–Dirac distribution

$$f(E) = \frac{1}{\exp\left(\frac{E-E_{\mathrm{F}}(T)}{k_{\mathrm{B}}T}\right)+1}, \tag{4.4}$$

whereas E_{F} is the Fermi energy (in intrinsic semiconductors at low temperatures usually close to the centre of the band gap). The number of excited electrons in a band thereby depends on the density of states and the distribution function $f(E)$.

During laser operation, a semiconductor gain medium is brought out of thermodynamic equilibrium, owing to the fact that for stimulated emission, a large number of electrons (holes) need to accumulate near the conduction (valence) band edge in order to be available for radiative recombination. This is achieved by a pumping process, which elevates particles into the excited state, whereas in such case, charge carriers within each band are in thermodynamic equilibrium and can be described by a quasi-Fermi distribution

$$f(E) = \frac{1}{\exp\left(\frac{E-Q_{\mathrm{CB/VB}}}{k_{\mathrm{B}}T}\right)+1}. \tag{4.5}$$

The quasi-Fermi level $Q_{\mathrm{CB/VB}}$ (given for either band) describes the virtual boundary between occupied and unoccupied states, and is named *imref* (Fermi spelt in reverse). Below Q_{CB}, no states in the conduction band are free at 0 K. At temperatures >0 K this boundary is smeared out with a width of the order of $8k_{\mathrm{B}}T$ as common for the Fermi distribution.

The scenario which comes closest to the term "inversion" for a semiconductor laser requires that the states in the conduction band close to the edge are strongly populated while those at the valence band edge are comparably depleted, i.e. the population of holes at the valence band edge is equally high in a charge-neutral device. As soon as $Q_{\text{CB}} - Q_{\text{VB}} > E_{\text{g}}$, transparency is achieved for photons with the energy $h\nu = E_{\text{g}}$ and all states/the majority of states (at zero/finite temperature) at the edge of the conduction (valence) band are occupied (vacant) and can contribute to stimulated emission via radiative recombination of electron–hole pairs, i.e. relaxation of electrons from conduction to valence band. This special non-equilibrium condition is achieved by charge-carrier injection into a diode (p-n) junction (or strong above-band optical excitation) and is referred to as Bernard–Duraffourg condition after Bernard and Duraffourg [27] (see Fig. 4.3c).

Without optical feedback, such a device with light emitting p-(i-)n junction will only act as a source of incoherent (non-amplified) light, thus act as LED, or superluminescence diode with amplified spontaneous emission (ASE).[2] This renders optical resonators basically indispensable. In fact, for edge emitters which are cleaved along the crystal-lattice planes on two opposite sides, a Fabry–Pérot resonator is naturally formed by two opposing parallel facets, giving enough feedback from the lossy resonator with slightly more than 30% reflectivity per mirror (merely given by the refractive index contrast between gain medium and air $n/n_0 \approx 3.5$). In contrast, QW or QD VCSELs require highly-reflective mirrors (e.g. DBRs) for sufficient gain.

4.2.4 *Similarities and Differences Between Polariton and Photon Lasers*

Having described the peculiarities of both types of devices, namely the polariton and the photon laser, it is important to understand similarities as well as differences between the two. While the terminology suggests that both lasers belong to the same family of devices, the working principles of both clearly unravels the differences. Indeed, both devices are considered very similar in terms of light output, as a polariton condensate is desired to achieve coherent emission of light from a stimulated system similar to the output of a conventional laser. Nevertheless, it is somehow misleading to label both devices as *lasers*, using an acronym which only represents the mechanism behind coherent light generation in conventional lasers.

The only link between polaritonic and photonic lasers is provided by a stimulation process: it leads to coherent emission of radiation. In both cases, the desired output occurs after an excitation process, which brings the system strongly out of equilibrium. However, for the sake of correctness, one has to clearly distinguish between the two mechanisms which succeed the pumping process and the consecutive

[2] ASE represents spontaneous emission experiencing single-pass gain for an inverted active medium. In a laser device, it represents a parasitic effect that depletes the gain (i.e. poses a loss channel) and causes heat by re-absorption in non-inverted (non-transparent) regions of the structure.

spontaneous relaxation of matter excitation towards a reservoir state of electron–hole pairs—something that both types of devices have in common. It is this process following the preparation of the reservoir state's heavy occupation, which is a stimulated process for both, though *not* stimulated emission in both cases.

In the case of a conventional laser, the out-of-equilibrium inverted population of electron–hole pairs (charge-carrier plasma in the proximity of their excitation minimum) is depleted by circulating cavity photons, resulting in stimulated emission by radiative recombination of electron–hole pairs. Thereby, relaxation of charge carriers from an incoherent bath (plasma) to the lasing ground state induced by an incident photon causes a clone photon and, consequently, a coherent beam of light. In contrast, electron–hole pairs in the polariton laser are in a low-density regime and form Coulomb-bound states (incoherent excitons). These are in strong interaction with the light field and relax from their reservoir as bosonic polaritons by a stimulated-scattering process efficiently into their ground state, forming a coherent (non-equilibrium) Bose condensate of exciton–polaritons. Out of this coherent state—the ground state of the polaritonic system—decay of photons from the optical cavity leads to a coherent stream of photons.

With this in mind concerning the working principle, spectral features and emission characteristics can be easily compared and further similarities and differences highlighted.

The pumping scheme for both devices can be very similar, given the structural similarities and the necessity to inject excited charge carriers into the active medium. Thereby, both systems are brought out-of-equilibrium and undergo several relaxation processes, one of which is a stimulated process. However, the polariton laser is necessarily operated below the Mott density in order to maintain the excitonic as well as bosonic character of the matter excitation for polaritonic emission as well as stimulated final-state scattering, whereas the photon laser requires gain from an inverted medium, i.e. feeds from uncorrelated charge carriers in an electron–hole plasma.

Directionality of emission is given for both types of lasers, owing to the stimulated emission process of the laser, which clones photons, and the ground-state emission from the condensate, which exhibits a narrow momentum distribution.

As a consequence of their working principles, although differing drastically, these devices both feature a monochromatic output of light. While this is achieved by stimulated emission from well-defined transitions in lasers, polariton condensates naturally exhibit a narrow energy distribution due to collective properties of the emitters (matter coherence) in a quantum degenerate state which is evidenced in the emission linewidth.

Coherence of the output signal is expected to be provided by both laser types, however, while a stable photon laser features high coherence and a Poissonian distribution of photons in the time domain, such result is not necessesarily obtained readily from dynamic polariton condensates. As polaritons are commonly out of equilibrium, not conserved (dynamic system) and prone to interactions with reservoir states, the temporal photon statistics can be spoiled and in many cases does not reach unity at particle densities reasonably far above the condensate threshold. Yet, the pronounced transition from a thermal light source (thermal state) towards a coherent state is typically obtained (see [17]), which is a significant indicator for spontaneous coherence build-up in the bosonic system. However, particularly good second-order temporal autocorrelation values comparable to that of a stable laser have only recently been demonstrated. In that case, a single-mode polariton laser, which features a condensate uncoupled from the excitonic reservoir as well as higher-energy polariton states, was used to obtain laser-like photon statistics and exhibited a high temporal coherence [81].

In summary, this comparison clearly shows that, although the concept of a polariton laser as considered by Imamoglu et al. in 1996 remains attractive with respect to the energy-efficient generation of coherent light, a *polariton laser* is not a laser in the real sense. The main features of both lasing regimes in terms of excitations

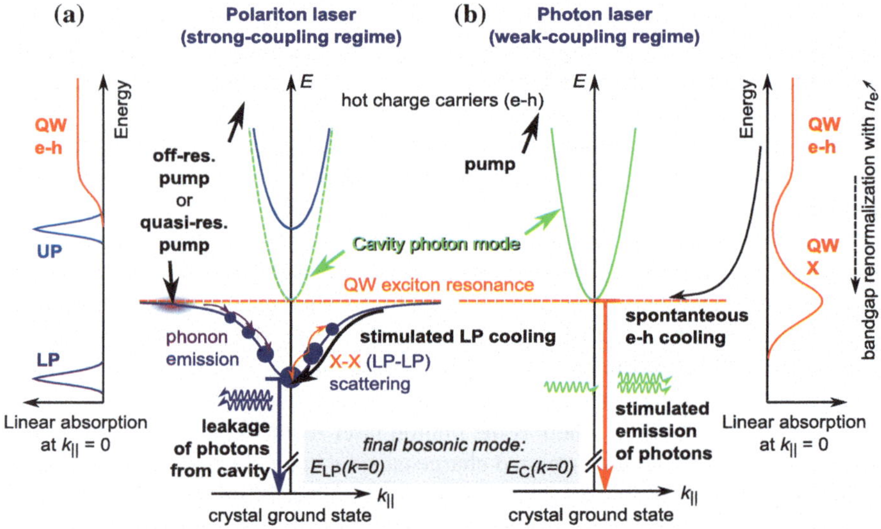

Fig. 4.4 Illustration of the similarities and differences between a polariton (**a**) and photon laser (**b**). Here, the dispersion for both the strong- and weak-coupling regime are sketched. The main difference arises from the underlying cooling process and the cause of coherent emission. Firstly, condensates are formed with the help of stimulated scattering. In contrast, the electronic system in a conventional laser cools down spontaneously to the relevant laser transition(s) after excitation. Secondly, a polariton laser employs a high-quality optical microresonator for coherent energy exchange between QW excitons in a low carrier-density regime with the light field to form hybrid bosonic quasi-particles. In contrast, a conventional laser uses a resonator for optical feedback to amplify cavity photon numbers (intensities) in the weak-coupling regime via the spectral overlap of resonator modes with the gain spectrum of the quantum-well system, which at high carrier densities is in an electron–hole (e–h) plasma regime. Ultimately, coherent light output from the polariton laser originates from the coherence of the exciton–polariton in the macroscopically-occupied LP ground state, and not from stimulated emission. Indeed, both systems provide light output due to photon leakage from the optical resonator owing to the finite reflectivity of the cavity mirrors, while there is a fundamental difference in the final bosonic mode: for a conventional laser and polariton laser this happens from the heavily occupied photon mode $E_C(0)$ and from the quantum-degenerate polariton state $E_{LP}(0)$, respectively. After [4, 7, 80]

and dynamics are compared in a schematic diagram in Fig. 4.4. Furthermore, polariton lasing brings up some drawbacks, such as density dependency, inferior output scalability and mendable coherence, while promising a fundamentally low operation threshold of the micro-sized semiconductor-laser-like system.

4.3 Identification of Polariton Lasing

This section is devoted to the differentiation of lasing in the strong-coupling regime, i.e. emission from a condensate of polaritons, and lasing in the weak-coupling regime, which represents classical lasing from the planar microresonator system, i.e. VCSEL.[3] The former is referred to as polariton lasing, the latter as *photon* (or *photonic*) lasing, for convenience, although polariton lasing has little to do with stimulated emission as mentioned above [4, 30, 33, 34, 82], while it exhibits many features of a laser [83, 84].

Unambiguous demonstration of a polariton laser remained a challenging task for a very long time given the strong similarities between both distinct laser regimes in microcavities. The question on how to identify a polariton laser remained subject of intense discussions in the literature [30, 33, 34]. Particularly, an unambiguous discrimination of the possible lasing regimes in the case of the occurrence of nonlinearities in the input–output characteristics was crucial for the claim of the first electrically-driven polariton laser (see [1]), and remains essential in general for any polariton-laser demonstration.

4.3.1 Prerequisites and the Signatures of a Polariton Condensate

A polariton condensate exhibits distinct signatures and certain criteria have to be fulfilled in order to observe such. Thus, it is essential to evidence these signatures and prove the fulfillment of necessary criteria in order to distinguish polariton lasing from conventional lasing. A list of signatures and criteria provided below give experimentalists a means of verification of a polariton regime above nonlinearity thresholds, for instance in the state of quantum degeneracy (cf. [12, 30, 33, 34]). Herewith, also signatures are described which fundamentally support the identification of a condensate regime (see Chap. 8).

1. Most importantly, strong light–matter coupling has to exist both below and above the ground-state nonlinearity threshold (polariton-laser/condensation threshold). If polaritons are evidenced at low densities, they have to be preserved even at densities around the threshold. A significant reduction of the exciton oscillator

[3] *Vertical-cavity surface-emitting laser.*

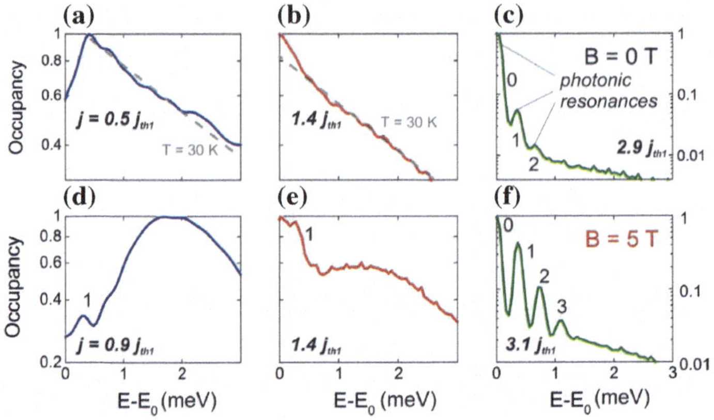

Fig. 4.5 Occupation functions for the three emission regimes of an electrically-driven polariton-laser device. **a** Maxwell–Boltzmann (MB) distribution implying a gas temperature of about 30 K in the linear regime, i.e. below the condensation threshold. **b** Deviation from a MB distribution of a thermalized polariton gas with Bose–Einstein-like distribution above the first nonlinearity threshold, for which the ground state shows a macroscopic occupation, whereas the higher-energy background remains thermalized. **c** Experimentally derived population indicates spectral signatures of the photonic resonances in the regime of weak-coupling lasing, corresponding to the fundamental mode and the higher (lateral) modes of a circular micropillar with a diameter of 20 μm. **d** to **f** show analog representations for the device in an external magnetic field of 5 T. Due to modifications in the relaxation efficiencies as a consequence of the lifted spin degeneracy in the external field, a pronounced bottleneck dominates the spectrum below threshold, occurring at approximately 1.7 meV above the weakly populated polariton ground state. Above both of the thresholds, the situation is comparable to the 0 T case, while **e** still indicates an uncondensed fraction at the bottleneck energies and **f** indicates a broader distribution of gain towards higher resonances than in **c**. Reused with permission. Reference [1] (Supplementary Information) Copyright 2013, Springer Nature

strength—and with it the coupling strength—at density levels shortly before a given phase transition has to be ruled out.

2. The phase transition from a low-density polariton gas to a condensate is accompanied by a nonlinearity in the input–output curve of the system with respect to ground-state emission. This represents stimulated scattering into the ground state, which becomes macroscopically occupied.
3. Above a critical excitation density, a strong narrowing of both the energy and k-space distribution has to be obtained for the polaritonic emission from the system, indicating macroscopic ground-state occupation and spontaneous formation of a coherent phase.
4. The preservation of a polaritonic dispersion above the observed threshold with indication of a pronounced ground-state occupancy can be seen as a signature of preserved strong coupling [6, 32, 86–89]. On the one hand, this is not readily clear in a pulsed excitation scheme, where time-integrated detection may represent emission from different regimes in the recorded output pattern, particularly if the expected energy level of the uncoupled photon mode is close to

the ground state emission (separation of the cavity and polariton ground-state $\Delta E_{cp} \ll \hbar\Omega$). On the other hand, this dispersion check becomes obsolete if interaction-dependent linearization of the polariton dispersion occurs owing to Bogoliubov excitations in the condensate [87], which is considered an evidence of itself for a non-photonic regime. However, for some polariton systems, other means of polariton verification are required, e.g. for discrete polaritonic modes in strong-confinement structures that do not exhibit the common dispersion charactertics of the unconfined 2D system.

5. The occurrence of a second nonlinear increase of the output intensity from the system above a first threshold at stronger excitation levels allows the distinct separation of three regimes. This can be used to attribute the low-density regime to incoherent and thermal polariton emission, the higher-density regime above a first threshold to polariton lasing, and the highest-density regime above a second threshold to photonic lasing (see [1, 12, 16–18]).
6. The exciton density around the condensation threshold must be significantly lower than the charge-carrier density, which is required to obtain the Bernard–Duraffourg condition (threshold for conventional lasing in the semiconductor structure) [27], and must be below the Mott density [90]. In some condensate systems, the exciton density may be of the same order of magnitude as the Mott density, which is not critical as long as the preservation of excitonic properties is evidenced. The Mott transition characterizes the bleaching of excitons that leads the system into an electron–hole plasma. Note, that for the appearance of polaritons, the following must hold true: $\tau_2 > t_{rt}$, that is, exciton dephasing time longer than the cavity photon's round-trip time (see [91, 92]). For an increased particle density υ or temperature T, the dephasing time shortens, e.g. through excitation-induced dephasing (EID) through exciton–exciton or electron–electron scattering or a Mott transition (see Chap. 17 of [93]). Thus, EID leads to a transition from a regime with normal-mode splitting to one with uncoupled resonances (emission originating from the bare cavity mode and excitons or plasma).
7. The linear and nonlinear regimes typically exhibit differences in the linewidth and energy of the emission modes. The laser transition is usually accompanied by a drop in the linewidth, which represents the degree of coherence for a laser. However, this does not necessarily hold true for microcavity systems where the transition to lasing can even lead to a linewidth broader than that of the LP mode. Moreover, even above the condensate threshold in the polaritonic regime, the behaviour of the mode does not follow a universal pattern. The interplay of different effects can lead to either broadening, narrowing, or even no noticable change, and may differ from structure to structure. For instance, an increased polariton–polariton interaction at higher densities can cause linewidth broadening [16, 17, 94], while at higher particle densities phase-space filling bleaches excitons (reducing the oscillator strength) and increases the photonic content of polaritons. Thus, the linewidth behaviour itself does not serve as a firm criterion, while the occurrence of a change is a good signature for the identification of a phase transition in the system.

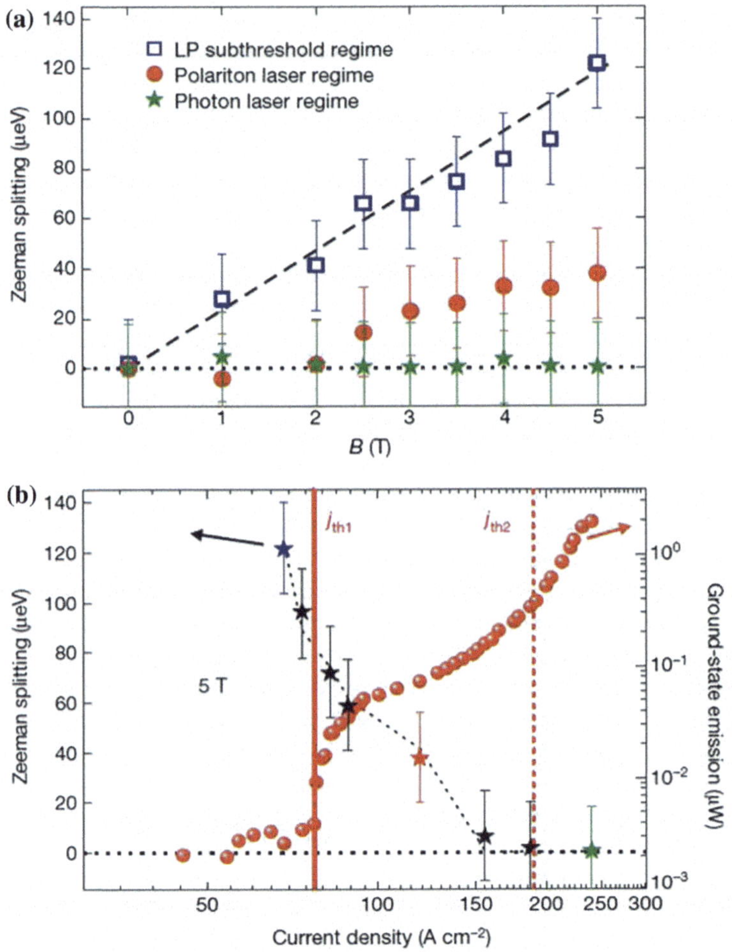

Fig. 4.6 Overview on the experimental procedure towards the unambiguous demonstration of polariton lasing in QW microcavities. By obtaining nonlinear emission signatures from the ground-state of a system, which exhibits a linear polariton regime at low particle densities, it is not readily provided that one can distinguish between polariton and photon laser operation, given the many similarities these regimes share. However, one can commonly characterize a two-threshold behaviour of the pump-density-dependent system in order to identify three regimes, with the first threshold for nonlinear emission out of the ground state being attributed to polariton lasing, and the final threshold to photon lasing above the Mott density. Fortunately, the verification of polariton lasing above the first threshold can be accomplished exploiting the interaction of the polariton's matter component with an external magnetic field, which can lead to Zeeman splitting of polaritonic modes even above the nonlinearity and thereby the establishment of spinor-condensates. To reveal the impact of increasing magnetic fields on the emission modes of a polariton-laser device in its three distinct operation regimes, the strength of an induced Zeeman splitting on the different modes for different density regimes was characterized (**a**). For emission between the two threshold densities (**b**), clearly a splitting was obtained at elevated fields, whereas the low-field response even indicated a spin-Meissner effect for spinor-polariton condensates (**a**), also investigated for optically-pumped condensates [85]. Reused with permission. Reference [1] Copyright 2013, Springer Nature

8. The energy level E_0 of the ground-state mode above a threshold can indicate the degree of excitonic bleaching and reveal a reduction in the coupling strength as E_0 shifts to higher energies (blue) towards the level of the resonance energy expected for an uncoupled cavity mode. The LP energy may experience a blue shift at the condensate threshold, which indicates a density-dependent effect due to macroscopic ground-state occupation. However, the energy position itself does not serve as a strong indication of a certain regime, as the photon laser mode can also be pinned to a level close to the polariton resonance. Nevertheless, the emission-energy level's behaviour may serve as a good signature for the identification of a phase transition, similar to the linewidth behaviour.
9. Dynamical BEC of polaritons requires spontaneous coherence build-up and is accompanied by an increased spatial phase correlation and long-range order. However, these features are strongly dependent on the coherence which the exciton gas in the polaritonic system reaches within itself, the magnitude of the condensate fraction and the degree of thermalization with the lattice. The demonstration of pronounced coherence effects clearly supports the claim of (quasi-)BEC. Temporal coherence is commonly evidenced in second-order temporal autocorrelation experiments as a change from a thermal towards a laser-like (coherent) emitter (see [5, 17, 22, 81, 95, 96]), and spatial coherence by interferometry (see [6, 20, 97–99]). Naturally, coherence is expected from the weak-coupling lasing regime as well. In the context of a BKT-like transition (named after Berezinskii, Kosterlitz and Thouless [100, 101]), also vortex–antivortex-pair formation (evidenced in spatial phase maps) plays a role which enable establishment of local phase order and a superfluid in a 2D system within the finite condensate spot [102, 103]. It shall be also noted that half-vortices [104, 105] cannot appear in the near-field pattern of a conventional laser for fundamental reasons, as stated in [34].
10. The distribution function of particles gives another clue whether a system has undergone Bose condensation. A thermal (Maxwell–Boltzmann) distribution of polaritons in their energy states typically represents an uncondensed system with a certain gas temperature, while above the critical particle density, condensate systems exhibit a distribution function clearly deviating from that of a thermal distribution. In this regime of macroscopic ground-state occupation, the system has to be described by a Bose–Einstein(-like) distribution. Since for most dynamic and non-equilibrium systems no chemical potential can be determined, a significant change from the Maxwell–Boltzmann distribution with strong ground-state occupancy is already a clear signature of dynamic condensation and polariton lasing (Fig. 4.5).
11. Finally, since the excitonic fraction of polaritons needs to be preserved above the threshold attributed to condensation (in contrast to photon lasing), the matter component of polaritons can be directly manipulated by an external magnetic field [106–108]. Thus, Zeeman splitting of the emission mode can provide unambiguous evidence of a polaritonic system in any density regime, even if the energy states are discrete [1, 108], provided that the "spin-Meissner" effect does not prevent the spinor-polariton system in the condensate regime to split up: Owing to

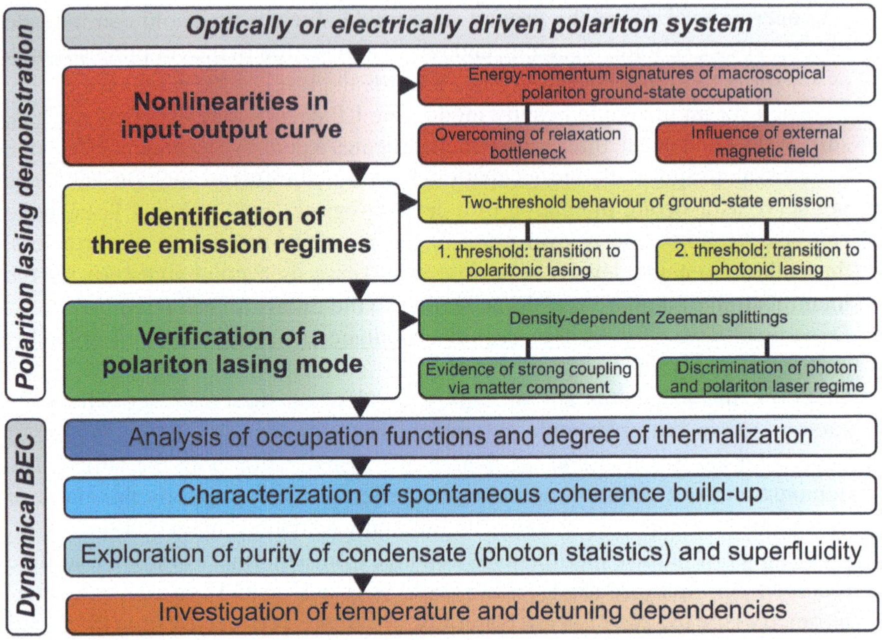

Fig. 4.7 Overview on the typical experimental procedure for the verification of an optically/electrically driven polariton laser by evidencing a two-threshold behaviour and by employing external magnetic fields to study the system's response in different excitation-density regimes. In addition, further steps concerning dynamical BEC studies are summarized, which help to deepen one's understanding of a polariton system's condensation behaviour and to determine its characteristics. The claim of a BEC-like regime is often linked to the demonstration of long-range order in the polariton gas through spatial correlation measurements. Moreover, the purity of a condensate and its interaction with reservoir states can be probed by photon statistics investigations. Adapted from [112] and extended. Courtesy of the author

interactions in the system in the regime of polariton lasing, the Zeeman splitting can be suppressed at magnetic fields below a critical field strength [109] (also see [85, 110, 111]). As the Zeeman splitting of polaritons scales with the excitonic fraction [108] and thereby with the coupling strength of the QW-microcavity system, even the transition to the weak-coupling regime (by bleaching of excitons, i.e. phase-space filling) can be observed in density-dependent measurements at the presence of an external magnetic field [1, 112] (see Fig. 4.6).

4.3.2 *Overview on the Typical Experimental Procedure*

Given the above described controversies in condensate identification, and the similarities to photon lasing, several typical experimental procedures can help to accomplish necessary characterization towards the unambiguous demonstration of polariton-laser operation. To achieve this goal, first of all, polaritonic emission from the respective structure needs to be evidenced. Then, naturally, nonlinearities in the ground-state emission characteristics should be obtained. The effect of the detuning and an external magnetic field on the luminescence features can be then addressed. For instance the occurrence of a bottleneck which is overcome at higher pump densities is a helpful indicator for macroscopic polariton ground-state occupation. Moreover, the influence of external fields on the dynamics and the strong-coupling regime provide further understanding of the nonlinearities in the system. Consecutively, the identification of three emission regimes by the observation of a two-threshold behaviour of the ground-state emission is key to the attribution of one threshold to polariton lasing and the second threshold in the microcavity system to conventional photon lasing, with particle density considerations supporting the transitions between these regimes. Finally, the verification of a polariton mode above the first threshold is crucial for the claim of polariton lasing. Pump-density-dependent Zeeman splitting of exciton–polaritons can be used to prove strong coupling at a given particle-density regime via the matter component of polaritons. Ultimately, the features and signatures obtained from these studies should then allow for a clear discrimation between polaritonic and photonic lasing in the considered QW microresonator, as has been done for the first unambiguous demonstration of an electrically-driven polariton laser both for quasi-DC (200-ns-pulsed) and DC current injection [1, 112] (also see follow-up reports about the same system [113–117]).

To provide an overview, this general procedure for the identification of polariton lasing is schematically summarized in Fig. 4.7 with an extended characterization roadmap for dynamic polariton BEC systems.

References

1. C. Schneider, A. Rahimi-Iman, N.Y. Kim, J. Fischer, I.G. Savenko, M. Amthor, M. Lermer, A. Wolf, L. Worschech, V.D. Kulakovskii, I.A. Shelykh, M. Kamp, S. Reitzenstein, A. Forchel, Y. Yamamoto, S. Höfling, An electrically pumped polariton laser. Nature **497**, 348 (2013)
2. G. Björk, A. Karlsson, Y. Yamamoto, Definition of a laser threshold. Phys. Rev. A **50**(2), 1675–1680 (1994)
3. J. McKeever, A. Boca, A.D. Boozer, J.R. Buck, H.J. Kimble, Experimental realization of a one-atom laser in the regime of strong coupling. Nature **425**(6955), 268–271 (2003)
4. A. Imamoglu, R.J. Ram, S. Pau, Y. Yamamoto, Nonequilibrium condensates and lasers without inversion: exciton-polariton lasers. Phys. Rev. A **53**(6), 4250 (1996)
5. H. Deng, G. Weihs, C. Santori, J. Bloch, Y. Yamamoto, Condensation of semiconductor microcavity exciton polaritons. Science **298**(5591), 199–202 (2002)
6. J. Kasprzak, M. Richard, S. Kundermann, A. Baas, P. Jeambrun, J.M.J. Keeling, F.M. Marchetti, M.H. Szymańska, R. André, J.L. Staehli, V. Savona, P.B. Littlewood, B. Deveaud, L.S. Dang, Bose–Einstein condensation of exciton polaritons. Nature **443**(7110), 409–414 (2006)
7. H. Deng, H. Haug, Y. Yamamoto, Exciton-polariton Bose–Einstein condensation. Rev. Mod. Phys. **82**(2), 1489 (2010)

8. K.B. Davis, M.O. Mewes, M.R. Andrews, N.J. van Druten, D.S. Durfee, D.M. Kurn, W. Ketterle, Bose–Einstein condensation in a gas of sodium atoms. Phys. Rev. Lett. **75**(22), 3969– (1995)
9. C.C. Bradley, C.A. Sackett, J.J. Tollett, R.G. Hulet, Evidence of Bose–Einstein condensation in an atomic gas with attractive interactions. Phys. Rev. Lett. **75**(9), 1687 (1995)
10. M.H. Anderson, J.R. Ensher, M.R. Matthews, C.E. Wieman, E.A. Cornell, Observation of Bose–Einstein condensation in a dilute atomic vapor. Science **269**(5221), 198–201 (1995)
11. Y. Yamamoto, F. Tassone, H. Cao, *Semiconductor Cavity Quantum Electrodynamics* (Springer, Berlin, 2000)
12. H. Deng, G. Weihs, D. Snoke, J. Bloch, Y. Yamamoto, Polariton lasing vs. photon lasing in a semiconductor microcavity. Proc. Natl. Acad. Sci. USA **100**(26), 15318–15323 (2003)
13. A. Das, J. Heo, M. Jankowski, W. Guo, L. Zhang, H. Deng, P. Bhattacharya, Room temperature ultralow threshold GaN nanowire polariton laser. Phys. Rev. Lett. **107**(6), 066405– (2011)
14. P. Bhattacharya, T. Frost, S. Deshpande, M.Z. Baten, A. Hazari, A. Das, Room temperature electrically injected polariton laser. Phys. Rev. Lett. **112**, 236802 (2014)
15. A. Bhattacharya, Md.Z. Baten, I. Iorsh, T. Frost, A. Kavokin, P. Bhattacharya, Output polarization characteristics of a GaN microcavity diode polariton laser. Phys. Rev. B **94**(3), 5203 (2016)
16. D. Bajoni, P. Senellart, E. Wertz, I. Sagnes, A. Miard, A. Lemaître, J. Bloch, Polariton laser using single micropillar GaAs-GaAlAs semiconductor cavities. Phys. Rev. Lett. **100**(4), 047401 (2008)
17. J.-S. Tempel, F. Veit, M. Aßmann, L.E. Kreilkamp, A. Rahimi-Iman, A. Löffler, S. Höfling, S. Reitzenstein, L. Worschech, A. Forchel, M. Bayer, Characterization of two-threshold behavior of the emission from a GaAs microcavity. Phys. Rev. B **85**(7), 075318 (2012)
18. P. Bhattacharya, B. Xiao, A. Das, S. Bhowmick, J. Heo, Solid state electrically injected exciton-polariton laser. Phys. Rev. Lett. **110**(20), 206403 (2013)
19. J. Klaers, J. Schmitt, F. Vewinger, M. Weitz, Bose–Einstein condensation of photons in an optical microcavity. Nature **468**(7323), 545–548 (2010)
20. H. Deng, G.S. Solomon, R. Hey, K.H. Ploog, Y. Yamamoto, Spatial coherence of a polariton condensate. Phys. Rev. Lett. **99**(12), 126403– (2007)
21. R. Balili, V. Hartwell, D. Snoke, L. Pfeiffer, K. West, Bose–Einstein condensation of microcavity polaritons in a trap. Science **316**(5827), 1007–1010 (2007)
22. J. Kasprzak, M. Richard, A. Baas, B. Deveaud, R. Andre, J.-Ph. Poizat, L.S. Dang, Second-order time correlations within a polariton Bose–Einstein condensate in a CdTe microcavity. Phys. Rev. Lett. **100**(6), 067402 (2008)
23. A. Kavokin, G. Malpuech, *Cavity Polaritons* (Academic Press, New York, 2003)
24. B. Deveaud, *The Physics of Semiconductor Microcavities* (WILEY-VCH Verlag, Weinheim, 2007)
25. D. Sanvitto, V. Timofeev, *Exciton Polaritons in Microcavities: New Frontiers* (Springer, Berlin, 2012)
26. A.V. Kavokin, J.J. Baumberg, G. Malpuech, F.P. Laussy, *Microcavities*, vol. 1 (Oxford University Press, Oxford, 2017)
27. M.G.A. Bernard, G. Duraffourg, Laser conditions in semiconductors. Phys. Stat. Sol. (b) **1**(7), 699–703 (1961)
28. S. Pau, H. Cao, J. Jacobson, G. Björk, Y. Yamamoto, A. Imamoglu, Observation of a laserlike transition in a microcavity exciton polariton system. Phys. Rev. A **54**(3), R1789– (1996)
29. R. Butté, G. Delalleau, A.I. Tartakovskii, M.S. Skolnick, V.N. Astratov, J.J. Baumberg, G. Malpuech, A. Di Carlo, A.V. Kavokin, J.S. Roberts, Transition from strong to weak coupling and the onset of lasing in semiconductor microcavities. Phys. Rev. B **65**(20), 205310 (2002)
30. D. Bajoni, P. lasers, Hybrid light-matter lasers without inversion. J. Phys. D Appl. Phys. **45**(31), 313001 (2012)
31. M. Richard, J. Kasprzak, R. Romestain, R. André, L.S. Dang, Spontaneous coherent phase transition of polaritons in CdTe microcavities. Phys. Rev. Lett. **94**(18), 187401 (2005)

32. M. Richard, J. Kasprzak, R. André, R. Romestain, L.S. Dang, G. Malpuech, A. Kavokin, Experimental evidence for nonequilibrium Bose condensation of exciton polaritons. Phys. Rev. B **72**(20), 201301 (2005)
33. B. Deveaud-Plédran, On the condensation of polaritons. J. Opt. Soc. Am. B **29**(2), A138–A145 (2012)
34. V.B. Timoffeev, Bose condensation of exciton polaritons in microcavities. Fizika i Tekhnika Poluprovodnikov **46**(7), 865–883 (2012). English: translation in Semiconductors **46**(7), S. 843, Springer (2012)
35. H. Deng, D. Press, S. Götzinger, G.S. Solomon, R. Hey, K.H. Ploog, Y. Yamamoto, Quantum degenerate exciton-polaritons in thermal equilibrium. Phys. Rev. Lett. **97**(14), 146402 (2006)
36. L.V. Butov, A.V. Kavokin, The behaviour of exciton-polaritons. Nat. Phot. **6**(1), 2 (2012)
37. B. Deveaud-Plédran, The behavior of exciton-polaritons. Nat. Phot. **6**, 205 (2012)
38. F. K. Kneubühl, M. W. Sigrist, Laser (Teubner, 2005)
39. J.P. Gordon, H.J. Zeiger, C.H. Townes, Molecular microwave oscillator and new hyperfine structure in the microwave spectrum of NH3. Phys. Rev. **95**(1), 282–284 (1954)
40. A.L. Schawlow, C.H. Townes, Infrared and optical masers. Phys. Rev. **112**(6), 1940–1949 (1958)
41. T.H. Maiman, Stimulated optical radiation in Ruby. Nature **187**(4736), 493–494 (1960)
42. K. Smith, R.M. Thomson, *Computer Modeling of Gas Lasers* (Springer, Boston, 1978)
43. W. Koechner, *Solid-State Laser Engineering* (Springer, New York, 2006)
44. G.P. Agrawal, N.K. Dutta, *Semiconductor Lasers* (Springer, New York, 1993)
45. T. Numai, *Fundamentals of Semiconductor Lasers* (Springer, Tokyo, 2015)
46. J.L. Hall, Nobel lecture: defining and measuring optical frequencies. Rev. Mod. Phys. **78**(4), 1279–1295 (2006)
47. T.W. Hänsch, Nobel lecture: passion for precision. Rev. Mod. Phys. **78**(4), 1297–1309 (2006)
48. A. Javan, W.R. Bennett, D.R. Herriott, Population inversion and continuous optical maser oscillation in a gas discharge containing a He-Ne mixture. Phys. Rev. Lett. **6**(3), 106–110 (1961)
49. R.N. Hall, G.E. Fenner, J.D. Kingsley, T.J. Soltys, R.O. Carlson, Coherent light emission from GaAs junctions. Phys. Rev. Lett. **9**(9), 366–368 (1962)
50. H. Kroemer, Nobel lecture: quasielectric fields and band offsets: teaching electrons new tricks. Rev. Mod. Phys. **73**(3), 783–793 (2001)
51. Z.I. Alferov, Nobel lecture: the double heterostructure concept and its applications in physics, electronics, and technology. Rev. Mod. Phys. **73**(3), 767–782 (2001)
52. R.J. Glauber, Photon correlations. Phys. Rev. Lett. **10**(3), 84–86 (1963)
53. M. Kuznetsov, F. Hakimi, R. Sprague, A. Mooradian, High-power (>0.5-w cw) diode-pumped vertical-external-cavity surface-emitting semiconductor lasers with circular tem$_{00}$ beams. IEEE Photonics Technol. Lett. **9**(8), 1063–1065 (1997)
54. U. Keller, A.C. Tropper, Passively modelocked surface-emitting semiconductor lasers. Phys. Rep. **429**, 67–120 (2006)
55. B. Heinen, T.L. Wang, M. Sparenberg, A. Weber, B. Kunert, J. Hader, S.W. Koch, J.V. Moloney, M. Koch, W. Stolz, 106 W continuous-wave output power from vertical-external-cavity surface-emitting laser. Electron. Lett. **48**(9), 516–517 (2012)
56. A. Rahimi-Iman, Recent advances in VECSELs. J. Opt. **18**(9), 093003 (2016)
57. M.A. Gaafar, A. Rahimi-Iman, K.A. Fedorova, W. Stolz, E.U. Rafailov, M. Koch, Mode-locked semiconductor disk lasers. Adv. Opt. Photon. **8**(3), 370 (2016)
58. M. Guina, A. Rantamäki, A. Härkönen, Optically pumped vecsels: review of technology and progress. J. Phys. D Appl. Phys. **50**(38), 383001 (2017)
59. M. Fujita, A. Sakai, T. Baba, Ultrasmall and ultralow threshold GaInAsP-InP microdisk injection lasers: design, fabrication, lasing characteristics, and spontaneous emission factor. IEEE J. Sel. Topics Quantum Electron. **5**(3), 673–681 (1999)
60. F. De Martini, G.R. Jacobovitz, Anomalous spontaneous-stimulated-decay phase transition and zero-threshold laser action in a microscopic cavity. Phys. Rev. Lett. **60**(17), 1711–1714 (1988)

61. J.-M. Gérard, Solid-state cavity-quantum electrodynamics with self-assembled quantum dots, in *Single Quantum Dots: Fundamentals, Applications, and New Concepts* (Springer, Berlin, 2003), pp. 269–314
62. G. Eisenstein, D. Bimberg, *Green Photonics and Electronics* (Springer, Cham, 2017)
63. K.J. Vahala, Optical microcavities. Nature **424**(6950), 839–846 (2003)
64. S. Reitzenstein, A. Forchel, Quantum dot micropillars. J. Phys. D Appl. Phys. **43**(3), 033001–(2010)
65. N.N. Ledentsov, The way to quantum dot lasers (Semiconductor News: http://wmw-magazine.ru/uploads/volumes/05/17.pdf) (2001)
66. S. Reitzenstein, T. Heindel, C. Kistner, A. Rahimi-Iman, C. Schneider, S. Höfling, A. Forchel, Low threshold electrically pumped quantum dot-micropillar lasers. Appl. Phys. Lett. **93**(6), 061104 (2008)
67. Z.G. Xie, S. Götzinger, W. Fang, H. Cao, G.S. Solomon, Influence of a single quantum dot state on the characteristics of a microdisk laser. Phys. Rev. Lett. **98**(11), 117401 (2007)
68. S. Noda, Seeking the ultimate nanolaser. Science **314**(5797), 260 (2006)
69. S. Strauf, F. Jahnke, Single quantum dot nanolaser. Laser Photon. Rev. **5**(5), 607–633 (2011)
70. P. Michler (ed.), *Single Quantum Dots: Fundamentals, Applications and New Concepts* (Springer, Berlin, 2003)
71. D.A. Nakdali, M.K. Shakfa, M. Gaafar, M. Butkus, K.A. Fedorova, M. Zulonas, M. Wichmann, F. Zhang, B. Heinen, A. Rahimi-Iman, W. Stolz, E.U. Rafailov, M. Koch, High-power quantum-dot vertical-external-cavity surface-emitting laser exceeding 8 W. IEEE Photon. Technolog. Lett. **26**(15), 1561–1564 (2014)
72. E.U. Rafailov, M.A. Cataluna, W. Sibbett, Mode-locked quantum-dot lasers. Nat. Photon. **1**(7), 395–401 (2007)
73. E.M. Purcell, Spontaneous emission probabilities at radio frequencies. Phys. Rev. **69**(11–12), 681 (1946)
74. A.A. Michelson, E.W. Morley, On the relative motion of the Earth and the luminiferous ether. Am. J. Sci. **s3–34**(203), 333–345 (1887)
75. R.J. Glauber, The quantum theory of optical coherence. Phys. Rev. **130**(6), 2529–2539 (1963)
76. R.J. Glauber, Coherent and incoherent states of the radiation field. Phys. Rev. **131**(6), 2766–2788 (1963)
77. R. Hanbury Brown, R.Q. Twiss, Correlation between photons in two coherent beams of light. *Nature* **177**, 27–29 (1956)
78. S.M. Ulrich, C. Gies, S. Ates, J. Wiersig, S. Reitzenstein, C. Hofmann, A. Löffler, A. Forchel, F. Jahnke, P. Michler, Photon statistics of semiconductor microcavity lasers. Phys. Rev. Lett. **98**(4), 043906– (2007)
79. M. Aßmann, F. Veit, M. Bayer, M. van der Poel, J.M. Hvam, Higher-order photon bunching in a semiconductor microcavity. Science **325**(5938), 297–300 (2009)
80. H. Deng, *Dynamic condensation of semiconductor microcavity polaritons*. Ph.D. thesis, Department of Applied Physics, Stanford University, 2006
81. S. Kim, B. Zhang, Z. Wang, J. Fischer, S. Brodbeck, M. Kamp, C. Schneider, S. Höfling, H. Deng, Coherent polariton laser. Phys. Rev. X **6**(1), 011026 (2016)
82. A. Kavokin, G. Malpuech, F.P. Laussy, Polariton laser and polariton superfluidity in microcavities. Phys. Lett. A **306**(4), 187–199 (2003)
83. D. Bajoni, P. Senellart, A. Lemaître, J. Bloch, Photon lasing in GaAs microcavity: similarities with a polariton condensate. Phys. Rev. B **76**(20), 201305(R) (2007)
84. H. Ohadi, E. Kammann, T.C.H. Liew, K.G. Lagoudakis, A.V. Kavokin, P.G. Lagoudakis, Spontaneous symmetry breaking in a polariton and photon laser. Phys. Rev. Lett. **109**(1), 016404 (2012)
85. J. Fischer, S. Brodbeck, A.V. Chernenko, I. Lederer, A. Rahimi-Iman, M. Amthor, V.D. Kulakovskii, L. Worschech, M. Kamp, M. Durnev, C. Schneider, A.V. Kavokin, S. Hofling, Anomalies of a nonequilibrium spinor polariton condensate in a magnetic field. Phys. Rev. Lett. **112**, 093902 (2014)

86. G. Christmann, R. Butté, E. Feltin, J.-F. Carlin, N. Grandjean, Room temperature polariton lasing in a GaN/AlGaN multiple quantum well microcavity. Appl. Phys. Lett. **93**(5), 051102 (2008)
87. S. Utsunomiya, L. Tian, G. Roumpos, C.W. Lai, N. Kumada, T. Fujisawa, M. Kuwata-Gonokami, A. Löffler, S. Höfling, A. Forchel, Y. Yamamoto, Observation of Bogoliubov excitations in exciton-polariton condensates. Nat. Phys. **4**(9), 700–705 (2008)
88. E. Wertz, L. Ferrier, D.D. Solnyshkov, P. Senellart, D. Bajoni, A. Miard, A. Lemaître, G. Malpuech, J. Bloch, Spontaneous formation of a polariton condensate in a planar GaAs microcavity. Appl. Phys. Lett. **95**(5), 051108 (2009)
89. A. Rahimi-Iman, A.V. Chernenko, J. Fischer, S. Brodbeck, M. Amthor, C. Schneider, A. Forchel, S. Höfling, S. Reitzenstein, M. Kamp, Coherence signatures and density-dependent interaction in a dynamical exciton-polariton condensate. Phys. Rev. B **86**(15), 155308 (2012)
90. N.F. Mott, *Metal Insulator Transitions*, 2nd edn. (Taylor & Francis, Abingdon, 1990)
91. V. Savona, L.C. Andreani, P. Schwendimann, A. Quattropani, Quantum well excitons in semiconductor microcavities: unified treatment of weak and strong coupling regimes. Solid State Commun. **93**(9), 733–739 (1995)
92. G. Panzarini, L. C. Andreani, Quantum theory of exciton polaritons in cylindrical semiconductor microcavities. Phys. Rev. B **60**(24), 16799–16806 (1999)
93. C.F. Klingshirn, *Semiconductor Optics* (Springer, Berlin, 2012)
94. D. Porras, C. Tejedor, Linewidth of a polariton laser: theoretical analysis of self-interaction effects. Phys. Rev. B **67**(16), 161310(R) (2003)
95. M. Aßmann, J.-S. Tempel, F. Veita, M. Bayer, A. Rahimi-Iman, A. Löffler, S. Höfling, S. Reitzenstein, L. Worschech, A. Forchel, From polariton condensates to highly photonic quantum degenerate states of bosonic matter. Proc. Natl. Acad. Sci. USA **108**, 1804–1809 (2011)
96. M. Amthor, S. Weißenseel, J. Fischer, M. Kamp, C. Schneider, S. Höfling, Electro-optical switching between polariton and cavity lasing in an InGaAs quantum well microcavity. Opt. Express **22**(25), 31146–31153 (2014)
97. C.W. Lai, N.Y. Kim, S. Utsunomiya, G. Roumpos, H. Deng, M.D. Fraser, T. Byrnes, P. Recher, N. Kumada, T. Fujisawa, Y. Yamamoto, Coherent zero-state and π-state in an exciton-polariton condensate array. Nature **450**(7169), 529–532 (2007)
98. D.N. Krizhanovskii, K.G. Lagoudakis, M. Wouters, B. Pietka, R.A. Bradley, K. Guda, D.M. Whittaker, M.S. Skolnick, B. Deveaud-Pl'edran, M. Richard, R. André, L.S. Dang, Coexisting nonequilibrium condensates with long-range spatial coherence in semiconductor microcavities. Phys. Rev. B **80**(4), 045317 (2009)
99. G. Roumpos, M. Lohse, W.H. Nitsche, J. Keeling, M.H. Szymańska, P.B. Littlewood, A. Löffler, S. Höfling, L. Worschech, A. Forchel, Y. Yamamoto, Power-law decay of the spatial correlation function in exciton-polariton condensates. Proc. Natl. Acad. Sci. USA (2012). https://doi.org/10.1073/pnas.1107970109
100. V.L. Berezinskii, Destruction of long-range order in one-dimensional and two-dimensional systems possessing a continuous symmetry group. II. Quantum systems. J. Exp. Theor. Phys. **34**, 610 (1972)
101. J.M. Kosterlitz, D.J. Thouless, Ordering, metastability and phase transitions in two-dimensional systems. J. Phys. C: Solid State Phys. **6**(7), 1181 (1973)
102. M.D. Fraser, G. Roumpos, Y. Yamamoto, Vortex-antivortex pair dynamics in an exciton-polariton condensate. New J. Phys. **11**, 113048 (2009)
103. G. Roumpos, M.D. Fraser, A. Löffler, S. Höfling, A. Forchel, Y. Yamamoto, Single vortex-antivortex pair in an exciton-polariton condensate. Nat. Phys. **7**(2), 129–133 (2011)
104. K.G. Lagoudakis, T. Ostatnický, A.V. Kavokin, Y.G. Rubo, R. André, B. Deveaud-Plédran, Observation of Half-Quantum Vortices in an Exciton-Polariton Condensate. Science **326**(5955), 974–976 (2009)
105. L. Dominici, G. Dagvadorj, J.M. Fellows, D. Ballarini, M. De Giorgi, F.M. Marchetti, B. Piccirillo, L. Marrucci, A. Bramati, G. Gigli, M.H. Szymañska, D. Sanvitto, Vortex and half-vortex dynamics in a nonlinear spinor quantum fluid. Sci. Adv. **1**(11), e1500807 (2015)

106. T.A. Fisher, A.M. Afshar, M.S. Skolnick, D.M. Whittaker, J.S. Roberts, Vacuum Rabi coupling enhancement and Zeeman splitting in semiconductor quantum microcavity structures in a high magnetic field. Phys. Rev. B **53**(16), R10469 (1996)
107. A. Armitage, T.A. Fisher, M.S. Skolnick, D.M. Whittaker, P. Kinsler, J.S. Roberts, Exciton polaritons in semiconductor quantum microcavities in a high magnetic field. Phys. Rev. B **55**(24), 16395 (1997)
108. A. Rahimi-Iman, C. Schneider, J. Fischer, S. Holzinger, M. Amthor, S. Höfling, S. Reitzenstein, L. Worschech, M. Kamp, A. Forchel, Zeeman splitting and diamagnetic shift of spatially confined quantum-well exciton polaritons in an external magnetic field. Phys. Rev. B **84**(16), 165325 (2011)
109. Y.G. Rubo, A.V. Kavokin, I.A. Shelykh, Suppression of superfluidity of exciton-polaritons by magnetic field. Phys. Lett. A **358**(3), 227–230 (2006)
110. A.V. Larionov, V.D. Kulakovskii, S. Höfling, C. Schneider, L. Worschech, A. Forchel, Polarized nonequilibrium Bose–Einstein condensates of spinor exciton polaritons in a magnetic field. Phys. Rev. Lett. **105**(25), 256401 (2010)
111. P. Walker, T.C.H. Liew, D. Sarkar, M. Durska, A.P.D. Love, M.S. Skolnick, J.S. Roberts, I.A. Shelykh, A.V. Kavokin, D.N. Krizhanovskii, Suppression of zeeman splitting of the energy levels of exciton-polariton condensates in semiconductor microcavities in an external magnetic field. Phys. Rev. Lett. **106**(25), 257401– (2011)
112. A. Rahimi-Iman, *Nichtlineare Eekte in III/V Quantenlm-Mikroresonatoren: Von dynamischer Bose–Einstein-Kondensation hin zum elektrisch betriebenen Polariton-Laser* (Cuvillier Verlag Göttingen, 2013)
113. A. Rahimi-Iman, C. Schneider, S. Höfling, Electrically driven polariton lasing. Opt. Photonics News **24**(12), 30 (2013)
114. A. Rahimi-Iman, C. Schneider, N.Y. Kim, J. Fischer, I.G. Savenko, M. Amthor, L. Worschech, V.D. Kulakovskii, I.A. Shelykh, M. Kamp, S. Reitzenstein, A. Forchel, Y. Yamamoto, S. Höfling, An electrically driven polariton laser, in *Asia Communications and Photonics Conference 2013*, OSA Technical Digest (online), Beijing, November 2013 (Optical Society of America, 2013), p. AW3B.6
115. C. Schneider, J. Fischer, M. Amthor, S. Brodbeck, I.G. Savenko, I.A. Shelykh, A. Chernenko, A. Rahimi-Iman, V.D. Kulakovskii, S. Reitzenstein, N.Y. Kim, M. Durnev, A.V. Kavokin, Y. Yamamoto, A. Forchel, M. Kamp, S. Höfling, Exciton-polariton lasers in magnetic fields, in *Proceedings of the SPIE, Quantum Sensing and Nanophotonic Devices XI*, vol. 8993 (SPIE, 2013), p. 899308
116. M. Amthor, J. Fischer, I.G. Savenko, I.A. Shelykh, A. Chernenko, A. Rahimi-Iman, V.D. Kulakovskii, S. Reitzenstein, N.Y. Kim, M. Durnev, A.V. Kavokin, Y. Yamamoto, A. Forchel, M. Kamp, C. Schneider, S. Höfling, Exciton-polariton laser diodes, in *Proceedings of the SPIE, Nanophotonics and Micro/Nano Optics II*, ed. by Z. Zhou, K. Wada (SPIE, 2014)
117. M. Amthor, C. Schneider, A. Rahimi-Iman, S. Reitzenstein, Y. Yamamoto, M. Kamp, S. Höfling, An exciton-polariton light source for low-power laser applications. SPIE Newsroom (2014)

Chapter 5
Optical Microcavities for Polariton Studies

Abstract In the past two decades, optical microcavities with very high optical quality and their rapid development substantially enabled the achievement of polariton condensation and the investigation of bosonic many-body phenomena such as superfluidity and Bose–Einstein condensation. Similarly, the demonstration of a polariton-laser device strongly relied on technological advances in the fabrication of multi-quantum-well (QW) microresonators. In this context, the general design and concept of optical structures for polariton physics will be summarized and the prominent example of a planar microcavity based on III/V semiconductors introduced. Beginning with the concept of planar Fabry–Pérot microcavities with an optical cavity sandwiched between Bragg mirrors, the principles of QW-based polariton structures will be explained. Thereafter, the resonator properties such as the transmission function, density of states and quality factor will be summarized which are relevant for the experimental realization of polaritons in practical structures.

A. Rahimi-Iman, *Polariton Physics*, Springer Series in Optical Sciences 229,
https://doi.org/10.1007/978-3-030-39333-5_5

5.1 Fabry–Pérot Microcavities

Optical cavities based on the principles of the interferometer developed by Charles Fabry and Alfred Pérot more than a century ago are widely used in microresonator structures for lasers and light–matter-coupled systems. In such Fabry–Pérot microresonators, light is reflected back and forth multiple times from two parallel opposing mirrors, which form the optical cavity (see e.g. resonator sketch in Fig. 2.5 in Sect. 2.2). This leads to an interference of the electro-magnetic waves and a standing-wave pattern of the light field between these mirrors. Details on the principle of a Fabry–Pérot interferometer can be found for example in [1].

For semiconductor systems, Fabry–Pérot cavities can be achieved by various means. In the simplest case, two opposing cleaved edges of a crystal act as a Fabry–Pérot resonator, but good reflectors can also be achieved by metallic surfaces, dielectric mirrors or photonic crystals. Interference-based mirrors such as one-dimensional photonic crystals (periodic sequence of layer pairs), which are referred to as Bragg reflectors [2], can result in highly-reflective mirrors. Such mirror type is commonly used to achieve planar (i.e. 2D) microresonator structures for various kinds of optical devices. One important advantage of Bragg mirrors, also referred to as distributed Bragg reflectors (DBRs), originates from the possibility of monolithic growth together with the active region of a microcavity structure. Moreover, by its design and composition, it can be well tailored for a certain wavelength range and reflectivity specification. A compact overview of different resonator types with their figures of merit is given for instance in [2].

5.1.1 *Distributed Bragg Reflectors*

Usually, in planar microresonator structures, an optical cavity sandwiched between two highly-reflective DBRs allows for a strong photon confinement based on the distributed feedback principle with quality factors representing several thousands of round trips for a photon prior to leakage through the naturally imperfect mirrors. The good confinement is used in lasers based on DBRs to achieve lower loss rates, i.e. to provide strong feedback to the active medium and to improve the device efficiency, such as in VCSELs (see for instance [3]).

DBRs provide out-of-plane (i.e. vertical) photon confinement in planar microcavities and are composed of many pairs of alternating $\lambda/4$-thick (quarter-wavelength) dielectric layers (DL1/DL2). As will be seen in the following, the number of layer pairs and the refractive index contrast of the employed materials determine the reflectivity of such a DBR structure. The thickness of each layer $d_{\text{lay}} = \lambda_{\text{res}}/4n_{\text{DL1/DL2}}$ is given by the resonance (design) wavelength λ_{res} and the effective refractive index $n_{\text{DL1/DL2}}$ of the respective layer material [4]. A stack of two layers DL1 and DL2 with different refractive index is referred to as mirror pair.

In the III/V material system, such pairs are typically made of AlAs and (Al)GaAs, a material system which provides very well controllable growth and a developed fabrication routine. In particular, the AlAs/GaAs system is very attractive owing to its perfect lattice matching of its compounds. While in binary DBRs, i.e. mirrors with binary semiconductors, the refractive index contrast is highest, one can fine-tune material properties such as lattice matching and the band structure of the system in ternary mirrors, which can be of importance for different material systems and devices. The refractive indices of AlAs and GaAs at room temperature and for a photon energy of $\approx 1.3\,\mathrm{eV}$ amount to $n_{\mathrm{AlAs}} \approx 2.95$ and $n_{\mathrm{GaAs}} \approx 3.54$ [5], respectively. In the following, GaAs will be considered as the mirror pair's layer material with higher refractive index, complemented by AlAs with correspondingly lower index.

Light incident on the layers of a DBR will be partially reflected at each interface. Does the reflection take place at the optically thicker medium (AlAs$\rightarrow$GaAs), the reflected part experiences a phase shift of π. This does not occur for reflection at the optically thinner medium (GaAs$\rightarrow$AlAs). Thus, destructive interference is obtained for the transmitted fraction of the beam for the $\lambda/4$-thick layer system, while constructive interference is obtained for the reflected fraction [3, 6]. Accordingly, the employment of a large number of mirror pairs increases the DBR's reflectivity, and an overall reflectivity of more than 99% can be achieved, even better than 99.9%. For a III/V semiconductor DBR, an exemplary calculation of the reflection spectra at normal incidence is shown in Fig. 5.1 for different numbers of mirror pairs. Here, it is clearly seen that such a mirror features a spectral region of high reflectivity, named photonic (energy) gap or stop band, with the highest reflectivity (for the given number of mirror pairs N)

$$R_{\mathrm{max}} = \frac{1 - \frac{n_{\mathrm{C}}}{n_{\mathrm{ext}}}\left(\frac{n_{\mathrm{H}}}{n_{\mathrm{L}}}\right)^{2N}}{1 + \frac{n_{\mathrm{C}}}{n_{\mathrm{ext}}}\left(\frac{n_{\mathrm{H}}}{n_{\mathrm{L}}}\right)^{2N}} \tag{5.1}$$

at its centre corresponding to the Bragg wavelength, whereas n_{C}, n_{ext}, and $n_{\mathrm{H/L}}$ are the refractive indices of the cavity medium, the external medium, and the higher/lower index material of the DBR, respectively. The stop band is surrounded by low reflectivity regions with fringes. The reflectivity drop occurs at wavelengths that do not fulfil the constructive interference conditions of the Bragg structure any more. The high-reflectivity band widens with an increasing number of mirror pairs, and simultaneously the maximum reflectivity increases for instance from $\approx 96\%$ to almost 100% when increasing N from about 10 to about 30 in the GaAs/AlAs system. This concept holds true for any spectral region, provided that non-absorbing layer materials with good refractive index contrast exist.

It is worth noting that one approach to maximize the contrast is to replace one of the layers by air (vacuum). Indeed, such structure referred to as "air-Bragg" can reach very high reflectivities with a comparably low number of pairs, but is correspondingly harder to produce and less rigid/robust. One example of an air-gap based DBR is given for example in [7], used in GaN/AlGaN microcavities for the study of room-temperature trapped exciton–polaritons.

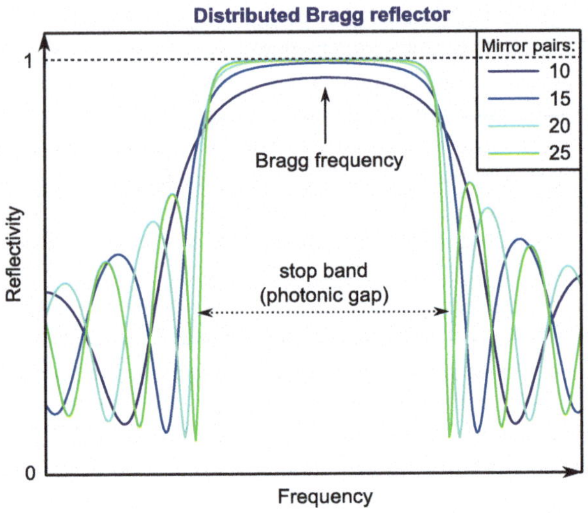

Fig. 5.1 Modelled reflection spectra at normal incidence for a III/V semiconductor DBR at room temperature as an example. The number of mirror pairs is varied between 10 and 25 to show the impact of the number on the high-reflectivity band

5.1.2 Planar Microresonator Structures

Monolithic Microcavities

With high-reflectivity DBRs at hand, the deterministic strong confinement of cavity photons can be achieved by the employment of a monolithic Fabry–Pérot microresonator, which comprises two opposing DBRs and a cavity spacer. As the DBR determines which wavelength λ_{res} experiences highest reflectivity, the design wavelength is thus referred to as the Bragg wavelength λ_{Bragg}. The thickness $d_{\mathrm{cav}} = i \times \lambda_{\mathrm{Bragg}}/2n_{\mathrm{C}}$ of the optical cavity with refractive index n_{C} has to be chosen as an integer number i of the halved Bragg wavelength [4]. The squared electric-field distribution of an optical resonator mode in representative cavities with $\lambda/2$ ($i = 1$) and λ ($i = 2$) thickness is presented in Fig. 5.2a and b, respectively. In these examples, QWs are incorporated into the structure as optically-active region inside the cavity. Such standing-wave field distribution can be calculated by using the well-known transfer-matrix method (e.g. summarized in [31]), or for instance with the cavity modelling framework (CAMFR) program [8]. A detailed description on how optical properties of thin-film stacks can be modelled one-dimensionally by the transfer-matrix method can be found in the literature, e.g. in [9].

For the mode penetrating the Bragg structure, the quadratic amplitude of the field exhibits an exponential decrease. This fact has to be considered when designing the semiconductor microcavity, particularly when it comes to the determination of standard Fabry–Pérot results, where the Fabry–Pérot cavity length has to be replaced with the effective cavity length (see [4, 10, 11]) determined by

$$L_{\mathrm{eff,C}} = L_{\mathrm{C}} + L_{\mathrm{DBR}} = L_{\mathrm{C}} + \frac{\lambda_{\mathrm{DBR}}}{2n_{\mathrm{C}}} \frac{n_{\mathrm{DL1}} n_{\mathrm{DL2}}}{|n_{\mathrm{DL2}} - n_{\mathrm{DL1}}|}, \tag{5.2}$$

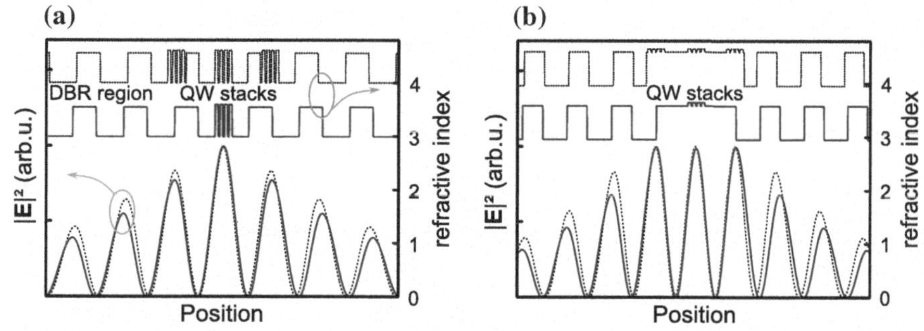

Fig. 5.2 Calculated cavity-mode fields (squared electric field, left axis) in microcavity structures with 12 (black dotted) and 4 QWs (blue solid line) together with refractive index profile (right scale) for **(a)** an AlGaAs $\lambda/2$ cavity and **(b)** a GaAs λ cavity. In comparison to the refractive index profile of the 4-QW structure, the profile for the 12-QW structure is arbitrarily offset by one. The central field maxima align well with the QW positions. Adapted from [12]. Courtesy of the author

whereas L_C and L_{DBR} denote the cavity spacer thickness and the penetration depth into the DBRs, respectively, with $n_{DL1} < n_{DL2}$.

Photonic Energy Gap and the Cavity Resonance

The wavelength-dependent reflectivity of DBRs and microresonators, and thus their stop band, can be calculated using the transfer-matrix method [3, 6]. For a Fabry–Pérot microcavity, the stop band features a pronounced reflectivity minimum (dip) at the resonance wavelength λ_{res}, corresponding to a high transmittance of the Fabry–Pérot structure at this wavelength, as can be seen for an example shown in Fig. 5.3. Here, a representative stop band of a microresonator with AlAs/GaAs DBRs and GaAs λ-thick cavity is shown. This example structure, which features a pronounced resonance in the centre of the high-reflectivity region, consists of 13 and 25.5 mirror pairs in the top and bottom DBR, respectively. In this λ-sized microcavity, the single visible resonant mode is the only Fabry–Pérot-resonator mode which can be accommodated within the stop band, given the large free spectral range (FSR) of the micro-size resonator. The quality of the resonator structure and the reflectivity of the mirrors determine the spectral width of this reflectivity dip, which can be hardly resolved spectrally for high-quality microcavities owing to the very sharp resonance with very narrow line width.

Strong Photon Confinement

Planar microresonator structures confine the photon field in the vertical direction, but the photons are free to propagate in the plane of the cavity spacer. This gives rise to a characteristic angle dependency of the resonator mode, and to the attribution of an effective mass to the cavity photons, as described in Chap. 2. However, cavity structures can be achieved which confine photons much stronger spatially. These structures are for instance one-dimensional photonic-wire cavities or even (quasi-) zero-dimensional cavities (photonic quantum boxes) which exhibit a discrete mode

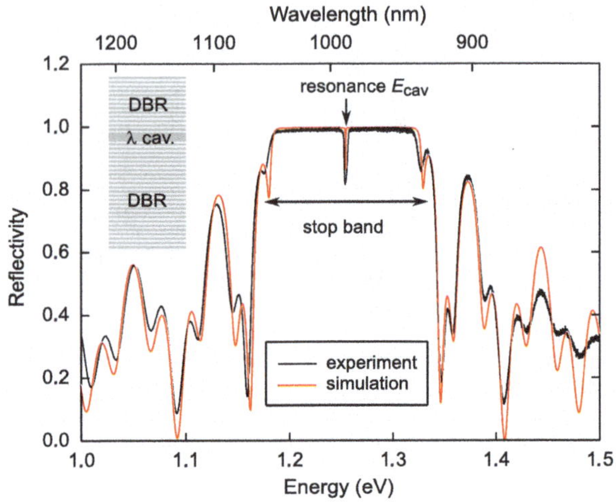

Fig. 5.3 Example diagram of a III/V microcavity's reflectivity spectrum with stop band and Fabry–Pérot resonance. Here, the theoretical behaviour (red) is plotted together with experimental data (black). Both spectra correspond to room temperature conditions. An inset indicates the structure with DBRs and cavity spacer. Adapted from [12]. Courtesy of the author

spectrum owing to the strong quantization effects obtained in such structures. Examples for such systems based on Fabry–Pérot microresonators will be described in the following chapters, where 0D micropillars [13–16], 0D buried mesas [17–20] and 1D structures [21, 22] are utilized to strongly confine photons and guide polaritons along one axis, respectively. It is also worth noting that optically-pumped polariton lasing has also been demonstrated with photonic crystal microcavities [23], which typically exhibit strong (0D spatial) confinement of light with high Q-factors.

Tunability of the Cavity Mode

Common cavity–polariton structures are monolithic in nature, i.e. they are achieved from the bottom to the top mirror in one go without physical break as the outcome of one and the same growth procedure. This does not mean a growth interruption and intermediate processing step cannot happen, as is the case for instance for buried photonic (quantum) boxes in planar microresonator structures (cf. [17, 24]). However, there is no air gap (e.g. as the cavity region) or differently deposited material system as spacer/buffer/cap layer, Bragg defect or active region found inside microresonators that were monolithically grown. In certain cases, the completion of microcavities can rely on different deposition methods and material systems when lattice matching conditions are widely softened, e.g. for 2D materials inbetween dielectric mirrors [25].

In aforementioned monolithical microcavity structures, a natural wedge obtained during wafer growth can be exploited for cavity mode tuning. On the one hand, this is very advantageous when a homogeneously-distributed gain region with good quality is spread over the relevant wafer region, which exhibits a tapered cavity profile. In such situation, the cavity–exciton detuning can be conveniently changed by moving along the wafer piece, i.e. the actual microcavity sample under investigation. In some cases, the quality of the sample hardly changes within centimetres of the grown structure and, if appropriately designed, the sample can give access to the so-to-say full detuning range from positive to negative detunings, which allows easy mapping of the anti-crossing behaviour by lateral displacement of the structure under the focal spot of the measurement system. This is not only fast, but gives a much more complete picture of the system's behaviour than if one had to look at each detuning system individually, which needed to be prepared sample by sample. On the other hand, propagation along the sample could lead to deviating behaviour when addressing different spots on one and the same sample due to possible inhomogeneities, defects and artefacts, as no spot equals the other—particularly in strained samples with strong lattice mismatch between components. Indeed, at a certain spot, hardly more than one cavity detuning configuration can be studied, while temperature tunable exciton modes can be shifted with respect to the cavity resonance. As thermal effects can have a detrimental influence on polariton gases, particularly with regard to condensation behaviour, it is understandable that this is not the ideal tuning knob. Most importantly, the tapered cavity profile cannot be used well for tuning purposes when the emitters' lateral dimensions are microscopic.

Besides monolithic cavities, tunable open microcavities can be conveniently used in connection with 2D, 1D and 0D excitonic systems. Such cavities not only allow to flexibly set the cavity length (commonly via piezo-electric actuators), they also give experimental access to the intra-cavity space which can be used for coupling experiments with nanosheets, colloidal quantum emitters, molecules, dispersed particles or particles in a polymer matrix, or even for coupling between the resonator-light field and two different emitter systems inside the cavity, to name but a few (cf. [26–28]).

5.2 Implementation of Quantum Wells

Once a planar microcavity can be established, the employment of optically active regions becomes crucial. For planar structures, quantum wells naturally serve well as active layers, in which they can host electron–hole pairs for interaction with the cavity light field. These exciton-confining layers can be with little effort incorporated into planar microresonator structures during growth of the cavity structure. In principle, single or multiple QWs can be employed and these wells are usually arranged as compact stacks inside the cavity region, as is indicated in Fig. 5.2 in the form of thin refractive index modulations that are arranged in stacks.

Each QW in such a stack is separated by its neighbours via a thin layer of cavity material, which also serves as the electronic barrier of the QW system. Indeed, the position of the QWs in the cavity in out-of-plane direction (growth direction) has to be chosen with care in order to fulfil resonance conditions for strong light–matter coupling in the microcavity system [30–32]: As the strength of light–matter interaction depends on the spatial overlap of the resonator-mode profile with the QWs, careful positioning of the QWs on the anti-nodes of the electric light field inside the cavity enables efficient polariton formation in such planar resonators. In Fig. 5.2, the QWs are clearly located at the maximum (maxima) of the intra-cavity intensity distribution of the transverse-electric (TE) mode of the resonator in the case of 4

(12) QWs, providing strong interaction between QW excitons and cavity photons. This also explains that stacks of multiple QWs cannot be arbitrarily expanded which limits the size of a stack and the number of its wells.

5.2.1 Distribution of Quantum Wells

In a QW-microcavity system, a homogeneous distribution of excitons in all quantum wells is desired in order to maximize the number of emitters coupled to the field, while the density of excitons per QW is dilute. A proper placement of a single compact stack is well achieved at the centre of the cavity for an optimal implementation into the microresonator. However, it becomes more challenging in the case of multiple stacks. In order to incorporate a considerable number of QWs into the microcavity, whereas the QWs cannot be well fitted into one stack, QWs have to be distributed onto several stacks located at available cavity-field anti-nodes. Typically, in addition to the central anti-node, the neighbouring anti-node positions can be used for the placement of additional stacks. This distribution of QWs on several stacks allows for all available QWs to strongly couple to the confined light field. In contrast, too expanded stacks would only enable strong coupling of the inner wells to the cavity mode. Furthermore, the separation of the QWs in these stacks should remain sufficient

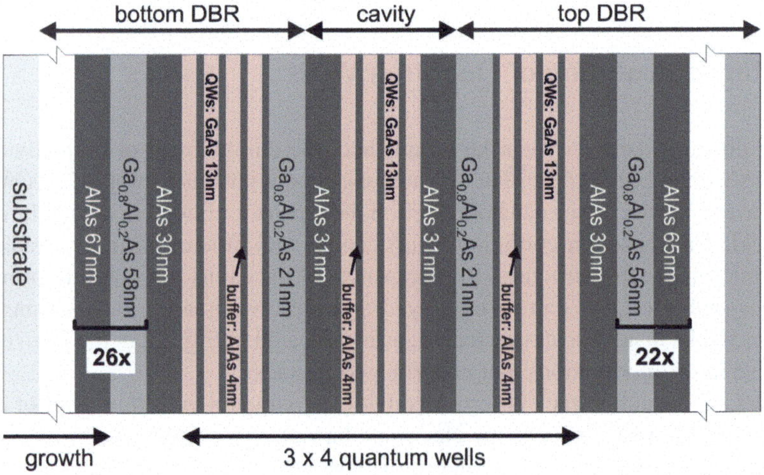

Fig. 5.4 Schematic drawing of an undoped microcavity structure with 12 QWs, which are located in the active region around the central lobes of the standing wave cavity field, after [12]. The top DBR is typically less reflective for directional out-coupling. The number of mirror pairs in each of the DBRs is indicated for the representative III–V planar Fabry–Pérot microresonator structure. This QW-microcavity design is representative for those polariton structures based on the GaAs/AlAs material system with which typical Rabi splittings of about 15 meV have been achieved at cryogenic temperatures. Such polariton structures have been MBE-grown for studies such as in [29]

in order to maintain exciton confinement: the choice of too narrow barriers would enhance tunnelling effects, which undermine separate confinement of excitons in each well that is desired for the achievement of high numbers of particles in a dilute gas.

In the case of 4 QWs, the employment of a single stack has been found to be a proper solution. In contrast, 12 QWs are commonly distributed on three stacks consisting each of 4 QWs and being located at the three central anti-node positions. It is understandable that for "standard" microcavity designs only the central cavity-mode anti-nodes are most pronounced. However, in short cavities, e.g. with $\lambda/2$ or λ thickness, the cavity mode's two neighbouring anti-nodes are already located in the region of the first mirror pair of each DBR (see Fig. 5.2). Nevertheless, this arrangement of QWs has been well established and is widely used in optically-pumped III/V polariton structures (cf. [33, 34]). Of course, there is no standard design for QW-microcavities and the number of QWs and their distribution may vary depending on the desired coupling strength, polariton densities and cavity parameters. One example of a III–V QW-microcavity is provided in Fig. 5.4.

5.2.2 *Number of Quantum Wells*

Besides the cavity design and the distribution of QWs, also the number of QWs in the microresonator structure is of importance for polariton studies. Since the coupling strength of an emitter–cavity system strongly depends on the number of QWs, a high number of QWs is desired with respect to a pronounced coherent energy exchange and a robust polariton system, which does not easily quit the strong-coupling regime [35, 36]. Furthermore, the achievable polariton densities of the system are strongly affected by the number of QWs. This is understandable, as the Mott transition can be postponed for the excitonic system at high polariton densities when the excitons, which form the polaritons, are distributed over a couple of QWs while being in a dilute gas in each of the QWs. Thus, the realization of a polariton BEC is facilitated by the number of QWs, which can provide a critical particle density for condensation [34, 37] before strong coupling vanishes (see Sect. 3.1.2), by avoiding saturation of the exciton density per QW over a large range of excitation powers. In simple words, the particle-density-dependent bleaching of strong coupling up to the final transition to weak coupling [38–40] can be delayed (see phase diagrams for polariton systems provided in the literature [41] and Fig. 3.1 of Chap. 3).

5.2.3 *Excitation Schemes*

In the field of polariton studies, QW excitons in microcavities are commonly excited optically, using either a resonant ($E_{\mathrm{pump}} = E_{\mathrm{LP/UP}}$) or non-resonant excitation scheme ($E_{\mathrm{pump}} > E_{\mathrm{LP/UP}}$), and either pulsed or continuous-wave pumping.

In contrast, electrical pumping of excitons in polariton structures by charge-carrier (electron and hole) injection from two poles via electrical contacts is only feasible non-resonantly. Therefore, similar to diode-laser structures, a p-i-n[1] diode design is chosen [40, 42, 43], with the cavity spacer being intrinsic in order to minimize absorption by donor and acceptor states. Electrical excitation not only requires doped structures and suitable contacts, but also a homogeneous injection of charge carriers in all QWs. This becomes particularly crucial as electrons and holes are injected into the structure from opposite sides in a p-i-n diode. Thus, the architecture of the structure plays a key role for efficient and effective electrical pumping. In this context, the QW arrangement in the microcavity structure is of importance.

From the technological point of view, a single stack of QWs is most easily realized and most homogeneously supplied with electrons and holes. For structures with multiple QWs distributed over different stacks one can also employ Esaki diodes [44] inside the structure, which facilitate distribution of charge carriers over the whole structure via the tunnelling effect exploited by the Esaki-diode structure [45]. In such a configuration, the designed band structure allows for efficient forwarding of charge carriers after each stack into the consecutive region with the next stack. An example of a polariton system with integrated tunneling junction operated at room temperature has been shown by Brodbeck et al. [46]. In addition, one can think of the electrical pumping of multi-stack QW structures to be achieved by efficient excitation of a single stack, which feeds the remaining stacks with excitons via optical coupling of all QWs with the resonator mode [47].

As there is no standard pumping scheme, it is also important to note that each pumping scheme has its advantages and disadvantages. This may be in parts further elaborated on in later chapters (also see Sect. 3.2).

5.3 Optical Properties of Resonators

As the microresonator system is key to the observation of polariton-condensation phenomena, it is worth discussing its properties in the following briefly. Understandably, the size and dimensionality affect the mode spectrum and the confinement potential determines the lifetime of photons in photonic structures. With figures of merit such as the cavity finesse F and the quality factor Q, the important aspect of resonator quality can be assessed.

---

[1] **P**ositive-**i**ntrinsic-**n**egative doping.

5.3.1 Free Spectral Range, Cavity Finesse, Photonic Density of States

The resonator mode spacing is commonly referred to as free spectral range (FSR). In external cavity lasers, the equally-spaced modes in the frequency domain can be so densily packed that it resembles a quasi-continuum of modes. This is well comparable with the theory of energy states in quantum boxes, for finite box lengths approaching very large values. In contrast, strong confinement potentials drastically reduce the number of resonator modes, e.g. standing waves between opposing mirrors. The mode spacing is determined by the cavity length (or in other words one full round trip of a photon) and is referred to as FSR

$$\delta\nu = \frac{c}{2nL}, \tag{5.3}$$

with n the refractive index of the medium, and c the speed of light in vacuum.

The filter function (transmission spectrum) for a Fabry–Pérot interferometer or resonator is characterized by the cavity finesse F. For a Fabry–Pérot resonator, the improvement of the mirror reflectivity leads to a reduction of the modes' linewidth which in turn leads to a higher finesse. The linewidth is defined as the full width at half maximum (FWHM) $\Delta\nu = 1/\tau$ (also given as decay rate times Planck's constant in units of energy, γ). From the linewidth and the FSR, the finesse can be derived

$$F = \frac{\delta\nu}{\Delta\nu}. \tag{5.4}$$

With the knowledge of the mode energy and mode linewidth, also the quality factor for a resonator mode is given, i.e. $Q = \nu/\Delta\nu$ (see (2.14) discussed in Sect. 2.2.2). As both F and Q depend on the linewidth, they increase with increasing reflectivity of the mirrors. For a half-wavelength cavity, the Q-factor of the fundamental cavity mode is equivalent to the finesse.

To complete this section, a brief look at the photonic density of states (c-DOS) is taken. For further details, the interested reader is referred to the literature [6, 48, 49]. In 3D with mode volume V, the photon density of states at the energy $\hbar\omega$ of an emitter is described by

$$\rho_{3D}(\omega) = \frac{\omega^2 V n^3}{\pi^2 c^3}. \tag{5.5}$$

Thus, with increasing frequency, the number of modes (i.e. the c-DOS) grows quadratically (see Fig. 5.5a). This also indicates the vast amount of free-space modes existing for photons. As mentioned earlier, the spontaneous emission rate $1/\tau_0$ of an emitter at frequency ω_e is directly linked to the c-DOS in its environment. In 3D vacuum, it is

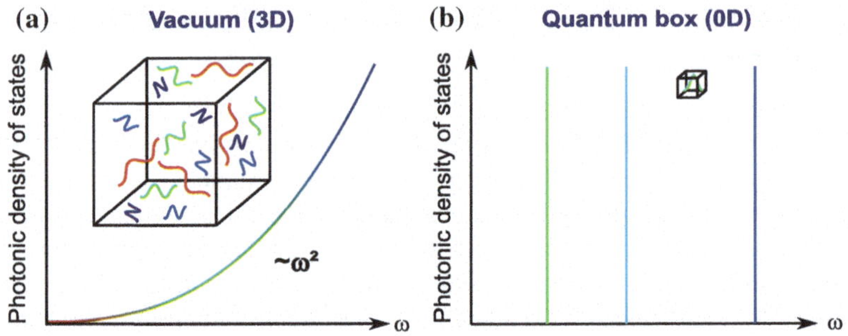

Fig. 5.5 Schematic diagram of the photonic density of states (c-DOS) in 3D (**a**) and 0D (**b**). The c-DOS has a significant effect on light–matter interaction, as it can enhance or suppress the emission rate of an emitter. 0D structures, so-called photonic quantum boxes, with their atomistic spectrum of discrete cavity resonances are therefore of particular interest for cavity quantum electrodynamics (cQED) studies involving single quantum emitters. Insets: Sketches of the spatial dimensionality

$$\frac{1}{\tau_0} = \frac{2\pi}{\hbar^2} M^2 \rho(\omega_e), \tag{5.6}$$

with $M = |\langle f | \mathbf{d} \cdot \mathbf{E} | i \rangle|$ the dipole transition-matrix element for the emitter–field system ((2.19) in Sect. 2.3.1), whereas $|f\rangle$ and $|i\rangle$ denote the final and initial state, respectively. $\rho(\omega_e)$ can be also seen as the density of final states ρ_f.

In contrast, for low-dimensional structures such as photonic quantum boxes, the density of states is drastically altered. In 0D cavities, the energy states of the electromagnetic field become discrete (see Fig. 5.5b). For a pillar structured in a planar cavity system with square-shaped in-plane confinement profile (lateral dimensions $d_x = d_y = d$) the photon energies are given by

$$\hbar\omega_{0D} = \frac{\hbar c}{n} \sqrt{k_\perp^2 + k_x^2 + k_y^2}, \tag{5.7}$$

with $k_i = (q_i + 1)\pi/d$ ($q_i = 0, 1, 2, \ldots$ are the transverse mode numbers, $i = x, y$) [50, 51]. The ground-state mode of a square-shaped micropillar is thereby at $E_{0,0,0} = (\hbar c/n)((2\pi/\lambda_C)^2 + 2(\pi/d)^2)^{1/2}$ ($k_\perp = 2\pi/\lambda_C$) with longitudinal mode number $q_\perp = 0$.

In a single-mode cavity with energy ω_C, the density of states becomes a Lorentzian

$$\rho_C(\omega) = \frac{2}{\pi} \frac{\Delta\omega}{4(\omega - \omega_C)^2 + \Delta\omega^2} \tag{5.8}$$

with resonance linewidth $\Delta\omega$ corresponding to the Q-factor of the microcavity [6]. Photonic quantum boxes are used for instance in strong-coupling experiments involving a single emitter and an empty cavity [52–54]. In fact, even photonic molecules [55] or photonic lattices [56] can be achieved from photonic quantum boxes.

5.3.2 Resonator Quality

The resonator quality is an important attribute and characterizes optical confinement in a resonator. It is represented by the quality factor $Q = E_C/\gamma_C$ (γ_C the linewidth of the cavity mode with energy E_C), which is a direct measure of the cavity quality and indicates the average time photons will stay inside the resonator, or in other words, the rate at which the optical energy in the cavity decays due to loss channels such as absorption or leakage. The intrinsic optical quality of a planar microcavity can be determined by the transfer-matrix method (e.g. see [31, 57, 58]), from which the cavity resonance and its linewidth can be obtained.

The theoretical quality factor of planar microcavities can also be calculated according to the following equation [59, 60]

$$Q = \frac{2L_{\text{eff}}}{\lambda_{\text{res}}} \frac{\pi}{1 - R_1 R_2}, \tag{5.9}$$

whereas L_{eff}, λ_{res}, and $R_{1/2}$ denote the effective cavity length (5.2), the design wavelength and the upper/lower mirror reflectivity (5.1), respectively. The theoretical Q-factor of such planar structures can range between 10^4 and 10^6 and the exact calculated value of a certain cavity design depends on the number of mirror pairs in the upper and lower DBRs, the material composition (cf. (5.1)), the cavity thickness and the resonance energy.

Single-emitter systems usually require a high Q-factor in order to exhibit strong coupling between the emitter and the cavity light field. For such systems, which usually employ strong optical confinement in addition to strong electronic confinement, a theoretical consideration of the Q-factor-limiting effects in micropillars is summarized for instance in [61]. Usually, for short photon lifetimes $\tau_C \propto 1/\gamma_C$, coherent energy exchange cannot be observed due to the fast energy dissipation out of the cavity in such systems, e.g. based on QD micropillars. Yet, for QW microcavities, the coupling rate exceeds the dissipative loss rates due to a high number of emitters in the system (see (2.21) and (2.22)), and strong coupling can already be observed for experimental $Q \approx 1000$. Thus, the relatively low Q and, correspondingly, the short photon lifetime do not affect the coupling significantly in contrast to single-emitter systems. Nevertheless, the cavity decay rate has another effect in polariton systems, which is described in the following.

Experimental Q-Factor

The experimental Q-factor of a microcavity is obtained by $Q = E_C/\gamma_C \equiv \lambda_C/\Delta\lambda_C$ from the experimentally observed figures for the cavity resonance, whereas γ_C and $\Delta\lambda_C$ denote the measured linewidth of the photon mode in energy and wavelength scale, respectively. Although this value typically undercuts the intrinsic Q-factor owing to loss channels in the optical system (cf. [59]), it serves well for the characterization of the resonator's quality, which is mainly considered in planar resonators for light confined under normal incidence with respect to the cavity mirrors, i.e. at $k_{\parallel} = 0$ in the planar microcavity system.

Commonly, losses in microresonator are dominated by the quality of the imperfect mirrors and absorption of cavity photons in the materials of the DBR structures and the active region. The former fraction of losses is referred to as intrinsic losses, which determine the intrinsic Q-factor Q_{int}, while the latter is labelled absorption losses, which set Q_{abs}. In addition, etched structures suffer optical losses via scattering at side walls in relation to the microresonator diameter which is crucial for thin micropillars, adding the term Q_{scat} to an equation for the actual Q-factor of a real microresonator that summarizes intrinsic and extrinsic components [62]:

$$\frac{1}{Q} = \frac{1}{Q_{\text{int}}} + \frac{1}{Q_{\text{ext}}} = \frac{1}{Q_{\text{int}}} \left(+ \frac{1}{Q_{\text{scat}}} \right) + \frac{1}{Q_{\text{abs,QW}}} + \frac{1}{Q_{\text{abs,DBR}}}, \tag{5.10}$$

whereas Q_{abs} is represented by the two components $Q_{\text{abs,QW}}$ and $Q_{\text{abs,DBR}}$ for absorption losses in the QWs and DBRs, respectively [63].

One serious drawback of the use of doped layers in DBRs for an improved electrical conductivity is the reduction of the resonator quality. Although the doping of the mirror layers has little impact on their refractive indices, the absorption in the DBRs due to the introduced impurities can reduce the Q-factor in comparison with undoped structures [63]. Thus, it is obvious that a suitable doping scheme has to be found for an electrically driven structure according to the required Q-factor. Studies concerning the Q-factor of QD-micropillars for single-photon generation and low-threshold lasing revealed that despite doping of the DBRs, these structures, which are indeed very sensitive to the quality of the microcavity, did not suffer significantly from doping. For pillar diameters smaller than 2 μm, high Q-factors on the order of undoped structures were obtained, when side-wall scattering losses pose the dominant loss channel of such doped micropillar structures [63]. Similar doping schemes have been applied for polariton devices and the achieved results show that electrical current injection is not obtained at the cost of noticeable resonator-quality reduction in well-designed structures (see [64]).

Consequences for Polaritons

In polariton structures, the Q-factor can have an important effect on the ground-state relaxation and the condensation behaviour of polaritons, since the lifetime of confined photons in the microcavity plays a considerable role in this regard. An increase of the resonator quality directly increases the preservation time of polaritons. As the Q-factor is also responsible for condensation effects via the accumulation of particles in the polariton-dispersion minimum at $k_{\parallel} = 0$, where the photon lifetime predominantly determines the polariton lifetime ($\tau_C \ll \tau_X$), it is natural that high quality structures are more preferable for condensation studies. If the polariton lifetime is longer than the relaxation time, polaritons can relax via polariton–polariton and polariton–phonon scattering from their energetically higher reservoir state into the ground state (cf. Figs. 4.4a, 3.3a in Chaps. 4 and 3, respectively), forming a condensate in the regime of stimulated final-state occupation.

The importance of the Q-factor is even more understandable in the context of the thermalization process. To reach thermal equilibrium, the relaxation time and

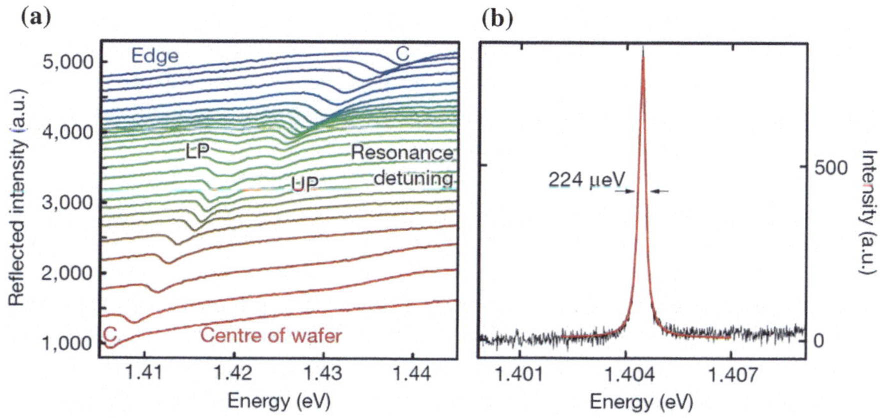

Fig. 5.6 Typical detuning map for a planar polariton system based on the wedge accross the cavity wafer. Here, the waterfall diagram of reflectivity spectra shows an anti-crossing behaviour for photonic (C) and exciton (X) resonances, forming lower (LP) and upper (UP) polaritons. **b** Micro-photoluminescence of a 2-μm micropillar with labelled linewidth corresponding to a Q-factor of more than 6300. Reused with permission. Reference [64] Copyright 2013, Springer Nature

thermalization time need to be much shorter than the cavity–polariton's lifetime in its ground state. Thus, the cavity's Q-factor determines—via the degree of thermalization reached in the polariton gas—which kind of condensate case can be finally obtained: *polariton laser*, *dynamical condensate* or *quasi-BEC* (see Sect. 4.1).

Determination of the Q-Factor

In practice, the linewidth of a cavity mode in planar microresonator structures can be extracted from photoluminescence or reflectivity measurements. However, these methods have some drawbacks. As soon as interaction with the embedded QWs is involved, the linewidth will be limited by absorption effects. Thus, the lower limit of the microcavity's experimental Q-factor can be best estimated by characterizing the nearly-uncoupled cavity mode at strong negative detunings, i.e. when $E_X \gg E_C$ and when the system at $k_{\|} = 0$ is almost purely photonic. This can be readily addressed for structures monolithically grown on wafers, for which a cavity wedge across the wafer cannot be circumvented owing to the growth technique. An example of a detuning map is given in Fig. 5.6a for the planar microcavity structure used in polariton diodes in [64], which shows that over a wide accessible detuning range, the pure photon mode can be hardly accessed since coupling effects hybridize the modes. Alternatively, the Q-factor of planar microresonator structures could be estimated from measurements with empty cavities, provided that the resonance is measurable and the resolution of the spectrometer is good enough.

Characterization of the cavity mode's linewidth in experiments is further complicated due to broadening of the mode when a spatially expanded region is probed by means of optical spectroscopy. Since structures grown by epitaxy commonly exhibit a tapered cavity thickness along the wafer radius, the detected signal might represent the sum over different detuning situations with little variation. Thus, a strong spatial selection needs to be applied to minimize this effect for planar resonators. Moreover, Q-factor measurements require angle-resolved detection in order to access the mode's linewidth at the photonic ground state at $k_{\|} = 0$, corresponding to normal incidence of light onto the cavity mirrors.

In this context, it is recommendable to use high-quality-processed microresonator pillars with small radii of approximately 2–4 μm in order to assess the Q-factor of planar microcavities as good as possible. This naturally requires that the fabrication of such pillars is technically achievable at the highest standards, allowing one to substitute the Q-factor for planar resonators with those of pillars etched from these planar structures. In such circular pillar structures, the strong optical confinement in three dimensions improves linewidth extraction from highly photonic modes and even facilitates the spectral isolation of the ground state from higher-energy cavity states. Figure 5.6b shows the micro-photoluminescence spectrum of such a micropillar processed from the planar (p-i-n-doped) microcavity structure used for the polariton-laser diode demonstrated in [64]. The isolated fundamental mode of a highly photonic micropillar cavity with a diameter of 2 μm exhibited a $Q \approx 6320$. Thereby, this 0D result indicates the lower limit of the planar microcavity's Q-factor.

References

1. J. Vaughan, *The Fabry-Perot interferometer* (Hilger, 1989)
2. J.K. Vahala, Optical microcavities. Nature **424**(6950), 839–846 (2003)
3. S. Yu, *Analysis and Design of Vertical Cavity Surface Emitting Lasers* (John Wiley & Sons, 2003)
4. M.S. Skolnick, T.A. Fisher, D.M. Whittaker, Strong coupling phenomena in quantum microcavity structures. Semicond. Sci. Technol. **13**, 645–669 (1998)
5. Ioffe-Institute. *Electronic Archive: New Semiconductor Materials. Characteristics and Properties* (St. Petersburg, 2012) http://www.ioffe.ru/SVA/NSM
6. A.V. Kavokin, J. Jeremy, G. Malpuech, F.P. Laussy, *Microcavities*. (Oxford University Press, Baumberg, 2007)
7. Renchun Tao, Kenji Kamide, Munetaka Arita, Satoshi Kako, Yasuhiko Arakawa, Room-temperature observation of trapped exciton-polariton emission in GaN/AlGaN microcavities with air-gap/III-nitride distributed bragg reflectors. ACS Photonics **3**(7), 1182–1187 (2016)
8. P. Bienstman, R. Baets, Optical modelling of photonic crystals and VCSELs using eigenmode expansion and perfectly matched layers. Opt. Quantum Electron. **33**(4–5), 327–341 (2001)
9. Z. Knittl, *Optics of Thin Films; an Optical Multilayer Theory* (Wiley, 1976)
10. V. Savona, L.C. Andreani, P. Schwendimann, A. Quattropani, Quantum well excitons in semiconductor microcavities: Unified treatment of weak and strong coupling regimes. Solid State Commun. **93**(9), 733–739 (1995)
11. Hui Deng, Hartmut Haug, Yoshihisa Yamamoto, Exciton-polariton Bose-Einstein condensation. Rev. Mod. Phys. **82**(2), 1489 (2010)
12. A. Rahimi-Iman, *Nichtlineare Eekte in III/V Quantenlm-Mikroresonatoren: Von dynamischer Bose–Einstein-Kondensation hin zum elektrisch betriebenen Polariton-Laser*. (Cuvillier Verlag Göttingen, 2013)
13. J.M. Gérard, D. Barrier, J.Y. Marzin, R. Kuszelewicz, L. Manin, E. Costard, V. Thierry Mieg, T. Rivera, Quantum boxes as active probes for photonic microstructures: the pillar microcavity case. Appl. Phys. Lett. **69**, 449 (1996)

14. J.P. Reithmaier, M. Röhner, H. Zull, F. Schäfer, A. Forchel, P.A. Knipp, T.L. Reinecke, Size dependence of confined optical modes in photonic quantum dots. Phys. Rev. Lett. **78**(2), 378–381 (1997)
15. Daniele Bajoni, Pascale Senellart, Esther Wertz, Isabelle Sagnes, Audrey Miard, Aristide Lemaître, Jacqueline Bloch, Polariton Laser Using Single Micropillar GaAs-GaAlAs Semiconductor Cavities, Phys. Rev. Lett. **100**(4), 047401 (2008)
16. S. Reitzenstein, A. Forchel, Quantum dot micropillars. J. Phys. D: Appl. Phys. **43**(3), 033001 (2010)
17. O. El Daïf, A. Baas, T. Guillet, J.-P. Brantut, R. Idrissi Kaitouni, J.L. Staehli, F. Morier-Genoud, B. Deveaud, Polariton quantum boxes in semiconductor microcavities. Appl. Phys. Lett. **88**(6), 061105 (2006)
18. Andreas Muller, Chih-Kang Shih, Lu Jaemin Ahn, Deepa Gazula Dingyuan, Dennis G. Deppe, High Q (33 000) all-epitaxial microcavity for quantum dot vertical-cavity surface-emitting lasers and quantum light sources. Appl. Phys. Lett. **88**, 031107 (2006)
19. P. Lugan, D. Sarchi, V. Savona, Theory of trapped polaritons in patterned microcavities. Phys. Stat. Sol. (c) **3**, 2428–2431 (2006)
20. S. Portolan, P. Hauke, V. Savona, Parametric photoluminescence of spatially confined polaritons in patterned microcavities. Phys. Stat. Sol. (b) **245**(6), 1089–1092 (2008)
21. E. Wertz, L. Ferrier, D.D. Solnyshkov, R. Johne, D. Sanvitto, A. Lemaître, I. Sagnes, R. Grousson, A.V. Kavokin, P. Senellart, G. Malpuech, J. Bloch, Spontaneous formation and optical manipulation of extended polariton condensates. Nat. Phys. **6**(11), 860–864 (2010)
22. J. Fischer, I.G. Savenko, M.D. Fraser, S. Holzinger, S. Brodbeck, M. Kamp, I.A. Shelykh, C. Schneider, S. Höfling, Spatial coherence properties of one dimensional exciton-polariton condensates. Phys. Rev. Lett. **113**(20), 203902 (2014)
23. S. Azzini, D. Gerace, M, Galli, I. Sagnes, R. Braive, A. Lemaître, J. Bloch, D. Bajoni, Ultra-low threshold polariton lasing in photonic crystal cavities. Appl. Phys. Lett. **99**(11), 111106 (2011)
24. A. Rahimi-Iman, C. Schneider, J. Fischer, S. Holzinger, M. Amthor, S. Höfling, S. Reitzenstein, L. Worschech, M. Kamp, A. Forchel, Zeeman splitting and diamagnetic shift of spatially confined quantum-well exciton polaritons in an external magnetic field. Phys. Rev. B **84**(16), 165325 (2011)
25. X. Liu, T. Galfsky, Z. Sun, F. Xia, E.C. Lin, Y.H. Lee, S. Kéna-Cohen, V.M. Menon, Strong light–matter coupling in two-dimensional atomic crystals. Nat. Photon. **9**(1), 30–34 (2014)
26. S. Dufferwiel, S. Schwarz, F. Withers, A.A.P. Trichet, F. Li, M. Sich, O. Del Pozo-Zamudio, C. Clark, A. Nalitov, D.D. Solnyshkov, G. Malpuech, K.S. Novoselov, J.M. Smith, M.S. Skolnick, D.N. Krizhanovskii, A.I. Tartakovskii, Exciton-polaritons in van der Waals heterostructures embedded in tunable microcavities. Nat. Commun. **6**, 8579 (2015)
27. L.C. Flatten, Z. He, D.M. Coles, A.A.P. Trichet, A.W. Powell, R.A. Taylor, J.H. Warner, J.M. Smith, Room-temperature exciton-polaritons with two-dimensional WS 2. Sci. Rep. **6**, 33134 (2016)
28. P. Qing, J. Gong, X. Lin, N. Yao, W. Shen, A. Rahimi-Iman, W. Fang, L. Tong, A simple approach to fiber-based tunable microcavity with high coupling efficiency. Appl. Phys. Lett. **114**, 021106 (2019)
29. A. Rahimi-Iman, A.V. Chernenko, J. Fischer, S. Brodbeck, M. Amthor, C. Schneider, A. Forchel, S. Höfling, S. Reitzenstein, M. Kamp, Coherence signatures and density-dependent interaction in a dynamical exciton-polariton condensate. Phys. Rev. B **86**(15), 155308 (2012)
30. Y. Yamamoto, F. Tassone, H. Cao, *Semiconductor Cavity Quantum Electrodynamics* (Springer-Verlag, 2000)
31. A. Kavokin, G. Malpuech, *Cavity Polaritons* (Academic Press, 2003)
32. B. Deveaud. *The Physics of Semiconductor Microcavities*. (WILEY-VCH Verlag, 2007)
33. J. Bloch, T. Freixanet, J.Y. Marzin, V. Thierry-Mieg, R. Planel, Giant Rabi splitting in a microcavity containing distributed quantum wells. Appl. Phys. Lett. **73**(12), 1694–1696 (1998)
34. H. Deng, G. Weihs, C. Santori, J. Bloch, Y. Yamamoto, Condensation of semiconductor microcavity exciton polaritons. Science **298**(5591), 199–202 (2002)

35. C. Weisbuch, M. Nishioka, A. Ishikawa, Y. Arakawa, Observation of the coupled exciton-photon mode splitting in a semiconductor quantum microcavity. Phys. Rev. Lett. **69**(23), 3314 (1992)
36. R. Houdré, R.P. Stanley, U. Oesterle, M. Ilegems, C. Weisbuch, Room-temperature cavity polaritons in a semiconductor microcavity. Phys. Rev. B **49**(23), 16761 (1994)
37. J. Kasprzak, M. Richard, S. Kundermann, A. Baas, P. Jeambrun, J.M.J. Keeling, F.M. Marchetti, M.H. Szymańska, R. André, J.L. Staehli, V. Savona, P. B. Littlewood, B. Deveaud, L. Si Dang. Bose-Einstein condensation of exciton polaritons. Nature **443**(7110), 409–414 (2006)
38. R. Houdré, J.L. Gibernon, P. Pellandini, R.P. Stanley, U. Oesterle, C. Weisbuch, J. O'Gorman, B. Roycroft, M. Ilegems, Saturation of the strong-coupling regime in a semiconductor microcavity: free-carrier bleaching of cavity polaritons. Phys. Rev. B **52**(11), 7810 (1995)
39. J. Bloch, B. Sermage, C. Jacquot, P. Senellart, V. Thierry-Mieg, Time-resolved measurement of stimulated polariton relaxation. Phys. Stat. Sol. (a) **190**, 827–831 (2002)
40. S.I. Tsintzos, N.T. Pelekanos, G. Konstantinidis, Z. Hatzopoulos, P.G. Savvidis, A GaAs polariton light-emitting diode operating near room temperature. Nature **453**(7193), 372–375 (2008)
41. Alexey Kavokin, Guillaume Malpuech, Fabrice P. Laussy, Polariton laser and polariton superfluidity in microcavities. Phys. Lett. A **306**(4), 187–199 (2003)
42. Daniele Bajoni, Elizaveta Semenova, Aristide Lemaître, Sophie Bouchoule, Esther Wertz, Pascale Senellart, Jacqueline Bloch, Polariton light-emitting diode in a GaAs-based microcavity. Phys. Rev. B **77**(11), 113303 (2008)
43. A.A. Khalifa, A.P.D. Love, D.N. Krizhanovskii, M.S. Skolnick, J.S. Roberts, Electroluminescence emission from polariton states in GaAs-based semiconductor microcavities. Appl. Phys. Lett. **92**(6), 061107 (2008)
44. Leo Esaki, New phenomenon in narrow germanium p-n junctions. Phys. Rev. **109**(2), 603–604 (1958)
45. T. Knödl, M. Golling, A. Straub, R. Jäger, R. Michalzik, K.J. Ebeling, Multistage bipolar cascade vertical-cavity surface-emitting lasers: theory and experiment. IEEE JSTQE **9**(5), 1406 (2003)
46. S. Brodbeck, J.-P. Jahn, A. Rahimi-Iman, J. Fischer, M. Amthor, S. Reitzenstein, M. Kamp, C. Schneider, S. Höfling, Room temperature polariton light emitting diode with integrated tunnel junction. Opt. Express **21**(25), 31098–31104 (2013)
47. T.C.H. Liew, I.A. Shelykh, G. Malpuech, Polaritonic devices. Phys. E: Low-Dimens. Syst. Nanostructures **43**(9), 1543–1568 (2011)
48. K. Vahala, *Optical microcavities*. (World Scientific, 2004)
49. A.V. Kavokin, J.J. Baumberg, G. Malpuech, F.P. Laussy, *Microcavities*, vol. 1. (Oxford University Press, 2017)
50. T. Gutbrod, M. Bayer, A. Forchel, J.P. Reithmaier, T.L. Reinecke, S. Rudin, P.A. Knipp, Weak and strong coupling of photons and excitons in photonic dots. Phys. Rev. B **57**(16), 9950 (1998)
51. C.F. Klingshirn, *Semiconductor Optics*. (Springer, 2012)
52. T. Yoshie, A. Scherer, J. Hendrickson, G. Khitrova, H.M. Gibbs, G. Rupper, C. Ell, O.B. Shchekin, D.G. Deppe, Vacuum Rabi splitting with a single quantum dot in a photonic crystal nanocavity. Nature **432**(7014), 200–203 (2004)
53. J.P. Reithmaier, G. Sek, A. Löffler, C. Hofmann, S. Kuhn, S. Reitzenstein, L.V. Keldysh, V.D. Kulakovskii, T.L. Reinecke, A. Forchel, Strong coupling in a single quantum dot-semiconductor microcavity system. Nature **432**(7014), 197–200 (2004)
54. E. Peter, P. Senellart, D. Martrou, A. Lemaître, J. Hours, J.M. Gérard, J. Bloch, Exciton-photon strong-coupling regime for a single quantum dot embedded in a microcavity. Phys. Rev. Lett. **95**(6), 067401 (2005)
55. M. Bayer, T. Gutbrod, J.P. Reithmaier, A. Forchel, T.L. Reinecke, P.A. Knipp, A.A. Dremin, V.D. Kulakovskii, Optical modes in photonic molecules. Phys. Rev. Lett. **81**(12), 2582–2585 (1998)
56. M. Bayer, T. Gutbrod, A. Forchel, T.L. Reinecke, P.A. Knipp, R. Werner, J.P. Reithmaier, Optical demonstration of a crystal band structure formation. Phys. Rev. Lett. **83**(25), 5374–5377 (1999)

57. M. Born, E. Wolf, *Principals of Optics* (Cambridge University Press, 1999)
58. C. Wilmsen, H. Temkin, L.A. Coldren, *Vertical-Cavity Surface-Emitting Lasers* (Cambridge University Press, 1999)
59. A. Löffler, J.P. Reithmaier, G. Sek, C. Hofmann, S. Reitzenstein, M. Kamp, A. Forchel, Semiconductor quantum dot microcavity pillars with high-quality factors and enlarged dot dimensions. Appl. Phys. Lett. **86**, 111105 (2005)
60. E. F. Schubert, *Light-Emitting Diodes*, 2nd edn. (Cambridge University Press, 2006)
61. M. Karl, B. Kettner, S. Burger, F. Schmidt, H. Kalt, M. Hetterich, Dependencies of micro-pillar cavity quality factors calculated with finite element methods. Opt. Express **17**(2), 1144–1158 (2009)
62. S. Reitzenstein, C. Hofmann, A. Gorbunov, M. Strauss, S.H. Kwon, C. Schneider, A. Löffler, S. Höfling, M. Kamp, A. Forchel, AlAs/GaAs micropillar cavities with quality factors exceeding 150.000. Appl. Phys. Lett. **90**(25), 251109 (2007)
63. C. Böckler, S. Reitzenstein, C. Kistner, R. Debusmann, A. Löffler, T. Kida, S. Höfling, A. Forchel, L. Grenouillet, J. Claudon, J.M. Gérard, Electrically driven high-Q quantum dot-micropillar cavities. Appl. Phys. Lett. **92**(9), 091107 (2008)
64. C. Schneider, A. Rahimi-Iman, N.Y. Kim, J. Fischer, I.G. Savenko, M. Amthor, M. Lermer, A. Wolf, L. Worschech, V.D. Kulakovskii, I.A. Shelykh, M. Kamp, S. Reitzenstein, A. Forchel, Y. Yamamoto, S. Hoefling, An electrically pumped polariton laser. Nature **497**, 348 (2013)

Chapter 6
Technological Realization of Polariton Systems

Abstract Driven by the achievement of strong light–matter coupling and the observation of exciton–polaritons in microcavities, different experimental platforms for the study of light–matter interactions have been utilized. Recent developments highly benefited from the strive to obtain polariton BEC, and even electrically-driven polariton light sources for practical applications have been envisaged. This has required technological advances in growth and patterning to enable fabrication of the desired microcavities with two-dimensional excitonic gases at its core for coupling experiments and for the formation of quantum gases and superfluids based on polaritons. In this context, growth and processing of microcavity devices will be covered, beginning with typical examples originating from III/V epitaxy. Furthermore, a prominent approach for polariton device achievement in the form of a vertical-emitting-laser-like diode structure is presented. Thereafter, microcavity systems achieved with alternative material systems are summarized that each feature their own legitimacy, attractiveness and challenges, mainly developed and studied targeting room-temperature observation of BEC.

A. Rahimi-Iman, *Polariton Physics*, Springer Series in Optical Sciences 229,
https://doi.org/10.1007/978-3-030-39333-5_6

6.1 Growth and Processing of Microcavity Devices

While there is no standard growth and processing technique for QW-microresonators, cavity–polariton structures based on different materials and designs share many similarities concerning their fabrication. However, to provide an example, microcavities based on the well-established and prominent III/V system will be the primary subject of this section in order to provide a general overview on the growth and structuring principles in the domain of cavity–polaritons.

6.1.1 Epitaxy of Multilayered Structures

Multilayered structures and thin-layer heterostructures are commonly grown by molecular beam epitaxy (MBE) [1, 2] and metal-organic vapour-phase epitaxy (MOVPE) [3]. The latter technique offers a higher throughput at comparably high structural qualities—typically used for growth of lasers and other optical structures—and offers industrial capabilities (as MOVPE is capable of growing multiple wafers simultaneously in special growth reactors). In contrast, owing to its high control and very low output rates, MBE is widely used in the field of research for scientific studies, for prototyping and for the growth of ultra-precise high-quality structures. Thus, historically and for practical reasons, polariton structures are predominantly grown by MBE. A typical cross-sectional view of a planar microresonator design which is commonly grown by III/V epitaxy has been schematically shown in Fig. 5.4 of Chap. 5.

In the III/V semiconductor system, crystalline GaAs substrates are used for epitaxy of heterostructures. For this purpose, a rotating wafer of the substrate material (for GaAs microcavities usually a (100) surface) is overgrown layer by layer under exposure by the directed molecular beam of Ga, As and Al, with the concentration of the materials depending on the composition of each layer. While rotation is used to prevent inhomogeneous exposure of the wafer by the growth material, the centre of the wafer usually growth thicker than the outer rims. This leads to a radial decrease of layer thicknesses towards the edge of the wafer which results in a tapered microcavity height profile and thereby to a radial detuning feature of planar microresonator structures (see sketch in Fig. 6.1). This is an interesting peculiarity that allows one to tune the resonator mode with respect to the QW-exciton energy, which remains nearly constant over the whole wafer area, simply by changing the spot on the wafer along the path from the centre to the edge. Typically, the resonator mode shifts around 25–30 meV towards higher energies, providing access to various detuning settings Δ for polariton studies. Indeed, processing of microstructures into these planar microcavities requires careful selection of the wafer piece with the desired cavity–emitter detuning, because once polariton diodes are fabricated at a certain spot of the wafer, flexibility to change the detuning and access to certain detuning regimes are strongly limited. Moreover, localization effects, which introduce energy

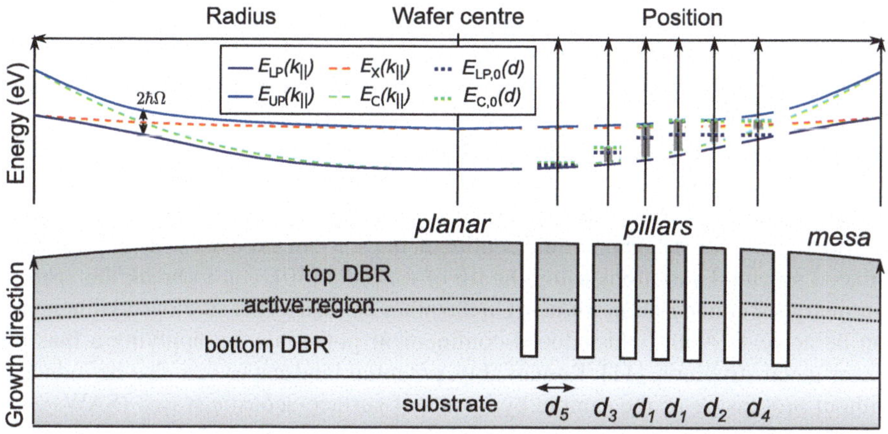

Fig. 6.1 Schematic drawing of a tapered microcavity growth profile for monolithic structures grown by epitaxy. The radial decrease of the grown layers' thicknesses from the centre to the edge of the wafer typically results in a radial detuning feature, which can be exploited for detuning studies on exciton–polaritons in such systems by addressing different positions on the same sample. This is indicated by a sketched energy–position (wafer radius) diagram above the cross-sectional view on the wafer. The position dependence has to be taken into account when processing confinement structures, which introduce additional energy shifts to the optical mode(s), such as for the ground state ($k = 0$). The quantization effect (up-shift indicated by grey bar for the corresponding cavity mode) is indicated on the right side for pillars of different radius at distinct positions on the wafer, with their ground-state energies for the corresponding bare cavity mode as well as the resulting polariton mode being discrete (dotted lines). Indices of the pillar diameter d are sorted with regard to size. Two scenarios are sketched: reduction of the size leading to increased confinement potentials (left part); increase of size arranged in a way that compensates the detuning for the cavity ground state due to tapering (right part). In contrast, for large enough mesa structures with negligible confinement, the radial dependence and the dispersion of the modes in phase space are preserved compared to the case in micropillars

shifts to the cavity mode due to optical confinement, have to be taken into account when etching low-dimensional structures such as micropillars (indicated in Fig. 6.1). These details are of great importance to the fabrication of polariton diodes, because current injection schemes favour the use of mesa and pillar structures, which laterally limit and vertically direct current flux through the structure.

6.1.2 Potential Landscapes and Polariton Boxes

While a strongly focussed laser beam acts as a natural source of localization with a gain-induced potential landscape [4], two separated pump-laser spots can even be used to form a parabolic harmonic-oscillator potential between them for condensate trapping [5]. It becomes obvious that playing with potential landscapes and tailoring quantum boxes for polaritons can strongly enrich the experimental possibilities regarding the observation and study of condensation phenomena. Various means of structuring and confinement have been explored for cavity–polariton so far, a selection of which shall be highlighted below.

A number of technological approaches exist for the lateral confinement of polaritons in so-called polariton traps. For instance, one can employ a local modulation of the cavity lengths (the spacer layer) through a growth interruption and consecutive etching process before overgrowth of the top DBR, leading to considerable quantization effects [6, 7]. Alternatively, a completed cavity can be etched into the shape of a micropillar for strong photonic confinement [8, 9], or exposed to local pressure-induced potential variations using the tip of a needle [10], e.g. to break the spatial symmetry for condensation studies. On the other hand, a rather shallow confinement can be achieved with field-induced confinement potentials, by applying a bias to planar metal structures [11]. Remarkably, potential landscapes can also be induced without processing of the sample by means of surface-acoustic waves (SAWs), as were used by Cerda-Méndez et al. to modify induced potentials on demand [12, 13]. Thereby, condensates formed in the landscape could be set to be between the unperturbed condensate regime and the fully confined regime, in which independent condensates form at the minima of the SAW potential with negligible particle tunneling between adjacent sites [13].

The low effective mass of polaritons ($\approx 10^{-4}$–10^{-5} times the effective exciton mass) allows one to observe 3D quantization effects already for trap diameters on the order of a few micrometers [14]. Through a lateral confinement of polaritons, it was proposed to design quantum emitters for quantum information processing applications, which for instance could employ confined polaritons as a source of indistinguishable single photons [15]. Indeed, theoretical considerations on the single polariton level have predicted single-photon emission from a photonic quantum dot structure with strong confinement potential for polaritons based on polariton quantum blockade [16]. In contrast to the low-density limit, for trap-polaritons in the high-density regime, condensates in excited polariton states were created by using a single trap potential naturally induced by an impurity or disorder [17]. Similarly, a micropillar confinement potential could be used for this purpose [18].

In addition to work with single quantum boxes, one can employ photonic molecule/lattice structures [19, 20] to study polariton molecules/crystals and the condensation within those coupled microcavities [21]. For a system of two coupled polariton micropillars, Abbarchi et al. have observed effects such as macroscopic quantum self-trapping and Josephson oscillations of exciton–polaritons [22]. Such Josephson oscillations were previously observed by Lagoudakis et al. in planar microcavities using two disorder traps [23]. For BECs, the discussion of the AC and DC Josephson effects took place in [24].

Furthermore, lattices of polariton structures can be used to study high-orbital bosonic condensates in the p- and d-orbital states in exciton–polariton systems. This can be readily done through the bottleneck condensation dynamics by controlling the polariton density [25]. According to Kim et al., a d-orbital condensate in an atom–optical lattice system has not yet been realized because of experimental difficulties. However, for polaritons, coherent anti-phased d-orbital condensates formed in a two-dimensional square-lattice potential has been achievable [25]. Similarly, anti-phased p-orbital polariton condensates were first demonstrated in a one-dimensional condensate array analogous to a Josephson π-junction array [26]. Exciton–polariton condensates were furthermore also reported near the Dirac point in a triangular lattice [27]. Due to the band formation in polariton lattices similar to that of atoms, certain crystal structures of (coupled) polaritonic traps can thereby be used to study exotic polariton band structures including those with Dirac cones [28].

A polariton condensate in a photonic crystal potential landscape comprising buried polariton traps was similarly reported [29], occuring at the M-point of the superlattice's Brillouin zone. Moreover, exploiting superlattices induced by surface-acoustic waves, exciton–polariton gap solitons in 2D lattices have been reported [30]. Another example of a 1D lattice of coupled polariton structures has dealt with polariton condensation and disorder-induced localization in a flat band [31]. Recently, excitation of localized condensates in the flat band of the exciton–polariton Lieb lattice has been proposed by theoretical work [32]. In fact, periodic arrays of polariton microstructures can also be employed for the study of exciton–polariton topological insulators [33, 34] and topological polariton condensates [35].

6.1.3 Doped Microresonators

For practical devices, electrical current injection into resonator structures is favoured over optical pumping schemes and is achieved via a p-i-n doping profile of the microcavity structure (cf. [36, 37]). However, such a diode-like doping scheme is well matched to the requirements of an optical microresonator. The features of a high-quality microcavity for LED-like operation will be described in the following.

In the case of a p-i-n structure, an n-doped GaAs (Si:GaAs) wafer with (100) surface is used as substrate, on which a similarly n-doped GaAs buffer layer (commonly a few hundred nanometers, e.g. $\approx$400 nm) is grown for a smooth epitaxy of the actual resonator structure. After completion of the microcavity growth, the bottom contact can be deposited (e.g. from the vapour phase) onto the reverse side of the wafer.

In the III/V system, n-doped DBRs typically employ silicon for doping, while p-doped DBRs are based on carbon doping of the mirror materials. In addition to the p-i-n design, also an opposite doping order can be achieved, i.e. in a n-i-p device. For such, a Zn:GaAs wafer can be used as substrate. Nevertheless, implementation of a p-i-n doping scheme for electrically-driven microcavities has been favoured so far.

In order to achieve as high as possible resonator qualities compared to undoped structures, p-i-n diodes typically employ gradually doped DBR structures. In such a tapered doping profile, the donor and acceptor densities in the top and bottom DBR, respectively, decrease from the one end of the DBR structure towards the cavity side. The cavity itself remains completely undoped, i.e. intrinsically doped. Thereby, an efficient current injection into the microresonator is achieved via the doped DBR structures (which are quite thick compared to the actual cavity spacer), while simultaneously the resonator quality is affected as little as possible by additionally introduced optical losses due to interaction with injected charge carriers and donor or acceptor states. This is particularly of relevance at the anti-node positions of the electric field standing wave pattern in the microcavity structure, where highest optical losses could be experienced as a consequence of doping. Furthermore, the gradient of the doping profile takes into account the unavoidable penetration of the light field into the first layers of the DBRs on the cavity side.

In the case of a pioneering polariton-laser structure known from the literature [37], the donor (Si) and acceptor (C) concentrations are raised from 1×10^{18} on the cavity side to $3 \times 10^{18}\,\mathrm{cm}^{-3}$ at the end of both the n-doped and p-doped DBR, respectively. For an optimum Ohmic contact from the top of the structure to the deposited metal contacts on it, the concentration of the two last doped mirror pairs of the p-doped mirror is further increased to $2 \times 10^{19}\,\mathrm{cm}^{-3}$. Moreover, such sophisticated diode structure also incorporates delta-doped layers in both the top and the bottom DBR for a further improved conductivity in the Bragg structure. In the aforementioned polariton-laser structure, the sheet density amounts to $10^{12}\,\mathrm{cm}^{-2}$. These additional doping sheets are systematically placed at every second material interface of the DBRs' mirror layers, where an optical node of the resonator mode is found, in order to minimize losses by absorption while achieving high carrier densities in the multi-layered system (also cf. [38] concerning the quality of doped micropillar structures based on this design principle).

6.1.4 Polariton Diodes

Having doped microcavities at hand, the next step towards the achievement of an operational polariton LED or laser diode requires homogeneous, local injection of charge carriers into the microresonator system for electrical excitation of excitons, from which polaritons are formed. However, since planar cavities—whether equipped with large-area or punctual contacts—are not suitable for achieving a well-defined directed current flow due to lateral charge-carrier diffusion, patterning of the planar microresonator structures becomes necessary.

Fortunately, semiconductor structures grown by epitaxy can be processed to yield pillar [37, 39, 40] and mesa structures [8, 19, 20, 28, 41, 42] of various shapes for different purposes via common etching techniques. Such techniques are wet-chemical etching and plasma etching [43] (dry etching, e.g. inductively coupled plasma or electron-cyclotron-resonance reactive-ion etching, used in [44, 45] towards high-quality micropillar fabrication). Highly-precise patterns with sub-micron resolution can be written into (either positive or negative) photoresists by electron-beam lithography (EBL) [46] or optical lithography [47, 48]. While EBL is known to produce very precise masks, it is also known to be relatively slow. In contrast, optical lithography is much faster, but exhibits a wavelength-limited resolution. Nevertheless, optical ultraviolet (UV) lithography is successfully employed in the field of microcavity processing to define rectangular or circular resonator posts (pillars/mesas). Yet, the quality of the processed structure is also dependent on the etching technique. While wet-chemical etching and certain methods of dry etching of micropillars can lead to under-etching of the post (e.g. as shown in [39, 44]), improved dry etching methods typically result in very clean and straight facets (such as in [45, 49]).

Etching of mesas and pillars offers a great advantage regarding concentrated current flow through the active region of the microcavity structure. Indeed, resonator posts of 100 µm diameter can be still considered a quasi-planar system, whereas small diameters below 20 µm cause a pronounced mode quantization and thereby the loss of the typical energy–momentum dispersion of planar cavity modes [8, 50]. However, a high ratio between the side-wall surface and the cavity volume at very small diameters leads to parasitic effects such as nonradiative recombination and dephasing effects at the etched surface, which result in a broadened excitonic line width [9, 51]. Nonetheless, pillar structures with a moderate diameter of 20 µm have been successfully employed for polariton-lasing studies [37]. Such relatively broad microscale posts combine the advantage of nearly planar cavities with a pronounced polaritonic dispersion feature and discretization of the photonic ground state due to the moderate confinement of the photon mode in such structure, while at the same time the role of the side wall is kept at a minimum.

For electrical contacting of the top facet, the micropillar landscape is commonly planarized by *benzocyclobutene* (BCB) [40], in order to provide a flat surface on which ring-shaped contact apertures connected to larger pads can be deposited. Those pads can later be conveniently linked to electrical pins of a connector via wire-bonding, while the ring aperture enables current injection into the system from the top and simultaneously allows for efficient light extraction from the top facet of the VCSEL-like structure (see Fig. 6.2a, [36]). Such top contacts, which attach to the upper layer of the DBR structure, can be achieved by Ti-Au contacts with inner and outer diameters of $d_i = 10\,\mu\text{m}$ and $d_a = 30\,\mu\text{m}$, respectively, as used in [37]. Moreover, current injection can be facilitated by a semi-transparent metal layer on the top facet for homogeneous current injection. On the surface of a p-doped DBR,

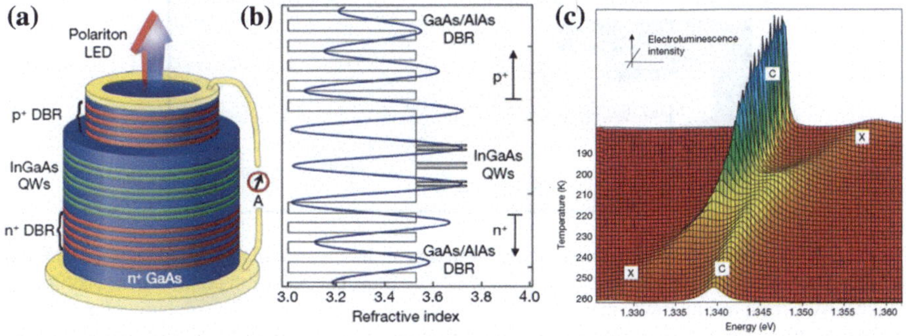

Fig. 6.2 **a** Schematic representation of a polariton LED device with electrical contacts. **b** The cavity field distribution with respect to the embedded QWs in the structure indicates the microcavity design. **c** From such GaAs-based LED, polariton modes were obtained in electroluminescence spectra for temperatures up to 260 K. Reused with permission. Reference [36] Copyright 2008, Springer Nature

a 3-nm-Ti/7-nm-Au compound layer has been used between DBR and ring contact to obtain an electrically-driven polariton laser [37]. The reverse-side contact of the very same device on the n-doped substrate has been achieved by a layer sequence of AuGe-Ni-Au. Although BCB is primarily employed for planarization, it also protects the structure with layers composed of aluminum from oxidation. Furthermore, BCB exhibits a refractive index of ≈ 1.5, which reduces the refractive index contrast from the cavity to its surroundings, and acts as an electrical insulator. A representative view on QW-microcavity polariton devices for electroluminescence studies is shown in Fig. 6.2, which depicts a sketch of the VCSEL-like structure (a) together with the cavity-field distribution with respect to the QW layers in the active region (b). From such a device, Tsintzos et al. demonstrated electroluminescence with anti-crossing at elevated temperatures close to 260 K [36] (see Fig. 6.2c).

Indeed, there is no standard design for a polariton LED, and different structures with different doping schemes, QW numbers and contact designs have been presented by various research groups in the early stage of polariton diode development [36, 53, 54]. These reportings included the demonstration of electroluminescence from polariton modes and led to the competitive investigation of power-dependent electroluminescence from such devices. Naturally, many optimization steps preceded the successful demonstration of an electrically-pumped polariton laser, while little is reported on the technical details of intermediate states towards the final structures which were relevant to reach the goal of that time: to achieve efficient current

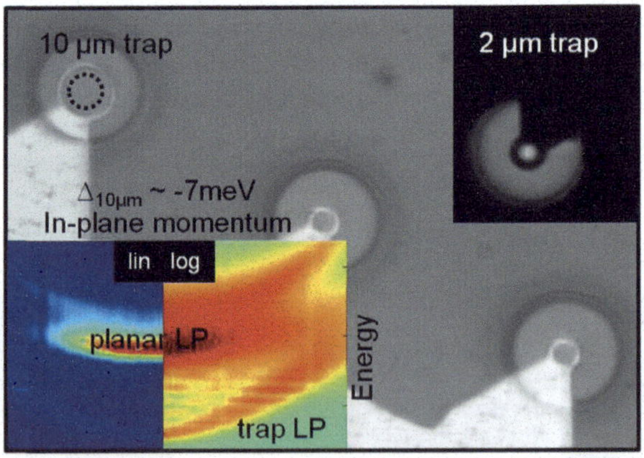

Fig. 6.3 Micrograph of a polariton-trap LED device, showing electrical top contacts on circular structures etched into a planar microcavity system. The traps for polariton confinement originating from cavity-height modulations in the spacer layer are not visible, since they are buried by the overgrown top DBR. The top-right inset shows electroluminescence from an LED with small polariton trap as recorded in a microscope image. For a similar but wider trap with weaker confinement effect, the electroluminescence spectrum from the polariton LED is shown in the bottom-left inset. Linearly as well as logarithmically plotted intensity profiles representing the energy–momentum dispersion from both the trap and the surrounding planar configuration are displayed. Adapted from [52]. Courtesy of the author

injection and to preserve polariton emission out of a VCSEL-like structure, i.e. light output which results from the strong-coupling regime in these QW-microresonators at sufficiently high current densities below the conventional (weak-coupling) laser threshold. Recently, polariton lasers with similar concept have been reported [37, 55], and also room-temperature polariton lasing from a differently designed polariton emitter claimed [56]. Alternatively, the development of novel polariton LED structures of various kind have been pursued, including ones with buried potential landscapes [57] (similar to those in Fig. 6.3) or Esaki diodes [58]. In addition, electrically-contacted structures enable the study of the influence of external electric fields on polaritons [59–61], commonly achieved by applying a reverse bias to the diode structure and affecting the exciton–cavity detuning of the system by the quantum-confined Stark effect (cf. [62]).

6.2 Microcavities for Different Material Systems

While the first experimental demonstration of a condensate of polaritons by H. Deng et al. in 2002 was achieved with a III/V microcavity structure [63], other material systems were successfully used to obtain similar results in the following years. The structure studied by Deng et al. consisted of an AlAs $\lambda/2$ cavity sandwiched by two DBRs, which were made of 16/20 alternating $Ga_{0.8}Al_{0.2}As$/AlAs $\lambda/4$ layer pairs (in the top/bottom DBR). In order to increase the exciton saturation limit, three stacks of QWs were incorporated into the structure at the three most central anti-node positions of the cavity-mode electric field, similar to the description of a 12 QW design of Sect. 5.2. These stacks consisted each of four 7-nm-thick GaAs QWs, which were separated by 3-nm-thick AlAs barriers. The binding energy of excitons in GaAs amounts to $\approx$10 meV, and the emission energy lies at around 1.61 eV. From this structure, a Rabi splitting of $\approx$15 meV was obtained, which was sufficient to observe strong coupling and condensation at cryogenic temperatures. However, for high temperature operation, materials with a strong binding energy of the excitons and thereby a large oscillator strength are favoured. Yet, alternative material systems are known that offer higher coupling strength than in the GaAs/AlAs system, and have been exploited for the demonstration of polariton condensation at both low and elevated temperatures up to room temperature. A few details about these material systems and the employed structures is given in the following to provide the reader with an overview on available microcavity designs and materials, and some of their peculiarities. For more information, the interested reader is referred to the cited articles within this section. A helpful overview chart from the literature regarding material systems is provided in Fig. 6.4 [64].

It is worth noting that not all polariton systems employ QW excitons as the active region and are also not necessarily in the small perturbation regime for the optical cavity. Commonly, few QWs with transparent barriers, which feature an almost constant background dielectric constant (or refractive index) in the spectral range of interest, are employed in III/V semiconductor microcavities. In those cases, the exciton resonance is considered a small perturbation of the cavity (compared to the

	Inorganic (large ε)	Inorganic (small ε)	Organic

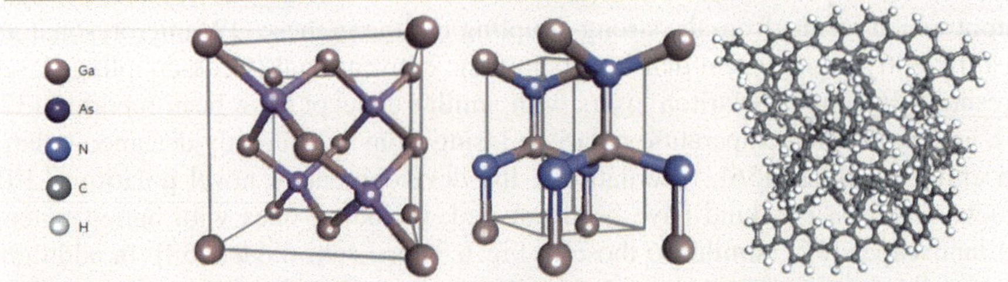

	Inorganic (large ε)	Inorganic (small ε)	Organic
Material	For example, GaAs, CdTe	For example, GaN, ZnO	For example, anthracene, MeLPPP, TDAF, cyanine dyes
Packing	Crystal (typically zincblende)	Crystal (typically wurtzite)	Amorphous, polycrystalline, single crystal (anthracene)
Exciton	Wannier-Mott	Wannier-Mott	Frenkel
Binding energy	5-25 meV	25-60 meV	0.5-1 eV
Linewidth	0.1-1 meV	5-10 meV	25-200 meV
Rabi splitting	3-25 meV	30-150 meV	0.1-1 eV
Dominant relaxation pathways	Acoustic phonons and exciton-exciton scattering	Acoustic, LO phonons and exciton-exciton scattering	Molecular vibrations and radiative
Source of exciton-exciton nonlinearity	Coulomb, saturation	Coulomb, saturation	Saturation

Fig. 6.4 Table on polariton semiconductor materials from [64] by D. Sanvitto and S. Kéna-Cohen: Three classes of semiconductor materials are compared. Inorganic semiconductors are divided according to ϵ, their relative permittivity, which determines the exciton Bohr radius and binding energy. [...] [The authors of this comparison draw attention to the fact that] most of the metrics scale accordingly. Materials in the two rightmost columns readily allow room-temperature operation. At low densities, the relaxation of inorganic polaritons along [their] dispersion occurs mainly via the emission of acoustic phonons. At higher densities, the polariton–polariton interaction due to the Coulomb potential between excitons plays a more dominant role. In most cases, the longitudinal optical (LO) phonon energy is too high to play an active role in relaxation, with ZnO being a notable exception. For organic polaritons, relaxation has been reported to occur via the emission of molecular phonons [66, 67] and radiatively [68]. Given the localized nature of Frenkel excitons, the Coulomb term is negligible for most organic materials. MeLPPP: methyl-substituted ladder-type poly(para-phenylene); TDAF: 2,7-bis[9,9-di(4-methylphenyl)-fluoren-2-yl]-9,9-di(4-methylphenyl)fluorene. Reused with permission. Reference [64] Copyright 2016, Springer Nature

empty cavity) due to the small optical density $\alpha(\omega_C)d \ll 1$ induced by the QW, whereas $\alpha(\omega)$ and d denote the frequency-dependent absorption and QW thickness, respectively. The cavity finesse hardly is affected and the effective refractive index of the cavity spacer remains basically the same compared to an empty cavity. In contrast, it has to be taken into consideration for bulk semiconductors used as spacer, such as ZnO, GaN or CdTe, that the material containing bulk-exciton resonances (actually 3D exciton–polariton resonances) does not act as a small perturbation to the cavity when employed as active medium. To give an example, for ZnO with about 0.3 μm spacer thickness L (cavity length), the optical density $\alpha(\omega_C)L \gg 1$ in the proximity of the matter resonance. This spoils the cavity effects due to the strong perturbation regime (defined by $\alpha(\omega)\lambda > 1$, with λ the wavelength in the medium) at the exciton resonance. Thus, cavity modes are only obtained in transparent regions, e.g. above/below that resonance, and only uncoupled exciton emission is expected at the exciton energy where no cavity mode can be formed. Moreover, if the refractive index drop from below to above exciton resonance is considerable, the next higher cavity mode evolving at those energies above resonance is not necessarily a continuation of the harmonic series of cavity modes established below the resonance. Attention has been drawn to these little-considered aspects when dealing with bulk spacers for instance in [65] in Chap. 17.

6.2.1 II/VI Microresonators

CdTe-Based Microcavities

II/VI semiconductor microcavities based on CdTe have been successfully utilized in the past two decades for the study of polaritons and remarkable investigations have contributed to the understanding of cavity–polaritons and their excitation-power dependent dynamics and characteristics at cryogenic temperatures. For instance, as an early work towards II/VI BEC studies, a spontaneous coherent phase transition of polaritons was observed in CdTe microcavities by Richard et al. [69], who shortly after also reported on the experimental evidence for non-equilibrium Bose condensation of exciton–polaritons [70]. Based on these activities, J. Kasprzak, M. Richard, S. Kundermann, A. Baas et al. were able to finally report the demonstration of BEC of exciton–polaritons in 2006 [71], which is considered the first complete investigation of polariton condensation able to link the observed condensation effect to BEC. Later, also ZnO microcavities (see below) were achieved in the II/VI material system which promise larger binding energies and access to polariton physics at elevated temperatures and in the visible spectral region.

CdTe microresonators grown by molecular beam epitaxy with $Mg_xCd_{1-x}Te$ and $Mn_xCd_{1-x}Te$ barrier and DBR layers [72, 73] can be regarded as an alternative to GaAs-based systems. They offer a few advantages, which were considered helpful for the initial performance of high-exciton density studies with the aim to see a nonlinear ground-state filling. Particularly, maintenance of strong coupling between the cavity mode and the QW excitons at high pump powers requires a material with rather large exciton binding energy. In CdTe quantum wells, the binding energy amounts to $\approx$25 meV, resulting in a higher vacuum Rabi splitting of $\approx$6 to 7 meV for a single QW compared to $\approx$4.2 to 4.4 meV for GaAs. The emission energy of about 1.67 eV is still close to that of GaAs QWs, though a bit higher energetic. Combined with a large number of QWs in the microresonator, a microcavity with 16 CdTe QWs in a 2λ CdMgTe cavity spacer can exhibit a Rabi splitting of 26 meV [69, 71] (for details on CdTe microcavity structures see [72, 73]). The Rabi splitting of a four-CdTe-QW structure with about 13 meV [74] almost reaches values of a 12-GaAs-QW structure (14 to 15 meV), which explains the initial interest in these structures. In addition, the larger binding energy and oscillator strength compensate the larger lattice mismatch in the II/VI structure, which renders the system more tolerant for inhomogeneous broadening. Moreover, less mirror pairs are needed in the DBRs to achieve a high-quality resonator owing to the larger refraction index contrast. Yet, growth of III/V materials still features advantages in quality and precision, while also electrical doping schemes can be more easily implemented in the GaAs/AlAs system.

6.2.2 *Inorganic Room-Temperature Polariton Systems*

GaN-Based Microcavities

With the pursuit of blue LEDs, (In)GaN-based structures had their début as semiconductor active media and enabled the era of LED-based white light—a successful endeavour for which three renowned researchers, Isamu Akasaki, Hiroshi Amano and Shuji Nakamura, respectively, were jointly awarded the Nobel prize in 2014. Besides the interesting emission wavelength, excitons in GaN exhibit a strong appeal when it comes to binding energies. This particular feature (with a Rabi splitting in corresponding microresonators typically exceeding 50 meV) encouraged scientists to think of a high-temperature polariton system based on GaN, which allows for the study of both polariton physics and condensation phenomena at elevated temperatures up to room temperature, similar to the ZnO system. Having developed microcavities in the footsteps of GaN VCSEL research, polariton lasers at room temperature have been reported since 2007 for optically-pumped GaN microcavities with different active-medium structures, such as bulk material [75], multi-QWs [76], or an embedded nanowire [77]. Recently, even electrically-driven bulk GaN polariton devices have been demonstrated [56]. Since the past decade, numerous reports on the design and characterization of polariton structures using this material system have been made, such as the one dealing with an all-dielectric GaN-based microcavity [78]. Here, a brief overview on some of the corresponding microresonator structures shall be provided.

In the first report on room-temperature polariton lasing, a bulk GaN microcavity was presented in the strong-coupling regime driven by pulsed non-resonant optical excitation, for which a transition to a coherent polariton state was claimed at increased pump densities one order of magnitude lower than the lowest reported value for optically-pumped InGaN QW VCSELs [75]. Grown on sapphire, the microresonator structure consisted of a central 210 nm $3\lambda/2$ bulk GaN cavity spacer sandwiched by two DBRs of different composition. The bottom DBR was composed of 35 $Al_{0.85}In_{0.15}N/Al_{0.2}Ga_{0.8}N$ grown on a buffer layer and was reported to facilitate reduction of built-in tensile strain. The top DBR complemented the cavity and consisted of a dielectric SiO_2/Si_3N_4 DBR with 10 mirror pairs [79]. Here, the bulk system, which emitted at about 360 nm, was preferred over QWs in order to circumvent the material-dependent broad QW linewidth and a pronounced quantum-confined Stark effect [80].

In the following year, room-temperature polariton lasing was reported for an optically-pumped hybrid AlInN/AlGaN multiple QW microcavity which contained an active region of multiple GaN/AlGaN QWs [76]. Non-resonant pulsed excitation led to a distinct pump-dependent transition into the energy ground-state of the emitting system, with the nonlinearity in the output intensity and the change of the angle-dependence of emission attributed to polariton lasing, i.e. coherent emission from the lower-polariton branch. The employed polariton system with $E_{\text{Rabi}} = 56$ meV was grown by MOVPE on (0001) sapphire and was composed of a 67 period GaN(1.2 nm)/$Al_{0.2}Ga_{0.8}N$(3.6 nm) multi-QW 3λ-cavity structure (described in more detail in [81]). Here, a small negative detuning ($\Delta \ll E_{Rabi}$) of the cavity mode with respect to the exciton resonance was desired, which was achieved by the insertion of a 67-nm-thick Si_3N_4 layer ($\approx 1.3\lambda/4$) before overgrowth of the cavity spacer by a dielectric SiO_2/Si_3N_4 top DBR with 13 mirror pairs.

Electrical pumping was later pursued with an in-plane Fabry–Pérot GaN based resonator, with electroluminescence and macroscopic ground-state occupation having been reported from such bulk-GaN pn-diode resonator structure recently [56]. The diode was grown as a 430 nm thick GaN p-n junction using plasma-assisted MBE. Prior to that, a 300 nm buffer layer of lattice-matched n-doped $In_{0.18}Al_{0.82}N$ layer was grown on a GaN-on-sapphire substrate. The optical cavity was designed to be an in-plane 5λ Fabry–Pérot resonator in the form of a 690 nm wide and 40 μm long ridge, patterned by reactive-ion etching and focused-ion-beam milling. The microcavity was completed by SiO_2/TiO_2 DBRs with 6/5 (left/right) mirror pairs on both opposing sides (facets of ridge). The fabrication of such a room-temperature polariton diode is described in more detail in [56].

These examples clearly illustrate the variety of approaches used in the GaN system to achieve optically and electrically pumped polariton systems. For more information on polaritons in this material system, the interested reader is referred to the original publications and the references therein.

ZnO-Based Microcavities

ZnO is a semiconductor with wurtzite structure. It has become particularly attractive for room-temperature polaritonics due to the strong exciton binding energies of around 60 meV and can directly act as the active region [82], i.e. in bulk form. Early reports of polariton condensates date back to 2010, when Sun et al. released their preprint on arXiv on their observation of a room-temperature one-dimensional polariton condensate in a ZnO microwire [83]. The structure employed was not a DBR-based microcavity, but a synthesized ZnO crystal forming a 1D photonic wire, in which polariton modes arise from the coupling of wire-photon modes with unconfined (bulk) ZnO excitons. Those polaritons can propagate freely along the wire and are strongly confined via the photon component in the perpendicular in-plane direction. Later work by other groups compared room-temperature polariton lasing versus photon lasing in a ZnO-based hybrid microcavity [84] or studied the tuning from an excitonic to photonic polariton condensate in a ZnO-based microcavity [85], showing growing interest in polaritonics in the ZnO material system. Room-temperature polariton lasing has been reported multiple times so far [85–87]. In some reports, closely spaced peaks occur in the polariton lasing regime [85, 88, 89] which could be indicative for strong spatial confinement effects in the potential landscape originating from inhomogeneities in the structure, or the pump spot.

Several different studies performed show the capability of employing ZnO for polaritonics, some of which promise operation well above room temperature [90]. For instance, exciton–polaritons in a ZnO-based half-wavelength microcavity were investigated with regard to polarization dependence and nonlinear occupation [82]. The cavity spacer was sandwiched between DBRs composed of yttria-stabilized zirconia (YSZ) and alumina, beginning and ending with YSZ in a 10.5-layer-pair configuration on both sides. The growth procedure for such structure is described in [91]. The same research group has reported on the observation of macroscopically coherent states of cavity–polaritons up to 250 K [89], attributing the signatures in the negative-detuning regime to a BEC, while for positive detunings, emission from an electron–hole plasma was reported. Moreover, the authors claimed ballistic propagation of the condensate in the pump-induced potential landscape [89].

6.2.3 *Organic Materials*

In contrast to their inorganic counterparts exhibiting Wannier–Mott excitons, organic semiconductors feature Frenkel excitons. Instead of being delocalized over several lattice sites with Bohr radii much larger than the unit cell of the lattice, Frenkel excitons are tightly-bound electron–hole pairs strongly localized at their host molecules which possess considerably higher binding energies ($\approx$1 eV) than their weakly-bound pendants (on the order of 10 meV), for which the strength of Coulomb interaction matters. This renders the latter prone to dissociation at elevated temperatures due to thermal activation, while Frenkel excitons are naturally highly stable at room temperature. Being affected by the details of the molecular structure and the organic semiconductor crystal, interactions of Frenkel excitons and polaritons in organic microcavities differ from that of Wannier–Mott excitons and are discussed for instance in [92].

Organic polariton lasing was first reported for microcavities containing anthracene single crystals in 2010 [94] after earlier demonstrations of strong coupling between Frenkel excitons and cavity photons [95, 96]. However, for anthracene no characteristic density-dependent blueshift was observed for the polaritonic ground-state mode which is expected for strongly interacting (e.g. Wannier–Mott-like) polaritons.

Later in 2014, nonlinear interactions in an organic polariton condensate were evidenced by Daskalakis et al. using a single film of thermally evaporated TDAF[1] sandwiched between two dielectric mirrors [93]. In that study, room-temperature polariton condensation was demonstrated in an organic microcavity exhibiting the (un)expected blue shift (for Frenkel excitons; attributed to exciton saturation) of the emission mode, which showed a superlinear input–output power dependence and long-range order. A schematic drawing of such an organic microcavity is shown in Fig. 6.5a. That study used microresonators made of TDAF layers with thickness in the range of 120–140 nm sandwiched between identical SiO_2/Ta_2O_5 DBRs with 6 mirror pairs (44/68 nm). Fabrication of these microcavities was achieved by radiofrequency magnetron sputtering of the bottom DBR on fused-silica substrates. After thermal evaporation of the TDAF layer, the top DBR was sputtered directly onto the

[1] 2,7-bis[9,9-di(4-methylphenyl)-fluoren-2-yl]-9,9-di(4-methylphenyl)fluorene.

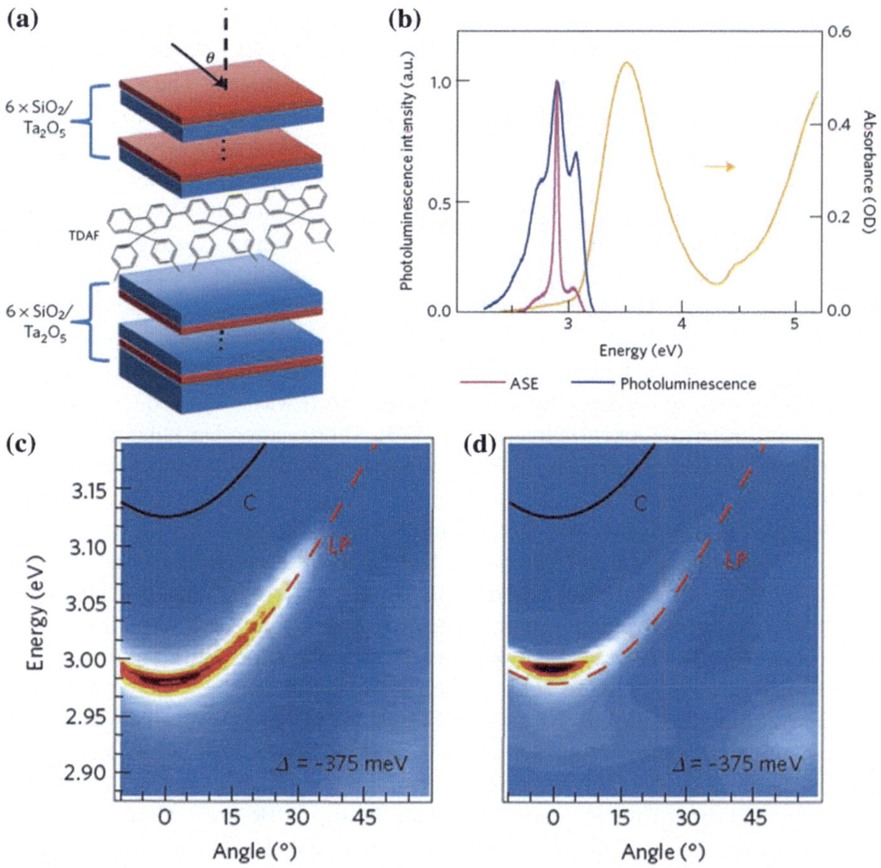

Fig. 6.5 A cavity with organic material, here a single film of TDAF, is schematically shown with alternating tantalum pentoxide (Ta_2O_5) and silicon dioxide (SiO_2) mirror pairs in (**a**). The molecular structure of the organic material is indicated. From a 120-nm-thick bare TDAF film, photoluminescence (blue) and its ASE spectrum (pink) are shown in (**b**). For PL, vibronic replicas are visible. The ASE peak occurs at the emission maximum and was obtained under pumping with an elongated (elliptical) pump spot for sufficient single-pass gain within the active medium. The absorption spectrum with pronounced inhomogeneous broadening of the exciton band at about 3.5 eV plotted with respect to the right scale was recorded for a 60-nm TDAF film. Density-dependent PL from a 120-nm-thick organic microcavity with angle resolution (exciton–cavity detuning $\Delta = -375\,\text{meV}$) at low (**c**) and high (**d**) excitation densities indicates the transition from a thermally distributed polariton gas to a macroscopically occupied polariton ground-state. The emission mode overcoming the threshold density attributed to polariton lasing shows typical signatures such as linewidth and momentum narrowing, and a density-dependent (not explicitly shown here) blueshift. Reused with permission. [93] Copyright 2014, Springer Nature

TDAF layer without breaking the vacuum. Those samples of the size 15×15 mm intrinsically featured detuning variations of the order of $\approx$50 meV due to thickness variations, which is beneficial for various polariton experiments. The Rabi splitting ($E_{\text{Rabi}} = 2\hbar\Omega = 2g$) in this structure at normal incidence reached $\approx$0.6 eV. The spectral signatures of the material system are summarized in Fig. 6.5b, with microcavity PL from the LP branch below and above a polariton condensation threshold shown in (c) and (d), respectively. Daskalakis et al. particularly discuss the observed density-dependent blueshift above the nonlinearity threshold as being a strong indicator of polariton condensation, as the peak in emission and amplified spontaneous emission (ASE) from bare TDAF were on the lower-energy side of the obtained polariton ground state. This would be in contrast to conventional laser frequency pulling which would have resulted in the emission being shifted towards the gain maximum, i.e. redshifted for their material system. Thus, the blueshift was attributed to polariton–polariton and polariton–exciton interactions [93]. Shortly after, Daskalakis et al. also reported on spatial coherence and stability in a disordered organic polariton condensate [97].

Basically simultaneous to the aforementioned work on organic polariton condensates, room-temperature non-equilibrium BEC of exciton–polaritons was reported in 2013 by Plumhof et al. in a polymer-filled microcavity [98]. In their study, the authors observed macroscopic ground-state occupation which exhibited long-range coherence above the critical excitation density and a behaviour different than that of conventional photon lasing. To achieve their results, a planar $\lambda/2$ microcavity with an effective thickness slightly larger than half of the polymer exciton's wavelength was grown with top and bottom SiO_2/Ta_2O_5 DBRs. The whole structure was grown on a fused-silica substrate by sputtering deposition, with the bottom DBR consisting of a higher number of mirror pairs (9.5) than the top DBR (6.5). Here, MeLPPP[2] was synthesized and a 35-nm layer of it was spin-coated onto the bottom-DBR's top dielectric layer (i.e. an SiO_2 layer, resembling a cavity spacer when completed by the top-DBR counterpart). Thus, MeLPPP was embedded in between two 50-nm SiO_2 spacer layers, which formed the cavity region encapsuled by the multiple mirror pairs of those surrounding DBRs. With this structure, a Rabi splitting and detuning of 116 and 77 meV (photon fraction $\cong$78%) was obtained, respectively. Peculiar for the quite rigid ladder-type conjugated polymer MeLPPP is the comparably very narrow linewidth of the excitonic resonance among organic semiconductors.

6.2.4 Perovskite-Based Exciton–Polariton Systems

Perovskites are a material class which has attracted considerable interest in the optoelectronics community, particularly with eyes towards solar cell applications. However, they are also of interest for lasing and strong-coupling studies. They exhibit the unique advantageous nature of, both, inorganic and organic materials. With large exciton binding energies and large oscillator strength they are promising for room-

[2]Methyl-substituted ladder-type poly(para-phenylene).

temperature polariton lasing as well. The inorganic and organic characteristics are for instance investigated in [99, 100].

Strong coupling has been studied in the past two decades by different groups [101, 102]. However, attempts to achieve polariton lasing were not successful. Following that, improved microcavities were grown, from which Q-factors of about 200–300 with monolithically grown opposing DBRs with perovskite spacer were achieved. One design with HfO_2/SiO_2 mirror pairs (7/13 for the top/bottom DBR) and $CsPbCl_3$ spacer material and active region is briefly highlighted, for which polariton lasing was investigated in recent years. With a detuning of −25 meV and a Rabi splitting of 265 meV, a clear polariton dispersion from the lower branch was reported in the polariton-lasing report by Su et al. in 2017 [103]:

CW excitation at about 3.5 eV (about 355 nm) led to the emission from the LP branch at 2.9 eV (about 430 nm) with clear flattening of the dispersion towards higher angles. Ultimately, 100-fs pulsed excitation at 375 nm with a spot size of about 25 μm showed a distinct nonlinearity in the input-output characteristics with typical spectral indicators of polariton lasing such as a linewidth reduction, spatial coherence build-up and drastical k-space narrowing.

6.2.5 Monolayer Transition-Metal Dichalcogenides

The material class of transition-metal dichalcogenides (TMDCs) has attracted considerable attention in recent years as it promises unique optoelectronic features in the monolayer limit—as the most prominent semiconducting representative of van-der-Waals materials, which consist of layered crystals that have weak out-of-plane bonds with no covalent contribution. Amongst these features are their spin–valley locking due to the three-fold lattice symmetry with pseudo-spins attributed to the two distinguishable momentum-space valleys at the corners of the Brillouin zone, namely at the K and K' points, the extraordinarily-high excitonic binding energies on the order of 0.5 eV and the high oscillator strength. These remarkable properties render them very appealing for light–matter interaction studies (see [104] and references therein).

The vast possibilities regarding out-of-plane stacking is particularly attractive (e.g. for 2D multi-well configurations, see Fig. 6.6, or indirect exciton systems for condensation experiments). Each of these heterostructure configurations opens up immense research possibilities. A sheer unlimited variability of layers in the sense of van-der-Waals epitaxy with band-gap engineering capabilities is given—taking also into account the impact of the stacking environment [106–109] and stacking angle [110–115] on the optical properties and charge-carrier dynamics in such multi-layered 2D-material systems.

Conveniently, TMDC in the monolayer regime are commonly direct-gap semiconductors and can be supported, separated, capped or encapsulated by the insulating 2D-material hexagonal boron nitride (hBN), which on the one hand helps preserving the material in ambient air when capped, and on the other hand can significantly improve the optical properties [116–120].

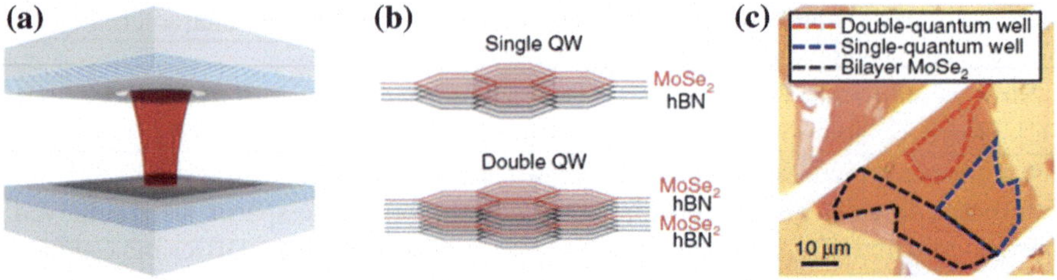

Fig. 6.6 **a** Schematic drawing of an open microcavity system used for light–matter interaction experiments with 2D semiconductors such as monolayers of $MoSe_2$, resembling ideal ultrathin quantum wells (QWs). **b** Two different types of active regions are displayed here: a single-QW-like configuration with a ML supported by hBN, and a double-QW-like configuration with two MLs incorporated between few-layer hBN sheets forming a stack. The corresponding stacks achieved by mechanical exfoliation and manual stacking are shown in (**c**), with distinct regions identified as single as well as double QW, and bilayer $MoSe_2$ (blue, red and black dashed lines, respectively). Reused with permission. Reference [105] (CC BY license) Copyright 2015, Springer Nature

Since room-temperature observation of strong light-matter coupling in practical emitter–resonator systems has required the use of large-band-gap materials to obtain correspondingly high binding energies of Wannier–Mott excitons for nonlinearity observations [76], the 2D excitons became promising for such studies. This is easily explained owing to their binding energies comparable to those of Frenkel excitons provided by organic materials, which are strongly localised and naturally deliver a high oscillator strength (see organic polariton condensate systems [93, 98]). In a monolayer, the excitons are literally confined in the out-of-plane direction onto an effectively atom-thick sheet. Thus, monolayer TMDCs are considered as nearly perfect quantum wells with exceptionally high binding energies of strongly confined excitons that have been studied in the literature in detail (see [121] and references therein). In 2014, Fogler et al. reported on the possibility to investigate high-temperature superfluidity with indirect excitons in van-der-Waals heterostructures [122].

In this context, 2D materials offer a new and unique testbed for light–matter coupling experiments at elevated temperatures, as pioneering work showed [123]. In terms of cavity–polariton research, Liu et al. have reported the first observation of exciton–polaritons employing 2D materials in optical microcavities at the end of 2014: In their strongly-coupled monolayer-MoS_2 microcavity, the 2D flake was sandwiched between two SiO_2-based DBRs, which were grown and terminated by a cavity-side SiO_2 layer. In this experiment at room temperature, a Q-factor of approximately 200, corresponding to an observed cavity mode's linewidth of 10 meV, was sufficient to see the single optical resonance split on the order of 50 meV when coupled to the MoS_2 flake's exciton mode (linewidth approximately 30 meV). This signature of polariton formation was obtained and characterized with angle-resolved reflection and photoluminscence spectroscopy.

These developments have led to further breakthroughs in this domain—using open and tunable cavity designs [105, 124] and showing large Rabi splittings (Fig. 6.7)—

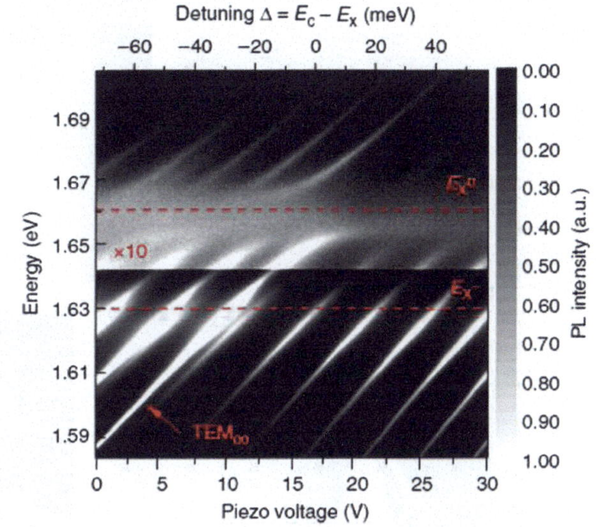

Fig. 6.7 Detuning map of microcavity PL with anti-crossings of the various longitudinal cavity modes with the TMDC monolayer exciton resonance (E_{X^0}). Here, an emission enhancement at the weakly-coupling ML trion's energy (E_{X^-}) is obtained for crossing cavity modes. Reused with permission. [105] (CC BY license) Copyright 2015, Springer Nature

which have kick-started exploration of this field in recent times [125–127]. Commonly, dielectric mirrors, i.e. DBRs based on SiO_2 are employed in such studies. Nevertheless, in some cases, metal mirrors can provide sufficient confinement to observe cavity–polaritons [124, 128] or Tamm-plasmon exciton–polaritons [125]. Wang et al. further compared strong coupling for both metal Fabry–Pérot and plasmonic optical cavities [126].

The control of the cavity–exciton detuning by the resonator length and the ability to displace (structured) cavity mirrors against each other in an open cavity design, as demonstrated by Dufferwiel et al. [105] or Flatten et al. [124], further increases the flexibility in the use of quantum materials. This allows one to find and establish resonance conditions after the construction of the sample and also facilitates alignment of the cavity to the active region, such as self-prepared stacks or grown flakes of TMDCs with their limited spatial (in-plane) dimensions. Early on, a normal-mode splitting was reported for neutral 2D excitons in $MoSe_2$, whereas the coupling strength of trions in the monolayer TMDC only led to weak-coupling behaviour for the charged-exciton resonance [105] (Fig. 6.7); in addition to coupling effects with single monolayers, Dufferwiel et al. also targeted a monolayer–hBN-stack-based double-well configuration for their investigations of strong coupling in their work (Fig. 6.6).

Beyond coupling experiments with a single-type emitter system, an experiment performed by Flatten et al. combined the coupling of the light-field with excitons in TMDC monolayers with the coupling with Frenkel excitons of organic molecules [128], using a similar open-cavity system as in their previous work on strong coupling at room temperature with monolayer WS_2 [124]. In their resonator structure with air gap between the opposing mirrors, a dielectric mirror featuring electrically-tunable WS_2-monolayer excitons on its surface was used together with an opposing mirror

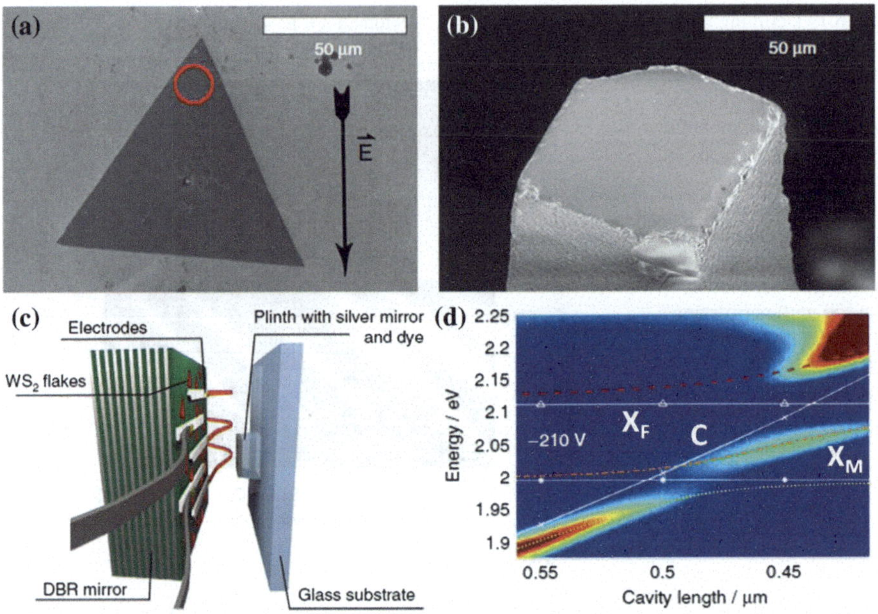

Fig. 6.8 Overview on the experiment with the idea to couple Mott–Wannier (M) excitons (X) in an inorganic 2D semiconductor and Frenkel (F) excitons in an organic material via the light field in a tunable microcavity to form hybrid polaritons. Here, monolayers of WS_2 tunable with electric fields were located on a dielectric mirror (**a, c**), whereas the organic material was deposited on top of a plinth with a silver mirror (**b, c**). The optical signatures of the hybrid Frenkel–Mott polaritons in transmission spectra are shown in (**d**) for the experimental case with lateral bias of −210 V, obtained by scanning over different cavity lengths in the tunable open microcavity. Reused with permission. Reference [128] (CC BY license) Copyright 2017, Springer Nature

made of silver, which was covered by the organic J-aggregated dye TDBC[3] for the dual-coupling experiment. Therefore, a polymer-dye layer of about 300 nm thickness was spin-coated onto the sub-mm-size silver mirror plinth (see Fig. 6.8). The typical anti-crossing between the tunable cavity mode and the Wannier–Mott as well as Frenkel exciton modes is displayed in the contour plot of Fig. 6.8d. The principle behind the experiment is comparable to that of Slootsky et al. [129], who demonstrated uniform Frenkel–(Wannier–Mott) hybridization of degenerate excitons that were hosted by two spatially separate layers within a strongly-coupled microcavity: the layer of nanoparticle ZnO provided Wannier–Mott excitons at E_{X1}, whereas the 3,4,7,8-naphthalene tetracarboxylic dianhydride (NTCDA) layers contributed Frenkel excitons at E_{X1} and E_{X2}.

Up-to-date, different cavity designs have been discussed and different TMDCs employed, ranging from WS_2 to $MoSe_2$. In fact, polariton studies with layered materials are not restricted to the class of TMDCs, as 2D-layered pervoskite materials

[3]A popular carbocyanine dye.

were used well before the rise of TMDCs for room-temperature polariton studies [102]. Particularly, the domain of valley polarization has become a hot topic for polariton research with 2D semiconductors with distinguishable K and K' valleys due to a time-reversal-symmetry break, given access to valley-selective excitation and detection schemes and the outlook for optical valleytronics (see for instance [130]).

Recent reports using TMDCs show pronounced circular polarization of 2D polaritons up to room temperature and explore optical control of polaritons in their microcavities [127, 131–134]. However, further optimisation of these systems is needed to obtain polariton condensates. Several of these approaches to perform valley-polarization studies in the strong-coupling regime involve monolithic (see [123, 125]) and open tunable cavities (see [105, 124]). As this field is still in motion, further achievements on 2D-materials-based microcavities are expected and the observation of polariton-condensation effects is imminent.

References

1. R. Dingle, W. Wiegmann, C.H. Henry, Quantum states of confined carriers in very thin $Al_xGa_{1-x}As$-GaAs-$Al_xGa_{1-x}As$ heterostructures. Phys. Rev. Lett. **33**, 827 (1974)
2. M.A. Herman, H. Sitter, *Molecular Beam Epitaxy: Fundamentals and Current Status* (Springer, Berlin, 1996)
3. G.B. Stringfellow, *Organometallic Vapor-Phase Epitaxy: Theory and Practice* (Elsevier, Amsterdam, 2012)
4. G. Roumpos, W.H. Nitsche, S. Höfling, A. Forchel, Y. Yamamoto, Gain-induced trapping of microcavity exciton polariton condensates. Phys. Rev. Lett. **104**(12), 126403 (2010)
5. G. Tosi, G. Christmann, N.G. Berloff, P. Tsotsis, T. Gao, Z. Hatzopoulos, P.G. Savvidis, J.J. Baumberg, Sculpting oscillators with light within a nonlinear quantum fluid. Nat. Phys. **8**, 190–194 (2012)
6. O. El Daïf, A. Baas, T. Guillet, J.-P. Brantut, R. Idrissi Kaitouni, J.L. Staehli, F. Morier-Genoud, B. Deveaud, Polariton quantum boxes in semiconductor microcavities. Appl. Phys. Lett. **88**(6), 061105–3 (2006)
7. R. Idrissi Kaitouni, O. El Daïf, A. Baas, M. Richard, T. Paraiso, P. Lugan, T. Guillet, F. Morier-Genoud, J.D. Ganière, J.L. Staehli, V. Savona, B. Deveaud. Engineering the spatial confinement of exciton polaritons in semiconductors. Phys. Rev. B. **74**(15), 155311 (2006)
8. T. Gutbrod, M. Bayer, A. Forchel, J.P. Reithmaier, T.L. Reinecke, S. Rudin, P.A. Knipp, Weak and strong coupling of photons and excitons in photonic dots. Phys. Rev. B **57**(16), 9950 (1998)
9. D. Bajoni, P. Senellart, E. Wertz, I. Sagnes, A. Miard, A. Lemaître, J. Bloch, Polariton laser using single micropillar GaAs-GaAlAs semiconductor cavities. Phys. Rev. Lett. **100**(4), 047401 (2008)
10. R. Balili, V. Hartwell, D. Snoke, L. Pfeiffer, K. West, Bose-Einstein condensation of microcavity polaritons in a trap. Science **316**(5827), 1007–1010 (2007)
11. S. Utsunomiya, L. Tian, G. Roumpos, C.W. Lai, N. Kumada, T. Fujisawa, M. Kuwata-Gonokami, A. Löffler, S. Höfling, A. Forchel, Y. Yamamoto, Observation of Bogoliubov excitations in exciton-polariton condensates. Nat. Phys. **4**(9), 700–705 (2008)
12. E.A. Cerda-Méndez, D. Krizhanovskii, K. Biermann, K. Guda, R. Bradley, R. Hey, P.V. Santos, M.S. Skolnick, One dimensional confinement of microcavity polaritons using non-piezoelectric surface acoustic waves. Phys. E: Low-Dimens. Syst. Nanostructures **42**(10), 2548–2551 (2010)

13. E.A. Cerda-Méndez, D.N. Krizhanovskii, K. Biermann, R. Hey, M.S. Skolnick, P.V. Santos, Wavefunction of polariton condensates in a tunable acoustic lattice. New J. Phys. **14**(7), 075011 (2012)
14. P. Lugan, D. Sarchi, V. Savona, Theory of trapped polaritons in patterned microcavities. Phys. Stat. Sol. C **3**, 2428–2431 (2006)
15. N. Na, Y. Yamamoto, Massive parallel generation of indistinguishable single photons via the polaritonic superfluid to Mott-insulator quantum phase transition. New J. Phys. **12**, 123001 (2010)
16. A. Verger, C. Ciuti, I. Carusotto, Polariton quantum blockade in a photonic dot. Phys. Rev. B **73**(19), 193306 (2006)
17. D. Sanvitto, A. Amo, L. Viña, R. André, D. Solnyshkov, G. Malpuech, Exciton-polariton condensation in a natural two-dimensional trap. Phys. Rev. B, **80**(4), 045301 (2009)
18. M. Maragkou, A.J.D. Grundy, E. Wertz, A. Lemaître, I. Sagnes, P. Senellart, J. Bloch, P.G. Lagoudakis, Spontaneous nonground state polariton condensation in pillar microcavities. Phys. Rev. B **81**(8), 081307 (2010)
19. M. Bayer, T. Gutbrod, J.P. Reithmaier, A. Forchel, T.L. Reinecke, P.A. Knipp, A.A. Dremin, V.D. Kulakovskii, Optical modes in photonic molecules. Phys. Rev. Lett. **81**(12), 2582–2585 (1998)
20. M. Bayer, T. Gutbrod, A. Forchel, T.L. Reinecke, P.A. Knipp, R. Werner, J.P. Reithmaier, Optical demonstration of a crystal band structure formation. Phys. Rev. Lett. **83**(25), 5374–5377 (1999)
21. M. Galbiati, L. Ferrier, D.D. Solnyshkov, D. Tanese, E. Wertz, A. Amo, M. Abbarchi, P. Senellart, I. Sagnes, A. Lemaître, E. Galopin, G. Malpuech, J. Bloch, Polariton condensation in photonic molecules. Phys. Rev. Lett. **108**, 126403 (2012)
22. M. Abbarchi, A. Amo, V.G. Sala, D.D. Solnyshkov, H. Flayac, L. Ferrier, I. Sagnes, E. Galopin, A. Lemaître, G. Malpuech, J. Bloch, Macroscopic quantum self-trapping and josephson oscillations of exciton polaritons. Nat. Phys. **9**, 275 (2013)
23. K.G. Lagoudakis, B. Pietka, M. Wouters, R. André, B. Deveaud-Plédran, Coherent oscillations in an exciton-polariton josephson junction. Phys. Rev. Lett. **105**(12), 120403 (2010)
24. S. Levy, E. Lahoud, I. Shomroni, J. Steinhauer, The a.c. and d.c. josephson effects in a Bose-Einstein condensate. Nature, **449**(7162), 579–583 (2007)
25. N.Y. Kim, K. Kusudo, C. Wu, N. Masumoto, A. Löffler, S. Höfling, N. Kumada, L. Worschech, A. Forchel, Y. Yamamoto, Dynamical d-wave condensation of exciton-polaritons in a two-dimensional square-lattice potential. Nat. Phys. **7**(9), 681–686 (2011)
26. C.W. Lai, N.Y. Kim, S. Utsunomiya, G. Roumpos, H. Deng, M.D. Fraser, T. Byrnes, P. Recher, N. Kumada, T. Fujisawa, Y. Yamamoto, Coherent zero-state and π-state in an exciton-polariton condensate array. Nature **450**(7169), 529–532 (2007)
27. N.Y. Kim, K. Kusudo, A. Löffler, S. Höfling, A. Forchel, Y. Yamamoto, Exciton-polariton condensates near the Dirac point in a triangular lattice. New J. Phys. **15**, 035032 (2013)
28. T. Jacqmin, I. Carusotto, I. Sagnes, M. Abbarchi, D.D. Solnyshkov, G. Malpuech, E. Galopin, A. Lemaître, J. Bloch, A. Amo, Direct observation of Dirac cones and a flatband in a honeycomb lattice for polaritons. Phys. Rev. Lett **112**(11), 116402 (2014)
29. K. Winkler, J. Fischer, A. Schade, M. Amthor, R. Dall, J. Geßler, M. Emmerling, E.A. Ostrovskaya, M. Kamp, C. Schneider, S. Höfling, A polariton condensate in a photonic crystal potential landscape. New J. Phys. **17**, 023001 (2015)
30. E.A. Cerda-Méndez, D. Sarkar, D.N. Krizhanovskii, S.S. Gavrilov, K. Biermann, M.S. Skolnick, P.V. Santos, Exciton-polariton gap solitons in two-dimensional lattices. Phys. Rev. Lett. **111**, 146401 (2013)
31. F. Baboux, L. Ge, T. Jacqmin, M. Biondi, E. Galopin, A. Lemaître, L. Le Gratiet, I. Sagnes, S. Schmidt, H.E. Türeci, A. Amo, J. Bloch, Bosonic condensation and disorder-induced localization in a flat band. Phys. Rev. Lett. **116**(6), 066402 (2016)
32. M. Sun, I.G. Savenko, S. Flach, Y.G. Rubo, Excitation of localized condensates in the flat band of the exciton-polariton Lieb lattice. Phy. Rev. B **98**(16), 161204 (2018)

33. T. Karzig, C.E. Bardyn, N.H. Lindner, G. Refael, Topological polaritons. Phys. Rev. X **5**(3), 031001 (2015)
34. C.E. Bardyn, T. Karzig, G. Refael, T.C.H. Liew, Topological polaritons and excitons in garden-variety systems. Phys. Rev. B **91**(16), 161413(R) (2015)
35. S. Klembt, T.H. Harder, O.A. Egorov, K. Winkler, R. Ge, M.A. Bandres, M. Emmerling, L. Worschech, T.C.H. Liew, M. Segev, C. Schneider, S. Höfling, Exciton-polariton topological insulator. Nature **562**(7728), 552–556 (2018)
36. S.I. Tsintzos, N.T. Pelekanos, G. Konstantinidis, Z. Hatzopoulos, P.G. Savvidis, A GaAs polariton light-emitting diode operating near room temperature. Nature **453**(7193), 372–375 (2008)
37. C. Schneider, A. Rahimi-Iman, N.Y. Kim, J. Fischer, I.G. Savenko, M. Amthor, M. Lermer, A. Wolf, L. Worschech, V.D. Kulakovskii, I.A. Shelykh, M. Kamp, S. Reitzenstein, A. Forchel, Y. Yamamoto, S. Hoefling, An electrically pumped polariton laser. Nature **497**, 348 (2013)
38. C. Böckler, S. Reitzenstein, C. Kistner, R. Debusmann, A. Löffler, T. Kida, S. Höfling, A. Forchel, L. Grenouillet, J. Claudon, J.M. Gérard, Electrically driven high-Q quantum dot-micropillar cavities. Appl. Phys. Lett. **92**(9), 091107 (2008)
39. J.M. Gérard, D. Barrier, J.Y. Marzin, R. Kuszelewicz, L. Manin, E. Costard, V. Thierry Mieg, T. Rivera, Quantum boxes as active probes for photonic microstructures: The pillar microcavity case. Appl. Phys. Lett. **69**, 449 (1996)
40. S. Reitzenstein, A. Forchel, Quantum dot micropillars. J. Phys. D: Appl. Phys. **43**(3), 033001 (2010)
41. E. Wertz, L. Ferrier, D.D. Solnyshkov, R. Johne, D. Sanvitto, A. Lemaître, I. Sagnes, R. Grousson, A.V. Kavokin, P. Senellart, G. Malpuech, J. Bloch, Spontaneous formation and optical manipulation of extended polariton condensates. Nat. Phys. **6**(11), 860–864 (2010)
42. A.A. Maksimov, I.I. Tartakovskii, E.V. Filatov, S.V. Lobanov, N.A. Gippius, S.G. Tikhodeev, C. Schneider, M. Kamp, S. Maier, S. Höfling, V.D. Kulakovskii, Circularly polarized light emission from chiral spatially-structured planar semiconductor microcavities. Phys. Rev. B **89**(4), 045316 (2014)
43. M. Sugawara. *Plasma Etching: Fundamentals and Applications* (Oxford Science Publications, Oxford, 1998)
44. A. Löffler, J.P. Reithmaier, G. Sek, C. Hofmann, S. Reitzenstein, M. Kamp, A. Forchel, Semiconductor quantum dot microcavity pillars with high-quality factors and enlarged dot dimensions. Appl. Phys. Lett. **86**, 111105 (2005)
45. S. Reitzenstein, C. Hofmann, A. Gorbunov, M. Strauss, S.H. Kwon, C. Schneider, A. Löffler, S. Höfling, M. Kamp, A. Forchel, AlAs/GaAs micropillar cavities with quality factors exceeding 150.000. Appl. Phys. Lett. **90**(25), 251109 (2007)
46. S. Reitzenstein, N. Gregersen, C. Kistner, M. Strauss, C. Schneider, L. Pan, T.R. Nielsen, S. Höfling, J. Mørk, A. Forchel, Oscillatory variations in the Q factors of high quality micropillar cavities. Appl. Phys. Lett. **94**(6), 061108 (2009)
47. A.J. Bennett, R.B. Patel, A.J. Shields, K. Cooper, P. Atkinson, C.A. Nicoll, D.A. Ritchie, Indistinguishable photons from a diode. Appl. Phys. Lett. **92**(19), 193503 (2008)
48. A. Dousse, L. Lanco, J. Suffczyński, E. Semenova, A. Miard, A. Lemaître, I. Sagnes, C. Roblin, J. Bloch, P. Senellart, Controlled light-matter coupling for a single quantum dot embedded in a pillar microcavity using far-field optical lithography. Phys. Rev. Lett. **101**(26), 267404 (2008)
49. J.P. Reithmaier, G. Sek, A. Löffler, C. Hofmann, S. Kuhn, S. Reitzenstein, L.V. Keldysh, V.D. Kulakovskii, T.L. Reinecke, A. Forchel, Strong coupling in a single quantum dot-semiconductor microcavity system. Nature **432**(7014), 197–200 (2004)
50. Giovanna Panzarini and Lucio Claudio Andreani, Quantum theory of exciton polaritons in cylindrical semiconductor microcavities. Phys. Rev. B **60**(24), 16799–16806 (1999)
51. D. Bajoni, Polariton lasers. Hybrid light-matter lasers without inversion. J. Phys. D: Appl. Phys. **45**(31), 313001 (2012)
52. A. Rahimi-Iman, *Nichtlineare Eekte in III/V Quantenlm-Mikroresonatoren: Von dynamischer Bose–Einstein-Kondensation hin zum elektrisch betriebenen Polariton-Laser* (Cuvillier Verlag Göttingen, Göttingen, 2013)

53. D. Bajoni, E. Semenova, A. Lemaître, S. Bouchoule, E. Wertz, P. Senellart, J. Bloch, Polariton light-emitting diode in a GaAs-based microcavity. Phys. Rev. B **77**(11), 113303 (2008)
54. A.A. Khalifa, A.P.D. Love, D.N. Krizhanovskii, M.S. Skolnick, J.S. Roberts, Electroluminescence emission from polariton states in GaAs-based semiconductor microcavities. Appl. Phys. Lett. **92**(6), 061107 (2008)
55. P. Bhattacharya, B. Xiao, A. Das, S. Bhowmick, J. Heo, Solid state electrically injected exciton-polariton laser. Phys. Rev. Lett. **110**(20), 206403 (2013)
56. P. Bhattacharya, T. Frost, S. Deshpande, M.Z. Baten, A. Hazari, A. Das, Room temperature electrically injected polariton laser. Phys. Rev. Lett. **112**, 236802 (2014)
57. K. Winkler, C. Schneider, J. Fischer, A. Rahimi-Iman, M. Amthor, A. Forchel, S. Reitzenstein, S. Höfling, M. Kamp, Electroluminescence from spatially confined exciton polaritons in a textured microcavity. Appl. Phys. Lett. **102**, 041101 (2013)
58. S. Brodbeck, J.-P. Jahn, A. Rahimi-Iman, J. Fischer, M. Amthor, S. Reitzenstein, M. Kamp, C. Schneider, S. Höfling, Room temperature polariton light emitting diode with integrated tunnel junction. Opt. Express **21**(25), 31098–31104 (2013)
59. D. Whittaker, T. Fisher, A. Afshar, M. Skolnick, P. Kinsler, J. Roberts, G. Hill, M. Pate, Vacuum rabi splitting in semiconductor microcavities with applied electric and magnetic fields. Il Nuovo Cimento D **17**(11), 1781–1786 (1995)
60. D. Bajoni, E. Semenova, A. Lemaître, S. Bouchoule, E. Wertz, P. Senellart, S. Barbay, R. Kuszelewicz, J. Bloch, Optical bistability in a GaAs-based polariton diode. Phys. Rev. Lett. **101**(26), 266402 (2008)
61. M. Amthor, S. Weißenseel, J. Fischer, M. Kamp, C. Schneider, S. Höfling, Electro-optical switching between polariton and cavity lasing in an InGaAs quantum well microcavity. Opt. Express **22**(25), 31146–31153 (2014)
62. C. Kistner, T. Heindel, C. Schneider, A. Rahimi-Iman, S. Reitzenstein, S. Höfling, A. Forchel, Demonstration of strong coupling via electro-optical tuning in high-quality QD-micropillar systems. Opt. Express **16**, 15006–15012 (2008)
63. H. Deng, G. Weihs, C. Santori, J. Bloch, Y. Yamamoto, Condensation of semiconductor microcavity exciton polaritons. Science **298**(5591), 199–202 (2002)
64. D. Sanvitto, S. Kéna-Cohen, The road towards polaritonic device. Nat. Mater. **15**, 1061 (2016)
65. C.F. Klingshirn, *Semiconductor Optics* (Springer, Berlin, 2012)
66. M. Litinskaya, P. Reineker, V.M. Agranovich, Fast polariton relaxation in strongly coupled organic microcavities. J. Lumin. **110**(4), 364–372 (2004)
67. D.M. Coles, P. Michetti, C. Clark, W.C. Tsoi, A.M. Adawi, J.S. Kim, D.G. Lidzey, Vibrationally assisted polariton-relaxation processes in strongly coupled organic-semiconductor microcavities. Adv. Funct. Mater. **21**(19), 3691–3696 (2011)
68. L. Mazza, L. Fontanesi, G.C. La Rocca, Organic-based microcavities with vibronic progressions: Photoluminescence. Phys. Rev. B **80**(23) (2009)
69. M. Richard, J. Kasprzak, R. Romestain, R. André, L.S. Dang, Spontaneous coherent phase transition of polaritons in CdTe microcavities. Phys. Rev. Lett. **94**(18), 187401 (2005)
70. M. Richard, J. Kasprzak, R. André, R. Romestain, L.S. Dang, G. Malpuech, A. Kavokin, Experimental evidence for nonequilibrium Bose condensation of exciton polaritons. Phys. Rev. B **72**(20), 201301 (2005)
71. J. Kasprzak, M. Richard, S. Kundermann, A. Baas, P. Jeambrun, J.M.J. Keeling, F.M. Marchetti, M.H. Szymańska, R. André, J.L. Staehli, V. Savona, P.B. Littlewood, B. Deveaud, L.S. Dang, Bose-Einstein condensation of exciton polaritons. Nature **443**(7110), 409–414 (2006)
72. R. André, D. Heger, L.S. Dang, Y.M. D'Aubigné, Spectroscopy of polaritons in CdTe-based microcavities. J. Cryst. Growth **184–185**, 758–762 (1998)
73. F. Boeuf, R. André, R. Romestain, L.S. Dang, E. Péronne, J.F. Lampin, D. Hulin, A. Alexandrou, Evidence of polariton stimulation in semiconductor microcavities. Phys. Rev. B **62**(4), R2279–R2282 (2000)
74. R. André, F. Boeuf, D. Heger, L.S. Dang, R. Romestain, J. Bleuse, M. Müller, Cavity-polariton effects in II-VI microcavities. Acta Phys. Pol. A **96**(5), 511–524 (1999)

75. S. Christopoulos, G.B.H. von Högersthal, A.J.D. Grundy, P.G. Lagoudakis, A.V. Kavokin, J.J. Baumberg, G. Christmann, R. Butté, E. Feltin, J.-F. Carlin, N. Grandjean, Room-temperature polariton lasing in semiconductor microcavities. Phys. Rev. Lett. **98**(12), 126405 (2007)
76. G. Christmann, R. Butté, E. Feltin, J.-F. Carlin, N. Grandjean, Room temperature polariton lasing in a GaN/AlGaN multiple quantum well microcavity. Appl. Phys. Lett. **93**(5), 051102 (2008)
77. A. Das, J. Heo, M. Jankowski, W. Guo, L. Zhang, H. Deng, P. Bhattacharya, Room temperature ultralow threshold GaN nanowire polariton laser. Phys. Rev. Lett. **107**(6), 066405 (2011)
78. K.S. Daskalakis, P.S. Eldridge, G. Christmann, E. Trichas, R. Murray, E. Iliopoulos, E. Monroy, N.T. Pelekanos, J.J. Baumberg, P.G. Savvidis, All-dielectric GaN microcavity: Strong coupling and lasing at room temperature. Appl. Phys. Lett. **102**(10), 101113 (2013)
79. R. Butté, G. Christmann, E. Feltin, J.F. Carlin, M. Mosca, M. Ilegems, N. Grandjean, Room-temperature polariton luminescence from a bulk GaN microcavity. Phys. Rev. B **73**(3), 033315 (2006)
80. M. Leroux, N. Grandjean, M. Laügt, J. Massies, B. Gil, P. Lefebvre, Quantum confined Stark effect due to built-in internal polarization fields in (Al, Ga)N/GaN quantum wells. Phys. Rev. B **58**(20), R13371–R13374 (1998)
81. G. Christmann, R. Butté, E. Feltin, A. Mouti, P.A. Stadelmann, A. Castiglia, J.F. Carlin, N. Grandjean, Large vacuum Rabi splitting in a multiple quantum well GaN-based microcavity in the strong-coupling regime. Phys. Rev. B **77**(8), 085310 (2008)
82. C. Sturm, H. Hilmer, R. Schmidt-Grund, M. Grundmann, Exciton-polaritons in a ZnO-based microcavity: polarization dependence and nonlinear occupation. New J. Phys. **13**(3), 033014 (2011)
83. L. Sun, S. Sun, H. Dong, W. Xie, M. Richard, L. Zhou, L.S. Dang, X. Shen, Z. Chen, Room temperature one-dimensional polariton condensate in a ZnO microwire. arXiv:1007.4686v1 (unpublished)
84. L. Tien-Chang, Y.-Y. Lai, Y.-P. Lan, S.-W. Huang, J.-R. Chen, W. Yung-Chi, W.-F. Hsieh, H. Deng, Room temperature polariton lasing vs photon lasing in a ZnO-based hybrid microcavity. Opt. Express **20**(5), 5530 (2012)
85. F. Li, L. Orosz, O. Kamoun, S. Bouchoule, C. Brimont, P. Disseix, T. Guillet, X. Lafosse, M. Leroux, J. Leymarie, M. Mexis, M. Mihailovic, G. Patriarche, F. Réveret, D. Solnyshkov, J. Zuniga-Perez, G. Malpuech, From excitonic to photonic polariton condensate in a ZnO-based microcavity. Phys. Rev. Lett. **110**(19), 196406 (2013)
86. W. Xie, H. Dong, S. Zhang, L. Sun, W. Zhou, Y. Ling, L. Jian, X. Shen, Z. Chen, Room-temperature polariton parametric scattering driven by a one-dimensional polariton condensate. Phys. Rev. Lett. **108**(16), 166401 (2012)
87. Q. Duan, X. Dan, W. Liu, L. Jian, L. Zhang, J. Wang, Y. Wang, G. Jie, H. Tao, W. Xie, X. Shen, Z. Chen, Polariton lasing of quasi-whispering gallery modes in a ZnO microwire. Appl. Phys. Lett. **103**(2), 022103 (2013)
88. T. Guillet, M. Mexis, J. Levrat, G. Rossbach, C. Brimont, T. Bretagnon, B. Gil, R. Butté, N. Grandjean, L. Orosz, F. Réveret, J. Leymarie, J. Zúñiga-Pérez, M. Leroux, F. Semond, S. Bouchoule, Polariton lasing in a hybrid bulk ZnO microcavity. Appl. Phys. Lett. **99**(16), 161104 (2011)
89. H. Franke, C. Sturm, R. Schmidt-Grund, G. Wagner, M. Grundmann, Ballistic propagation of exciton-polariton condensates in a ZnO-based microcavity. New J. Phys. **14**, 013037 (2012)
90. C. Sturm, H. Hilmer, R. Schmidt-Grund, M. Grundmann, Observation of strong exciton-photon coupling at temperatures up to 410 k. New J. Phys. **11**(7), 073044 (2009)
91. H. Hilmer, C. Sturm, R. Schmidt-Grund, B. Rheinländer, M. Grundmann, Observation of strong light-matter coupling by spectroscopic ellipsometry. Superlattices Microstruct. **47**(1), 19–23 (2010)
92. M. Litinskaya, Exciton polariton kinematic interaction in crystalline organic microcavities. Phys. Rev. B **77**(15), 155325 (2008)
93. K.S. Daskalakis, S.A. Maier, R. Murray, S. Kéna-Cohen, Nonlinear interactions in an organic polariton condensate. Nat. Mater. **13**(3), 271–278 (2014)

94. S. Kena-Cohen, S.R. Forrest, Room-temperature polariton lasing in an organic single-crystal microcavity. Nat. Phot. **4**(6), 371–375 (2010)
95. D.G. Lidzey, D.D.C. Bradley, M.S. Skolnick, T. Virgili, S. Walker, D.M. Whittaker, Strong exciton-photon coupling in an organic semiconductor microcavity. Nature **395**(6697), 53–55 (1998)
96. S. Kéna-Cohen, M. Davanço, S.R. Forrest, Strong exciton-photon coupling in an organic single crystal microcavity. Phys. Rev. Lett. **101**(11), 116401 (2008)
97. K.S. Daskalakis, S.A. Maier, S. Kéna-Cohen, Spatial coherence and stability in a disordered organic polariton condensate. Phys. Rev. Lett. **115**, 035301 (2015)
98. J.D. Plumhof, T. Stöferle, L. Mai, U. Scherf, R.F. Mahrt, Room-temperature Bose–Einstein condensation of cavity exciton-polaritons in a polymer. Nat. Mater. **13**, 247 (2014). Published Online Dec 2013
99. Q. Zhang, S.T. Ha, X. Liu, T.C. Sum, Q. Xiong, Room-temperature near-infrared high-Q perovskite whispering-gallery planar nanolasers. Nano Lett. **14**(10), 5995–6001 (2014)
100. Q. Zhang, R. Su, X. Liu, J. Xing, T.C. Sum, Q. Xiong, High-quality whispering-gallery-mode lasing from cesium lead halide perovskite nanoplatelets. Adv. Funct. Mater. **26**(34), 6238–6245 (2016)
101. T. Fujita, Y. Sato, T. Kuitani, T. Ishihara, Tunable polariton absorption of distributed feedback microcavities at room temperature. Phys. Rev. B **57**(19), 12428–12434 (1998)
102. G. Lanty, A. Bréhier, R. Parashkov, J.S. Lauret, E. Deleporte, Strong exciton-photon coupling at room temperature in microcavities containing two-dimensional layered perovskite compounds. New J. Phys. **10**(6), 065007 (2008)
103. R. Su, C. Diederichs, J. Wang, T.C.H. Liew, J. Zhao, S. Liu, W. Xu, Z. Chen, Q. Xiong, Room-temperature polariton lasing in all-inorganic perovskite nanoplatelets. Nano Lett. **17**(6), 3982–3988 (2017)
104. K. F. Mak and J. Shan, Photonics and optoelectronics of 2D semiconductor transition metal dichalcogenides. Nat. Photon. **10**(4), 216–226 (2016)
105. S. Dufferwiel, S. Schwarz, F. Withers, A.A.P. Trichet, F. Li, M. Sich, O. Del Pozo-Zamudio, C. Clark, A. Nalitov, D.D. Solnyshkov, G. Malpuech, K.S. Novoselov, J.M. Smith, M.S. Skolnick, D.N. Krizhanovskii, A.I. Tartakovskii, Exciton-polaritons in van der Waals heterostructures embedded in tunable microcavities. Nat. Commun. **6**, 8579 (2015)
106. Y. Yu, Y. Yu, C. Xu, Y.Q. Cai, L. Su, Y. Zhang, Y.W. Zhang, K. Gundogdu, L. Cao, Engineering substrate interactions for high luminescence efficiency of transition-metal dichalcogenide monolayers. Adv. Funct. Mater. **26**(26), 4733–4739 (2016)
107. A.V. Stier, N.P. Wilson, G. Clark, X. Xiaodong, S.A. Crooker, Probing the influence of dielectric environment on excitons in monolayer WSe2: Insight from high magnetic fields. Nano Lett. **16**(11), 7054–7060 (2016)
108. S. Lippert, L.M. Schneider, D. Renaud, K.N. Kang, O. Ajayi, J. Kuhnert, M.-U. Halbich, O.M. Abdulmunem, X. Lin, K. Hassoon, S. Edalati-Boostan, Y.D. Kim, W. Heimbrodt, E.-H. Yang, J.C. Hone, A. Rahimi-Iman, Influence of the substrate material on the optical properties of tungsten diselenide monolayers. 2D Mater. **4**(2), 025045 (2017)
109. L.M. Schneider, S. Lippert, J. Kuhnert, O. Ajayi, D. Renaud, S. Firoozabadi, Q. Ngo, R. Guo, Y.D. Kim, W. Heimbrodt, J.C. Hone, A. Rahimi-Iman, The influence of the environment on monolayer tungsten diselenide photoluminescence. Nano-Struct. Nano-Objects **15**, 84–97 (2018)
110. A.K. Geim, I.V. Grigorieva, Van der waals heterostructures. Nature **499**, 419 (2013)
111. K.S. Novoselov, A. Mishchenko, A. Carvalho, A.H. Castro Neto, 2d materials and van der waals heterostructures. Science **353**(6298), aac9439 (2016)
112. K. Liu, L. Zhang, T. Cao, C. Jin, D. Qiu, Q. Zhou, A. Zettl, P. Yang, S.G. Louie, F. Wang, Evolution of interlayer coupling in twisted molybdenum disulfide bilayers. Nat. Commun. **5**(1), 4966 (2014)
113. H. Zhu, J. Wang, Z. Gong, Y.D. Kim, J. Hone, X.-Y. Zhu, Interfacial charge transfer circumventing momentum mismatch at two-dimensional van der waals heterojunctions. Nano Lett. **17**(6), 3591–3598 (2017)

114. Y. Cao, V. Fatemi, A. Demir, S. Fang, S.L. Tomarken, J.Y. Luo, J.D. Sanchez-Yamagishi, K. Watanabe, T. Taniguchi, E. Kaxiras, R.C. Ashoori, P. Jarillo-Herrero, Correlated insulator behaviour at half-filling in magic-angle graphene superlattices. Nature **556**, 80 (2018)
115. L.M. Schneider, J. Kuhnert, S. Schmitt, W. Heimbrodt, U. Huttner, L. Meckbach, T. Stroucken, S.W. Koch, S. Fu, X. Wang, K. Kang, E.-H. Yang, A. Rahimi-Iman, Spin-layer and spin-valley locking in CVD-grown AA'- and AB-stacked tungsten-disulfide bilayers. J. Phys. Chem. C **123**(35), 21813–21821 (2019)
116. M. Selig, G. Berghäuser, A. Raja, P. Nagler, C. Schüller, T.F. Heinz, T. Korn, A. Chernikov, E. Malic, A. Knorr, Excitonic linewidth and coherence lifetime in monolayer transition metal dichalcogenides. Nat. Commun. **7**(1), 13279 (2016)
117. O.A. Ajayi, J.V. Ardelean, G.D. Shepard, J. Wang, A. Antony, T. Taniguchi, K. Watanabe, T.F. Heinz, S. Strauf, X.Y. Zhu, J.C. Hone, Approaching the intrinsic photoluminescence linewidth in transition metal dichalcogenide monolayers. 2D Mater. **4**(3), 031011 (2017)
118. F. Cadiz, E. Courtade, C. Robert, G. Wang, Y. Shen, H. Cai, T. Taniguchi, K. Watanabe, H. Carrere, D. Lagarde, M. Manca, T. Amand, P. Renucci, S. Tongay, X. Marie, B. Urbaszek, Excitonic linewidth approaching the homogeneous limit in MoS_2-based van der Waals heterostructures. Phys. Rev. X **7**(2), 021026 (2017)
119. J. Wierzbowski, J. Klein, F. Sigger, C. Straubinger, M. Kremser, T. Taniguchi, K. Watanabe, U. Wurstbauer, A.W. Holleitner, M. Kaniber, K. Müller, J.J. Finley, Direct exciton emission from atomically thin transition metal dichalcogenide heterostructures near the lifetime limit. Sci. Rep. **7**(1), 12383 (2017)
120. D. Edelberg, D. Rhodes, A. Kerelsky, B. Kim, J. Wang, A. Zangiabadi, C. Kim, A. Abhinandan, J. Ardelean, M. Scully, D. Scullion, L. Embon, I. Zhang, R. Zu, E.J.G. Santos, L. Balicas, C. Marianetti, K. Barmak, X.Y. Zhu, J. Hone, A.N. Pasupathy, Hundredfold enhancement of light emission via defect control in monolayer transition-metal dichalcogenides (2018), arXiv:1805.00127
121. G. Wang, A. Chernikov, M.M. Glazov, T.F. Heinz, X. Marie, T. Amand, B. Urbaszek, Excitons in atomically thin transition metal dichalcogenides. Rev. Mod. Phys. **90**(2), 21001 (2018)
122. M.M. Fogler, L.V. Butov, K.S. Novoselov, High-temperature superfluidity with indirect excitons in van der Waals heterostructures. Nat. Commun. **5**, 4555 (2014)
123. X. Liu, T. Galfsky, Z. Sun, F. Xia, E.C. Lin, Y.H. Lee, S. Kéna-Cohen, V.M. Menon, Strong light–matter coupling in two-dimensional atomic crystals. Nat. Photon. **9**(1), 30–34 (2014)
124. L.C. Flatten, Z. He, D.M. Coles, A.A.P. Trichet, A.W. Powell, R.A. Taylor, J.H. Warner, J.M. Smith, Room-temperature exciton-polaritons with two-dimensional WS2. Sci. Rep. **6**, 33134 (2016)
125. N. Lundt, S. Klembt, E. Cherotchenko, S. Betzold, O. Iff, A.V. Nalitov, M. Klaas, C.P. Dietrich, A.V. Kavokin, S. Höfling, C. Schneider, Room-temperature Tamm-plasmon exciton-polaritons with a WSe2 monolayer. Nat. Commun. **7**, 13328 (2016)
126. S. Wang, S. Li, T. Chervy, A. Shalabney, S. Azzini, E. Orgiu, J.A Hutchison, C. Genet, P. Samorì, T.W. Ebbesen, Coherent coupling of WS2 monolayers with metallic photonic nanostructures at room temperature. Nano Lett. **16**(7), 4368–4374 (2016)
127. Y.J. Chen, J.D. Cain, T.K. Stanev, V.P. Dravid, N.P. Stern, Valley-polarized exciton-polaritons in a monolayer semiconductor. Nat. Photon. **11**(7), 431–435 (2017)
128. L.C. Flatten, D.M. Coles, Z. He, D.G. Lidzey, R.A. Taylor, J.H. Warner, J.M. Smith, Electrically tunable organic-inorganic hybrid polaritons with monolayer WS2. Nat. Commun. **8**, 14097 (2017)
129. M. Slootsky, X. Liu, V.M. Menon, S.R. Forrest, Room temperature Frenkel-Wannier-Mott hybridization of degenerate excitons in a strongly coupled microcavity. Phys. Rev. Lett. **112**(7), 07640 (2014)
130. K.F. Mak, D. Xiao, J. Shan, Light–valley interactions in 2D semiconductors. Nat. Photon. **12**(8), 451–460 (2018)
131. S. Dufferwiel, T.P. Lyons, D.D. Solnyshkov, A.A.P. Trichet, F. Withers, S. Schwarz, G. Malpuech, J.M. Smith, K.S. Novoselov, M.S. Skolnick, D.N. Krizhanovskii, A.I. Tartakovskii, Valley-addressable polaritons in atomically thin semiconductors. Nat. Photon. **11**(8), 497–501 (2017)

132. Z. Sun, G. Jie, A. Ghazaryan, Z. Shotan, C.R. Considine, M. Dollar, B. Chakraborty, X. Liu, P. Ghaemi, S. Kéna-Cohen, V.M. Menon, Optical control of roomerature valley polaritons. Nat. Photon. **11**(8), 491–496 (2017)
133. N. Lundt, C. Schneider, P. Nagler, A. Nalitov, S. Klembt, M. Wurdack, S. Stoll, T.H. Harder, S. Betzold, V. Baumann, A.V. Kavokin, C. Schüller, T. Korn, S. Höfling, Valley polarized relaxation and upconversion luminescence from Tamm-plasmon trion-polaritons with a MoSe2 monolayer. 2D Mater. **4**(2), 025096 (2017)
134. N. Lundt, S. Stoll, P. Nagler, A. Nalitov, S. Klembt, S. Betzold, J. Goddard, E. Frieling, A.V. Kavokin, C. Schüller, T. Korn, S. Höfling, C. Schneider, Observation of macroscopic valley-polarized monolayer exciton-polaritons at room temperature. Phys. Rev. B **96**(24), 241403 (2017)

Chapter 7
Spectroscopy Techniques for Polariton Research

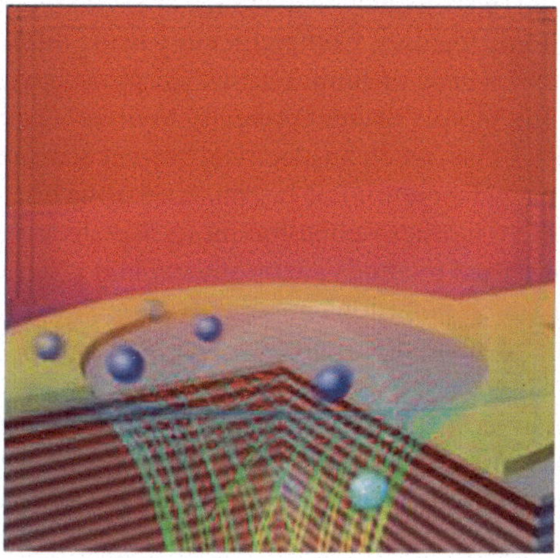

Abstract The emission from the exciton–polariton state that is formed in a strongly-coupled QW-microcavity system is correlated to the composite quasi-particles' state. Thus, a direct experimental access to the properties of polariton gases in solids is provided by means of optical spectroscopy. In fact, the light originating from the decay process of the polaritons inside a cavity is emitted under that angle relative to the cavity axis that corresponds to the in-plane momentum component of the cavity mode coupled to the QW exciton mode due to momentum conservation. This renders angle-resolved spectroscopy an indispensable tool for gathering information about the polaritonic system, which is strongly represented by its energy–momentum dispersion and the anti-crossing behaviour of the coupled optical resonances. Therefore, one can conveniently extract characteristics of the system by means of spectroscopy, which give insight into occupation numbers, effective masses and statistical distributions of particles in the system. In addition, time-resolved spectroscopy increases the ability to characterize polariton gases drastically, as the available and established methods shed light on those systems' dynamics on the most-relevant time scales. Many of the essential techniques will be briefly summarized here and the relevant terminology introduced.

A. Rahimi-Iman, *Polariton Physics*, Springer Series in Optical Sciences 229,
https://doi.org/10.1007/978-3-030-39333-5_7

7.1 Optical Spectroscopy

Optical spectroscopy is a widely used tool for probing the optical properties of matter, particularly with regard to resonances in the electronic system relevant for the considered spectral range. Indeed, it is at the core of light–matter interaction studies and utilizes the response of an electronic system, such as electrons in a solid, on the optical light field irradiated upon that matter. For instance, both, reflection and transmission of spectrally-broad light can provide an overview on the energies of possible optical transitions, or in more general terms, on the dielectric properties of an optical medium. Particularly, semiconductors are predestined for such optical investigations due to their energy (band) gaps located in the range of UV to IR, and due to their typically pronounced excitonic features. Complementary to the information obtained via absorption studies, fluorescence or luminescence spectroscopy allows one to study emitter properties. Is the sample excited optically to drive its emission, we refer to photoluminescence (PL), else, when an electrical current scheme is used, to electroluminescence (EL). Spatial resolution can be enhanced using microscopy, resulting in the use of the prefix micro in micro-photoluminescence/electroluminescence (μPL, μEL). These commonly used methods are briefly explained here. It should be noted that most systems, in which polaritons are clearly observed and in which condensation phenomena can be studied, are investigated at cryogenic temperatures. This deems additional efforts necessary to keep the sample in an evacuated environment at around liquid Helium temperatures. Special acquisition modes such as the real-space and Fourier-space resolving imaging/spectroscopy as well as time-resolved spectroscopy techniques will be addressed later in the chapter.

7.1.1 Reflection and Transmission Measurements

By looking at the back-reflected portion (R) of the incident light on a sample, one can already learn much about the optical absorption, although to fully deduce the absorption spectrum (A), a simultaneously acquired transmission spectrum (T) for the same spot and same irradiance is needed, as

$$A = 1 - T - R. \tag{7.1}$$

Reflection spectra (R) become particularly useful when it is about the characterization of a planar microcavity's stop-band and resonances formed by the opposing DBRs. Naturally, the stop-band of an individual DBR with its typical fringe pattern to its sides can be probed similarly, but a cavity-resonance dip or a number of resonance dips can only be obtained for the complete microcavity. For high-quality microresonators, the cavity resonances may become too sharp to be resolved in the spectrum or will be hardly seen. In addition, high-reflectance mirrors make it difficult to probe the interior of a structure as light in the range of the stop-band can hardly penetrate it.

For common QW-microcavities in the regime of strong coupling between an exciton mode and a cavity resonance, two distinct modes should be seen around the centre wavelength of the resonator, or in the vicinity of the uncoupled exciton and cavity mode's energetic position. Similar observation is expected for transmission measurements, provided that the substrate of the microcavity is not opaque in the relevant spectral region. In some material systems, the substrate must be etched away to be able to record transmission spectra.

A double dip (or peak in transmission) with clear anti-crossing behaviour is very indicative of the strong coupling regime. To see the expected normal-mode splitting occurring when the oscillators are coupled, the optical mode can be for instance tuned through the exciton resonance via changes of the incidence angle, the temperature or electric/magnetic fields. Please note that each tuning nob may have its advantages and disadvantages, and is neither equivalently nor equally addressing the observable modes. For instance, temperature typically shifts the exciton more than the cavity mode, whereas the angle practically only changes the cavity resonance energy, and a typical DC (direct current) electric field solely manipulates the exciton. Moreover, temperature changes may be detrimental to the observation of condensation phenomena, not to mention their impact on exciton dephasing/linewidth, binding energy/oscillator strength and thereby the coupling strength. Similarly, external fields change exciton properties, e.g. due to an E-field induced quantum-confined Stark effect (QCSE) (see Sect. 9.1.1) or a B-field magnetic confinement (see Sect. 9.1.3). These property changes, beyond a mere influence on the energy position (see QCSE-induced transition energy shift or diamagnetic shift), can arise for instance by an altered electron–hole wave-functions overlap in external fields, which can lead to a modified oscillator strength, or a spin-degeneracy lifting (see Zeeman spin splitting).

Commonly, basic information about a polariton system is obtained via aforementioned simple and effective optical characterization tools, amended by complementary luminescence experiments. Usually, acquisition of different spectral data is done within the same setup, which employs the light sources (e.g. thermal white light, pump lasers), all relevant optics for beam guidance, projection, imaging and signal acquisition (e.g. achromatic lenses, mirrors, beam splitters), and the dispersive element (e.g. a spectrometer such as a grating-based monochromator) with detector system (e.g. photodetector, charge-coupled device, i.e. CCD camera, Streak camera system). The literature has come up with a vast amount of setup descriptions and setup variations that serve specific measurement purposes, which the interested reader is encouraged to analyze and compare. Nevertheless, the following simplified example provided here for the sake of completeness can be regarded as a typical platform for measurements on microcavity systems (see Fig. 7.1). The optics and the beam alignment are drawn in a simplified and summarizing manner. Different angle-resolving measurement schemes used to probe the polaritons' dispersion are explained later.

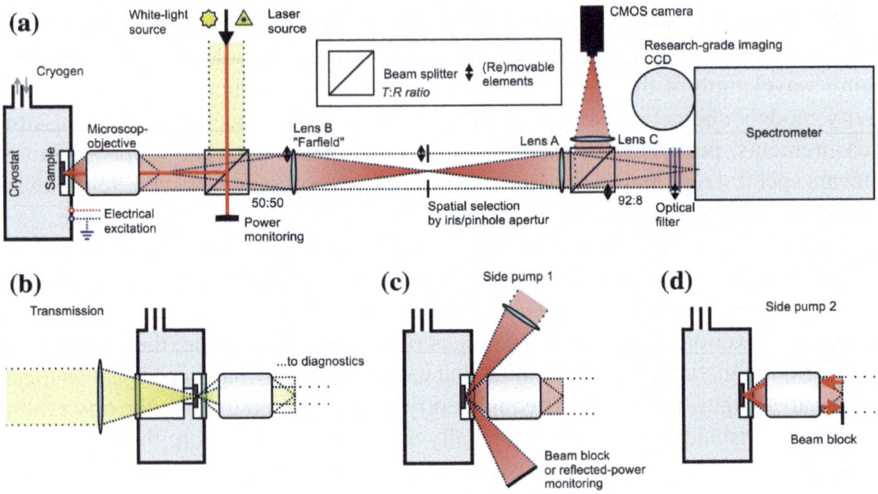

Fig. 7.1 **a** Schematic drawing of a polariton spectroscopy setup showing in-coupling of light via optics and detection of signal for spectroscopy by means of a monochromator and an attached research-grade imaging CCD camera. The optical axis links the sample plane via relay lenses with the imaging plane of the spectrometer at the entrance slit position. White-light can be used to monitor the sample position or to record reflectivity spectra. For photoluminescence studies, different types of lasers can be employed for optical excitation. **b** Example of a white-light transmission setup. Similarly, rear-side optical pumping of transparent samples can be performed with such geometry. Two different optical side-pump configurations are sketched in **c** and **d**

7.1.2 Micro-Photoluminescence Experiments

Light emitted by QW-microcavities carries essential information about the system's properties. Photoluminescence studies are widely used to investigate semiconductor quantum structures, such as quantum wells and quantum dots. It provides access to the output intensities and their spectral/spatial/angular distribution, the emission linewidth—for homogeneously-broadened emitters also directly the lifetime—and indicates the quality of the structure. Furthermore, time-resolving techniques are used to obtain transient luminescence signal for decay-time (lifetime) evaluations.

Charge-carrier excitation of matter in μPL experiments takes place via optical pumping (e.g. with a continuous-wave or pulsed laser) either through the same objective, which is employed for signal collection, or via an alternative optical path (e.g. from behind, from the side, see Fig. 7.1). Here, with regard to the excitation energy, one distinguishes between resonant, quasi-resonant and off-resonant excitation.

Excitation above band is not selective and typically floods the system with hot carriers that need to relax to their ground state. Getting rid of excess energy involves numerous phonon-emission steps and electronic-scattering processes. For a discussion of the dynamics after excitation, please see Sect. 3.2. The main advantages of off-resonant excitation, i.e. above-band excitation, are the less stringent requirements set for the pump source, the broadly-occupied energy states of the system and the easy-to-achieve luminescence measurements. Moreover, polaritons formed originate from incoherent relaxed excitons that are not correlated anymore to the pump light. However, the temperature distribution in the excited gas/sample and spectral-linewidth results may not represent the ideal scenario when polariton formation takes place in the presence of many uncoupled hot QW excitons as well as a considerable fraction of electron–hole plasma. Excess-energy pumping is also not ideal when a significant demand for effective/continuous heat dissipation exists.

Quasi-resonant excitation refers to the case when the excitons are not resonantly pumped into their ground state, but the energy of the input light is not sufficient to overcome the electronic band gap. This allows addressing individual polariton branches and creating populations at certain momenta or energies, while preventing a strong presence of hot carriers. To inject polaritons above their respective branches' ground state, or excitons with non-zero momentum, the incidence angle on the sample with respect to the normal can be varied. However, side-pumping requires optical access from the side (e.g. in terms of the working distance of the microscope objective, the window of a cryostat, the availability of a proper sample surface) when irradiating the sample under an incidence angle. In certain cases, even using the microscope objective can lead to excitation under an angle when the in-coupled laser beam is displaced from the optical axis. And since angle corresponds to in-plane momentum, such variations are of great interest for resonant excitation schemes that are used for the injection of polaritons/excitons with finite momentum.

If polaritons were directly excited in their ground-state of the respective dispersion (LP/UP), resonant excitation took place. Yet, quasi-resonant and resonant excitation imprint the properties of the pump light onto the polariton system, forming coherent excitons that feature the excitation light's polarization and are not uncorrelated. However, for a number of experiments, it is desirable to inject polaritons resonantly, as relevant studies in the literature outline.

7.1.3 Micro-Electroluminescence Studies

In order to study electrically-pumped polariton devices, various structures and contacting schemes have been proposed in the research community. The general aim is to enable condensation in a polariton gas, which is electrically driven. In such a system, light is generated due to current injection after charge-carrier recombination similar to μPL, but with an electrical bias applied, thus leading to electroluminescence (μEL). Electrical excitation remains a non-resonant pumping scheme, unless very special structures are achieved for selective excitation of individual quantum wells (e.g. potentially via built-in Esaki tunnelling diodes [1]). In p-i-n diodes, polariton electroluminescence occurs typically under flat-band conditions in forward bias, i.e. above the characteristic diode voltage of a given device. Signal is collected in μEL experiments in a similar fashion than in μPL experiments, and, usually, the same setup is employed together with contacted samples inside a cryostat with proper wiring.

Electroluminescence studies are typically performed in the continuous-wave (cw) mode with a DC bias applied, which injects constantly charge carriers into the structure and compensates for polariton decay through spontaneous emission. This operation mode bears the advantage that the studied system is usually thermalized and quasi-equilibrium conditions can be assumed. However, a high background of free charge carriers can spoil the properties of exciton–polaritons, reduce the coupling strength and the lifetime. In addition, heat dissipation can become a problem and the outflow of current-induced heat is limited to the structure's cross section and heat conductivity. Thus, high-current operation generally—even at cryogenic temperatures—poses a risk and can lead to structure and material deterioration which represents the loss of the device due to the irreversible nature of the degradation process. For VCSEL-like structures, thermal management, conductivity and sample quality need to be targeted to guarantee operation at elevated current densities. In DC mode, the diode (LED) bias can be regulated using a pre-resistor (a defined load) in a circuit with serial device connection, also allowing for the sample-current determination.

Electrically-contacted structures can also be operated under a high-frequency (HF) pulsed current in a quasi-DC mode, in order to avoid heating at high current densities. Hundreds of nanoseconds rectangular pulses generated by an HF-voltage source with repetition rates in the kHz range provide conditions similar to DC operation with temporal gaps long enough to allow excess power in the form of heat to be dissipated, keeping the sample cool, while injecting temporally-limited plateau-like current densities that mimic DC excitation. HF-pulsed excitation generally requires impedance matching of the input circuit.

Experiments with polariton diodes have shown that quasi-DC pumping can be well accomplished [2]. Owing to the picoseconds-short polariton lifetime and the ultrafast charge-carrier dynamics in semiconductors, $\geq$100 ns HF rectangular current pulses can be regarded as quasi-DC current injection into a polariton device. Indeed, the performance between quasi-DC [2, 3] and DC [3, 4] operation is well comparable.

7.2 Imaging and Real-Space Spectroscopy

Given the use of a set of lenses for light collection and focussing, it is natural to implement an imaging scheme into a spectroscopy setup resembling that of a regular microscope.

The typically self-built optical microscope comprises a microscope objective (MO) and a lens A in the primary optical path between sample and detector (screen). The MO is set to collect and collimate the light reflected/emitted from a sample, while lens A acts as focussing lens placed in the collimated beam path. The focal spot can be placed on a camera, monochromator entrance slit or other detectors. With MO and lens A, the sample surface (or the object in the focus of the MO) will be projected with certain magnification depending on the optics onto the backfocal plane of lens A. This plane is referred to as near-field (NF) or real-space plane. The projection of the NF plane can be useful for many measurements, where spatial information and signal strength matters.

Spectroscopy in the NF projection mode is very sensitive to positioning and allows one to investigate strongly-localized emitters with weak signals and without any momentum-space dispersion characteristics. These can be quantum dots, but also microresonators with strong 3D confinement and subsequently discrete mode energies. Owing to the (magnified) projection of the NF plane onto the spectrometer, a lot of signal from a microscopic area can be coupled in effectively. However, the lack of angle sensitivity leads to a k-space integrated intensity acquisition. If NF spectra are recorded as integrated line spectra (i.e. integrating over one axis of the 2D detector chip), spectra only represent the overall emission for the in-coupled projected sample area. Thus, spatial information is disregarded and lost. Nevertheless, it is the most common acquisition mode in μPL (or μEL) spectroscopy. Particularly when the sample excitation spot is also limited to a few micrometers, one typically obtains a projected spot size of a few pixels on the monochromator's CCD, unless a very high magnification is used.

In order to explain the advantages and the use of NF projection, two main examples are provided in the following.

7.2.1 Sample Imaging for Position Monitoring or Interferometry

The NF projection onto an imaging camera (e.g. webcam, research-grade CCD camera, CMOS camera), can give direct access to sample pictures as obtained under a microscope. This is particularly useful for in situ sample monitoring and position control during spectroscopic measurements. Imaging of the sample's surface can also be used to map the luminescence intensity in the field of view. In addition, magnified images of microscopic samples can also be used in combination with a Michelson interferometer, in order to obtain phase diagrams of the sample's emission from its surface facet (see Sect. 8.2.3). Such diagrams can be used to evaluate the spatial correlations of an emitting area such as polariton condensates [5, 6] or to observe vortices in polariton fluids [7–11] (see Sect. 8.3). Similarly, the NF plane can be projected onto a double slit to perform Young's double-slit experiments with polariton condensates [12, 13] (see Sect. 8.2.3).

Due to the projection of the NF plane onto the backfocal plane of aforementioned lens A, an aperture or pinhole in that plane can serve as a means of spatial selection if the setup is extended by additional projection optics that re-project the spatially-filtered aperture plane (NF plane) onto the detector plane of the respective acquisition instrument. Thus, many imaging spectroscopy setups make use of the confocal-microscopy configuration with an aperture in the joint focal plane of consecutive relay optics (e.g. between lenses A and B in Fig. 7.1a).

Indeed, the monochromator entrance slit with NF plane projected upon acts as a 1D spatial filter, which allows transmission of light in one orientation and truncates the image in the other direction perpendicular to that. With open slit, an imaging

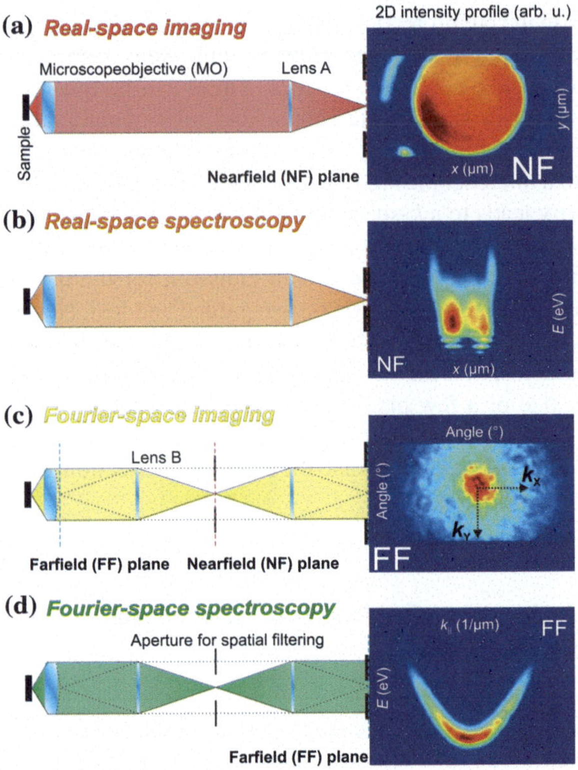

Fig. 7.2 **a** Schematic drawing of a microscope configuration for signal acquisition from the sample's real-space projection with an imaging spectrometer. The near-field (NF) image is projected onto the open entrance slit which allows recordings of intensity profiles with spatial resolution. **b** Analogue detection scheme with dispersive element in the spectrometer, and nearly-closed entrance slit. Spectra are given as intensity profiles with NF resolution in one spatial direction. **c** Similar to the NF acquisition scheme, a far-field (FF) acquisition scheme enables recordings of intensity profiles with momentum/angle resolution. **d** With dispersive element, spectra are given as intensity profiles with Fourier-space resolution in one momentum direction. Adapted from [3]. Courtesy of the author

monochromator's CCD can be like-wise used for μPL/μEL imaging, as an example false-colour image in Fig. 7.2a shows. This acquisition mode allows one to map the spatial distribution of emission (x–y NF intensity profile) from an excited sample area. In contrast, with nearly-closed entrance slit for high-resolution spectroscopy, the axis perpendicular to the wavelength/energy scale carries spatial information and can be read out from a 2D spectrum recorded by the imaging-monochromator's CCD, as described below.

7.2.2 Spatially-Resolved Spectra

In an imaging monochromator (typically a Cherny–Turner-type monochromator), spatial information of the signal projected onto the spectrometer entrance is preserved. The adjustable slit truncates that projection plane (here NF plane) in one direction based on the slit's closure. However, the direction parallel to the entrance slit is projected onto the imaging CCD of the monochromator. Thereby, the transmitted light is passing the dispersive element (commonly a grating). Thus, in a 2D spectrum (x–E intensity plot), each pixel line represents the spectrum for one particular sample position according to the obtainable spatial resolution (example shown in Fig. 7.2b). Conveniently, one can even scan the NF plane by displacing the projection laterally with regard to the position of the entrance slit using the last focussing lens in the imaging apparatus facing the monochromator, as sketched in Fig. 7.3. This is feasible owing to the imaging optics installed in a typical confocal microscope setup and the negligible imaging distortions caused. Thereby, signal from different sample positions within the MO-collected NF frame can be probed without physically moving the sample.

In case one integrates over the real-space axis of the 2D spectrum, for instance whenever spatial information can be disregarded, NF line spectra without any spatial and angular information are obtained. This is particularly useful when quantum dot samples (with the aim of signal maximisation) or spatially homogeneous samples such as high-quality quantum wells (when position dependencies on the microscale are of no relevance) are investigated with a very tightly focused excitation. In those cases, the exposure of the imaging CCD of a few pixels in the spatial axis orientation is not enough to obtain any spatial resolution.

However, is the excitation spot large and significantly projects onto the CCD, imaging capabilities may be very attractive. In such cases, spatially-resolved spectra can provide insight into the mode confinement and excitation spectrum in microresonators such as (buried or patterned) photonic quantum boxes. For polariton traps, the different discrete modes and their spatial distribution was revealed in such recordings by Idrissi-Kaitouni et al. [14] (structural trapping) and Roumpos et al. [15] (gain-induced trapping). Similarly, this technique is useful to monitor the PL along a correspondingly aligned nanowire or 1D polariton system [16] (Fig. 7.4; spatial information can also be combined with signal acquisition in the time domain, such as in [17]). In fact, imaging can be well used to see e.g. how a launched polariton gas propagates [18] (see Fig. 7.4) or to study ultrafast coherent switches with polariton condensates [19].

Recently, the identification of 2D materials in the monolayer, bilayer or heterostructure configuration has raised considerable interest among researchers that aim at the employment of such ultra-thin quantum materials in optical structures, such as microcavities. Spatially-resolved μPL is very useful for distinguishing monolayer parts from multilayer regions in microscopically small 2D-semiconductor flakes, or to identify the overlapping regions in stacked monolayer–monolayer heterostructures.

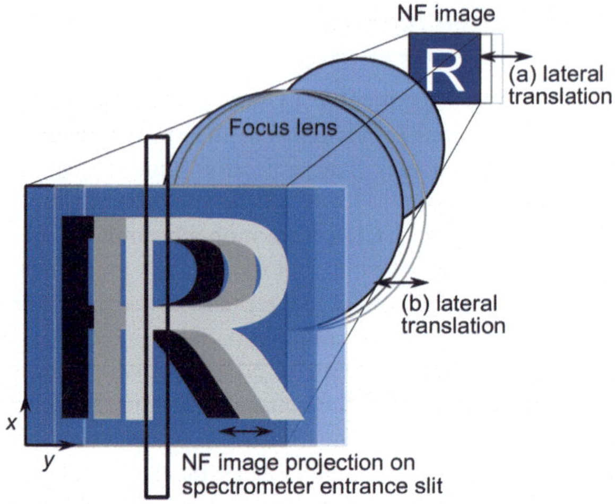

Fig. 7.3 Sketch of the real-space scanning possibilities based on (**a**) an in-plane sample translation for a fixed detection pathway and (**b**) lateral focus-lens translation for a fixed sample position. Owing to the monochromator's entrance slit, different positions projected onto the entrance plane (image plane) are transmitted into the monochromator. The direction parallel to the slit can be directly used for spatially-resolved spectroscopy with information about one spatial direction (e.g. the x direction), whereas information about the other direction (e.g. y) can be obtained by correspondingly one-directional scanning. For most optical microscopy setups with relatively large focus lens diameter, large setup magnification and negligible projection errors, moving the projected image (here magnified, schematic, real-space sample image, letter R) in the detection plane is a reasonably good option for NF scanning. This can be helpful when the measurement is very sensitive to sample displacements. However, any change of the focusing lens in front of the monochromator may also affect Fourier-space detection schemes incorporated into the same setup. In contrast, if the sample position in the collection plane of the imaging optics is displaced, it in turn leads to a translated sample image on the detection plane, but leaves the FF projection unaffected. In fact, a usually impractical but imaginable approach would be to move the spectrograph apparatus with respect to a fixed optical axis of the setup or vice versa, not indicated here. Alternatively, the fixed sample is often scanned (locally) in spectroscopic experiments by translating the microscope objective for an else fixed detection path, not shown here, but this comes at the cost of a projection displacement for both the NF and the FF configuration, which needs to be readjusted correspondingly

7.3 Fourier-Space-Resolved Spectroscopy

Angle-resolved spectroscopy is the key tool for the characterization of polariton modes in planar microresonator systems. Since the angle of photon emission out of a planar system (angle with respect to the sample normal) directly corresponds to the in-plane momentum of its emitter therein (see (2.9) and (2.10) in Chap. 2), angle-resolved spectroscopy provides a unique access to the dispersion of cavity–polariton modes. To obtain angular resolution and to show polariton branches in an energy–momentum diagram, different approaches have been established and used for data acquisition. Three prominent methods are summarized below.

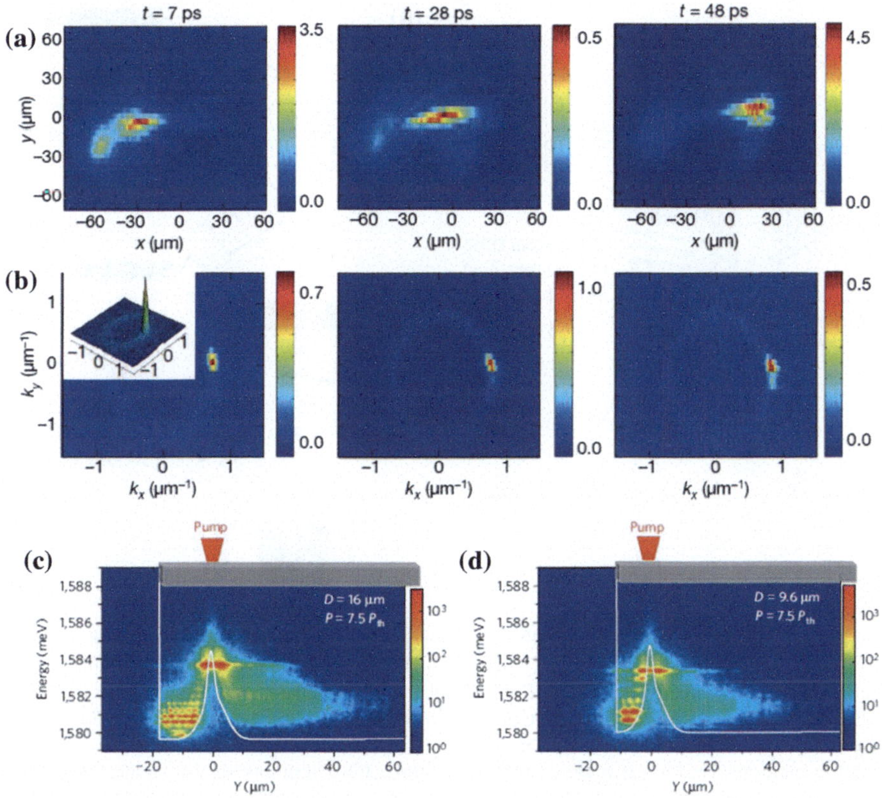

Fig. 7.4 **a** With real-space imaging techniques, even the 2D polariton gas propagation in a sample plane can be imaged by time-resolved PL recordings with microscopic spatial resolution. In this example, the gas set in motion by a short optical pulse is displayed at different times (from left to right) on the ps time scale [18]. **b** Corresponding Fourier-space images show a narrow k-space distribution at a finite momentum representing the polariton gas in motion. **c** With real-space spectroscopy, also a 1D polariton gas propagating along a wire-like planar microcavity system can be analyzed. In this second example, the trapping of polariton condensates is investigated when approaching the pump-laser spot to the wire end [16]. Thereby, a trap potential is caused, with a white solid line indicating the excitation-induced Coulomb potential barrier. **a** and **b** reused with permission. Reference [18] Copyright 2009, Springer Nature. **c** and **d** reused with permission. Reference [16] Copyright 2010, Springer Nature

7.3.1 Goniometer-Like Technique

The incidence angle of pump light can often be independently adjusted with respect to the acquisition direction. However, if the sample needs to be rotated under the collection optics with the pump spot and detection spot constantly maintained on the same sample position, one way to do so is to employ a goniometer-like setup (Fig. 7.5). For reflection measurements, incidence and reflection angle are equal and

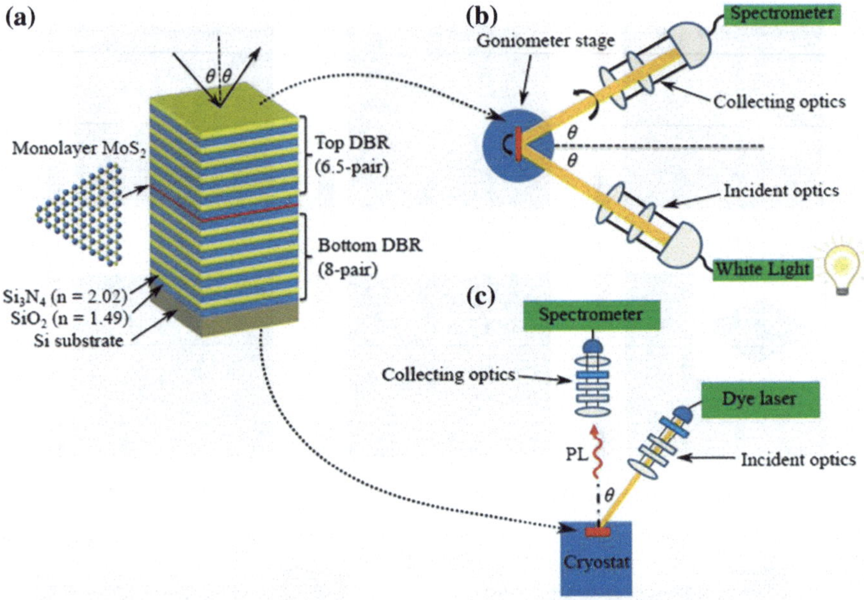

Fig. 7.5 **a** A microcavity structure with incorporated 2D semiconductor monolayer shown as a representative sample, which is typically investigated both with white-light reflection and photoluminescence spectroscopy. **b** Sketch of an optical setup for angle-resolved reflectivity measurements. Both, the excitation and detection path are configured as goniometer arms, with the sample mounted on the rotation stage in the path of the white light. This goniometer scheme is widely used for angle-resolved experiments. For practicability, fibre coupling of the incident light and the detected signal can be employed. **c** Sketch of an optical measurement setup for PL measurements, here with additional optics for polarization control. Reused with permission. Reference [20] (Supplementary Information) Copyright 2017, Springer Nature

thus both optical arms of the setup (input/output) have to be adjusted with respect to the sample surface. Therefore, a goniometer can be used to achieve this. For optical pumping, the angle is usually fixed with respect to the surface, corresponding to a certain in-plane momentum, while the detection path needs to be adjusted anglewise to scan different angles. In such a setup, a low NA of the collection optics actually improves the angular resolution. Truncation techniques with regard to transmitted angles could further help to enable sufficient discrimination between different detection angles in angle-resolved spectra, if the angle steps are finer than the NA coverage.

While this method is still found in recent literature due to its straight-forwardness, it is obvious that the limitations imposed by such a setup configuration and scanning method let one favour the employment of a more effective, easy-to-use and precise method to image the Fourier space of the microcavity emission and to reveal the polaritonic dispersion relation—ideally in a single-shot fashion. The manual (or automatized) scan over different angles is mainly used, when a Fourier-space projection technique is not available, e.g. due to the lack of an imaging monochromator or other restrictions imposed by the setup or experiment. Yet, the use of this goniometer-like angle-scanning technique also imposes restrictions on the measurement capabilities and the setup itself. Not every setup offers the necessary flexibility to perform sample rotations while changing at least one optical arm (e.g. in-coupling). To facilitate the use of this technique, fibre coupling of light to and from the rotatable sample can be used.

Motorization of the goniometer arm can serve sequential spectrum acquisition purposes. After a series of angles have been measured, their spectra are either evaluated individually for the extraction of peak energies as a function of the angle or stitched together digitally to achieve an energy–angle contour plot. It is worth noting that the obtained dispersion curves are good enough to indicate a polariton branch. In contrast, Fourier-space-resolved spectroscopy using a below described projection technique can provide higher angular resolution and much faster acquisition times owing to the single-shot detection via a CCD camera's pixel arrays of imaging spectrometers. Thereby, the signal from the sample of one particular moment ("moment" mainly with regard to timing and duration, but also position and measurement parameters) is instantaneously provided whereas being dispersed over energy and angle. Nevertheless, it is worth taking a look at another practical angle scanning approach, which can also be employed whenever imaging capabilities are not available or not needed, that is the pinhole translation method described briefly in the following.

7.3.2 Pinhole Translation Method

Getting access to the spectrum of a certain reflection/transmission or luminescence angle doesn't need to be that tricky compared to the goniometer-arm rotation technique or the Fourier-space projection method (of course, tricky in a different sense). In any Fourier-space projection plane available in the optical setup, one can quite simply insert a movable pinhole on a 1D (or even 2D) stage and manually or automatically scan along one in-plane momentum direction (or through that Fourier-space plane) in which the angle information is contained. The back-focal plane of the microscope objective directly provides a bull's-eye-like distribution of emission angles, as every set of parallel rays of light collected by the objective is focused onto the same spot in the back-focal plane, constructing the Fourier-space plane (see Figs. 7.2 and 7.6). Naturally, any proper projection of this plane in the optical setup can be used to image, filter and probe the signal in the Fourier space (see Figs. 7.7 and 7.8).

In some experiments, suppression of the signal from the ground state or the higher-momentum states is desired. This can be for instance achieved by masking the Fourier-space plane with a pin-board pin's head (or any other solid mask), an aperture (also applicable for angle-dependent photocurrent measurements, see [22]), or a programmable liquid-crystal screen or spatial light modulator (also applicable for beam shaping in the excitation path, see [23]).

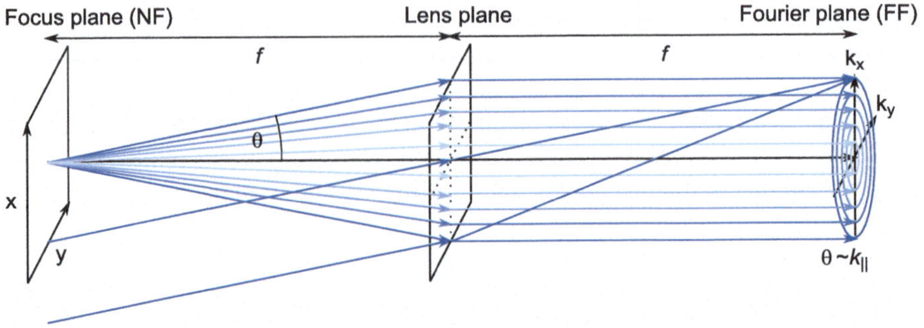

Fig. 7.6 Sketch of the Fourier-space construction in the backfocal plane of a lens with focal length f. The lens directly links the real space, i.e. near-field (NF) image in the frontal focal plane (with spatial directions x and y), with the momentum space, i.e. far-field (FF) image in the backfocal plane (with phase-space directions k_x and k_y), in a Fourier transformation. As indicated, parallel rays of light passing through the lens are focussed onto the same position in the backfocal plane of that lens. Incident rays collected by the lens under different angles with respect to the optical axis end up in different locations in that Fourier-space plane. The bull's-eye-like distribution of collected angles (coloured lines, from light to dark blue with increasing angle) in the FF image can be used for direct imaging of the k space, with emission angles θ being connected to the in-plane component of the photon wave-vector $k_{\|}$. For a radial-symmetric phase space, the in-plane momentum $k_{\|}$ corresponds to the distance from the centre (i.e. the radius) of the FF image (0 angle, black central line). Whenever the k-space has reduced symmetry due to effects of super lattices or sample anisotropies, the orientation of detection matters and information in k_x may considerably differ from that in k_y

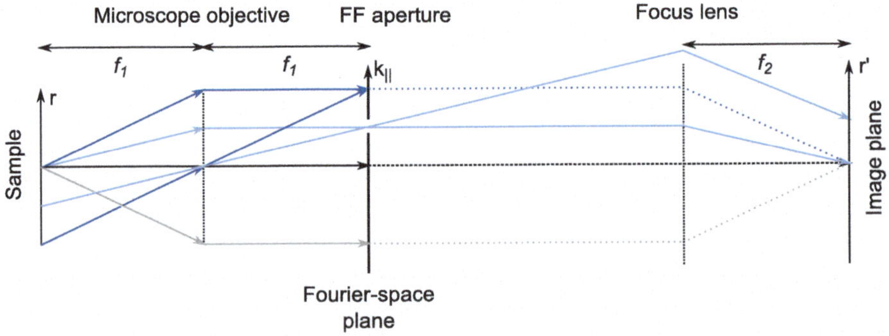

Fig. 7.7 Schematically drawn optical pathways indicated for an imaging configuration, with angle-resolution obtained in the backfocal plane of the collection optics (microscope objective). A (movable) pinhole/iris aperture inserted into this backfocal plane, i.e. the Fourier plane, can be used to transmit certain angles. By translating the aperture with the help of a mechanical stage, (signal from) different angles can be scanned serially to retrieve FF information. The image of the sample is then projected onto the detector (spectrometer) via the backfocal plane of a focus lens for acquisition. Behind this image projection plane, a Fourier-plane image occurs which can be used for Fourier-space spectroscopy as in [21]

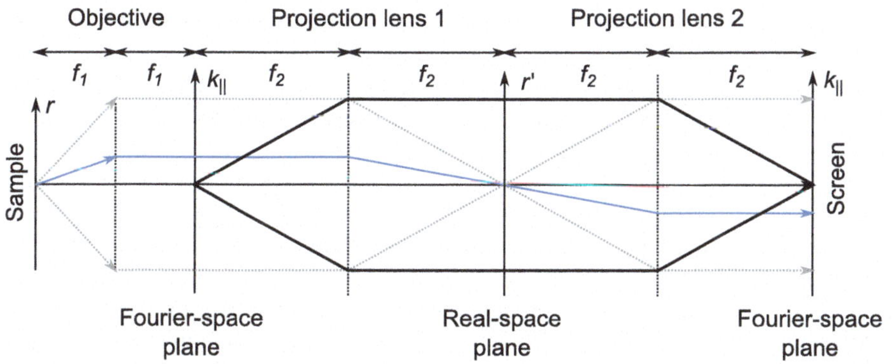

Fig. 7.8 Sketch of a $4f$ FF projection setup. Without the extra lens (FF lens) in-between microscope objective and the lens in front of the screen (focus lens), the setup resembles a real-space projection setup with microscopic resolution of the sample image. With two equivalent relay lenses used for the FF projection, a one-to-one image of the Fourier-space plane can be created in the plane of the screen. With every set of two equal lenses, the signal can be further relayed, whereas the magnification of images remains unchanged in an $n \times 4f$ configuration with equal f for all lenses of the n relay lens pairs. By removing the last FF lens of the relay configuration, the NF projection on the screen is restored. The intermediate NF projection plane can be used both for spatial filtering purposes and for correlation experiments based on spatial interference as performed in [13]. Any intermediate FF projections can be used equivalently for k-space filtering purposes

7.3.3 Single-Shot Angle-Resolved Acquisition

Fourier-space spectroscopy utilizes the angle information contained in the far-field (FF) projection of a sample's emission (Fig. 7.6). In contrast to NF spectroscopy, the FF plane in the microscopy setup is projected onto the imaging monochromator's CCD (i.e. screen, see Fig. 7.8) and enables the characterization of intensity profiles in the Fourier space. Similar to NF microscopy, which images the NF of a sample, as described above, FF microscopy images the FF. Ideally, FF projection provides a radial-symmetric intensity profile on the screen, if such symmetry is justified by the real-space symmetry (example profile shown in Fig. 7.2c). This is for instance typical for planar structures with no preference of orientation. Schematically, it can be compared to a bull's eye, with the centre of the collected FF circle representing 0°, while the radial direction represents a corresponding orientation of $\mathbf{k}_{\|} = k_x + k_y$. The edge of the circle represents the maximum collected angle of the MO's light cone as defined by the numerical aperture (NA). In all other cases where a peculiar directionality exists, it is important to align structures reasonably with respect to the slit orientation (here chosen to be x) in order to obtain suitable projections in k_x. Fourier-space intensity profiles can be readily obtained in the FF configuration when the monochromator's entrance slit is open and the collected signal is mirrored onto the CCD (entrance-slit plane imaged onto CCD), such as sketched in Fig. 7.2c.

In combination with a dispersive element (monochromator grating)—and when the entrance slit is nearly closed—2D angle-resolved spectra can be recorded in which each pixel line contains spectral information for a corresponding angle, i.e. in-plane momentum value. In other words, the axis perpendicular to the energy/wavelength scale on the monochromator's CCD represents collection angles. The resolution of the momentum space depends on the optics used and can be on the order of $\approx 0.05\,\mu\mathrm{m}^{-1}$ ($\approx 0.5°$). An example spectrum is shown in Fig. 7.2d for a lower-polariton (LP) branch in PL at low excitation densities, clearly revealing the energy–momentum dispersion features expected for this mode. While a cavity mode features a parabolic dispersion in the measured angle range, the LP's dispersion flattens to the sides owing to the anti-crossing behaviour when approaching the exciton mode at higher angles. This method of visualizing polariton modes is widely used in the literature and different setup configurations for FF spectroscopy have been depicted in publications on this subject [13, 21].

Prerequisite for a correct measurement of $k_{\|}$ (in-plane momentum, radial direction of the projected momentum-space circle) is the corresponding selection of the central cut through the Fourier-space plane by the entrance slit (down to the exposure of about 2 pixel columns on the CCD for very good E–k resolution). Only a central narrow cut of a well-projected (well-centred) FF plane guarantees conditions as close as possible to $k_y = 0$ and thereby $k_x = k_{\|}$ ($x \parallel$ to slit orientation), which allows coverage of the ground state of polariton modes in FF spectra. A near-infrared MO with 0.4 NA, which covers an angle range of approximately $\pm 22°$, enables investigations of the momentum space for wave-vectors of $\approx (0 \pm 2.2)\,\mu\mathrm{m}^{-1}$ according to (2.9) in the photon energy region of about 1.5–1.6 eV.

Owing to the lack of spatial sensitivity, real-space-integrated intensities are recorded in the FF configuration. Weak emitters with strong angular distribution of their emission will be recorded with very little (hardly detectable) intensities. This particularly happens, since most of the signal is cancelled off by the entrance slit, which in fact only transmits a tiny fraction of the FF plane's signal for correct FF spectroscopy. In contrast, a considerable amount of the light cone collected by the MO from a micro-sized spot is forwarded into the monochromator in NF spectroscopy. However, if the signal collected with the MO's NA is widely distributed in the FF plane, a narrow cut through the plane represents a major intensity loss.

Typically, a lens combination in the main optical path between sample and detection system is used to obtain a one-to-one projection of the Fourier-space plane from the back-focal plane of the MO to a point of interest. This point can be the forward-focal plane of the imaging monochromator, i.e. the entrance-slit plane. In the NF configuration, lens A focuses the collimated beam onto the monochromator entrance and projects the NF plane there. Adding an equivalent lens B between MO and lens A in a certain distance configuration allows one to project the FF plane (unmagnified) instead of the NF plane onto that entrance. For convenience, this lens B can be referred to as FF lens and can be introduced into the setup (or removed) with precision magnet mounts on demand. For an unchanged projection of the Fourier-space image, the distance between the back-focal plane of the MO and the monochromator entrance has to be four times the focal length of the lenses ($f_A = f_B = f_{A/B}$). The distance between the lenses in this $4f$ geometry becomes $2f_{A/B}$ and the distance to the MO's back-focal plane or monochromator's entrance slit equals $f_{A/B}$.

At the same time, this configuration offers a magnified real-space image in the joint focal plane (NF plane) between the equal-type lenses which can be used for spatial filtering, e.g. via an iris aperture or pinhole. This can be particularly useful for reflection measurements of samples, where signal from the surroundings must be suppressed in order to enable spectrum acquisition for an area of interest. Similarly, μPL acquisition can benefit from background suppression or spatial filtering when the pump spot is larger than the region of interest. Naturally, the smaller the pinhole, the higher the spatial selectivity. However, filtering comes at the cost of acquisition intensities due to the (intended) reduced spatial integration.

The same spectroscopy setup can be extended and modified for other measurements. For instance, magneto-optical studies require strong magnetic fields in specially-prepared cryostats, which set certain requirements for the experimental setup, such as a safety distance of ferromagnetic setup components from the magnetic

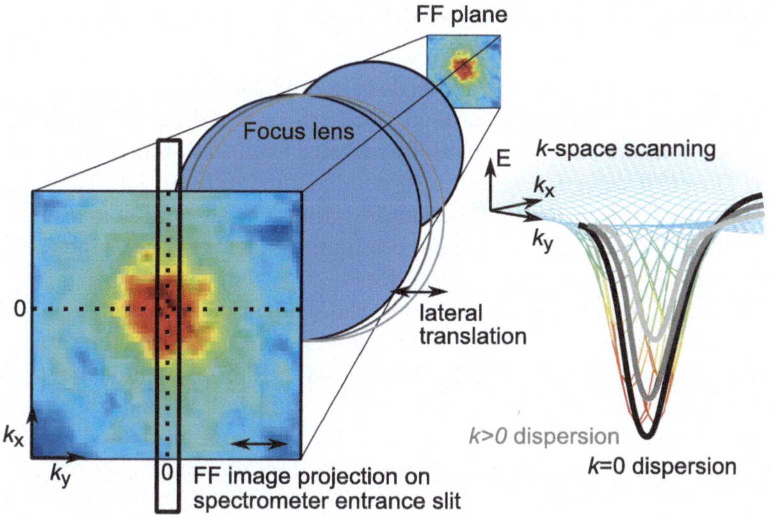

Fig. 7.9 Sketch of a Fourier-space scanning approach based on imaging optics similar to the case in Fig. 7.3, method type (b). A lateral focus-lens translation in the FF configuration allows one to project different slices of the phase-space plane onto an imaging monochromator's entrance slit, which transmits that signal slice into the spectrometer. The direction parallel to the slit can be directly used for momentum-resolved spectroscopy along one phase-space direction (e.g. k_x), which would be equivalent to $k_{\|}$ when the FF projection circle is centred on the slit (i.e. when $k_y = 0$, $k_x = k_{\|}$). Again, projection errors are typically negligible when translating the projected image (here FF plane) in the detection plane by the focus lens for scanning purposes in the direction orthogonal to the slit orientation. However, any change of the focusing lens in front of the monochromator may also affect real-space detection schemes incorporated into the same setup, which requires readjustment of the NF configuration after FF scanning. Here, the displayed image in the projection sketch is an example of a FF emission profile detectable in the Fourier plane (from red to blue with decreasing intensity). The mesh on the right represents a lower-polariton dispersion in the k_x-k_y plane (from red to blue with increasing quasi-particle energy). Black to grey lines represent different polariton dispersions in k_x along the scanned phase-space direction k_y, as indicated in the lateral-translation sketch on the left-hand side

fields. In some cases, an existing setup simply needs to be expanded to reach another apparatus. Therefore, the main optical path can be prolonged by relay lenses in combination with relay mirrors. For an unmagnified bridging projection (if desired), a lens pair with equal focus length is employed. For simplicity, such relay lenses are not depicted in setup overview sketches such as Figs. 7.1 and 7.2.

Similar to the aforementioned NF plane scanning approach, which used translation of a lens in the setup in front of the monochromator's entrance to displace the NF projection with regard to the slit's transmission window, the Fourier-space projection method can be used to finely scan different slices of the k-space plane with the very same lens translation. In this case, the setup is configured to project the FF image onto the slit, and this image is projected by the same focus lens, which projects the NF image onto the slit in the NF configuration. This is shown schematically in Fig. 7.9.

7.4 Time-Resolved Spectroscopy

While steady-state measurements are primarily used for the investigation of luminescence properties of polariton systems due to the spectral information obtained, the dynamics of an emitting system is only revealed by means of time-resolved spectroscopy. Different detection schemes in combination with pulsed excitation may serve the purpose of transient-luminescence or transient-absorption measurements. However, the temporal resolution can be a limiting factor if decay rates, transfer dynamics or the formation times are too fast.

A big pool of literature exists utilizing and explaining such techniques. Here, a brief overview shall be given on some essential approaches for time-trace measurements that can be used for instance to observe Rabi oscillations or to obtain quasi-particle lifetime information. Two distinct ultra-fast techniques are summarized, the streak-camera detection scheme and the pump–probe method, which have been established to provide ps and sub-ps temporal resolution.

7.4.1 Streak-Camera Measurements

Lifetime measurements for quantum emitters are commonly used to characterize properties of structures and materials. Fluorescence lifetimes can be extracted on different time scales, whereas the temporal resolution mainly depends on the excitation pulse shape as well as duration and the detector speed that together determine the instrument response function. Highly-sensitive and ultra-fast photodetectors such as avalanche photodetectors (APDs) can be used to measure very weak signals with temporal resolution in the sub-nanosecond range down to tens of picoseconds in combination with ps-pulsed excitation by ultra-short-pulse lasers and photon counters. Thereby, histograms of the emission decay are built up from incident photons of different excitation cycles, using a time-tagging mode. The overall time-trace acquisition speed depends on both the signal intensity and the repetition rate of the laser. However, processes such as exciton formation, polariton decay or Rabi oscillations can be still too fast. This does not imply that the alternatively used streak-camera measurements can temporally resolve all features, but the resolution can practically reach down to about 1 ps depending on the operation mode of the device, or even about one order of magnitude shorter times for newest device models. Thus, such streak-camera units have provided important opportunities to the optical characterization of charge-carrier dynamics in many material systems and quantum structures.

The Streak-Camera Unit

The advantage of a streak-camera unit placed behind a monochromator is its ability to resolve incident light both spectrally (or spatially, in imaging configuration) as well as temporally. Photons of different energy and time are projected onto a detector screen differently after passing the streak unit, resulting in a spatially-resolved signal for which one axis corresponds to time and the other to wavelength/energy (without dispersive element simply a spatial direction determined by an entrance slit). Incident photons coupled in through the spectrometer's entrance slit are spectrally resolved on the exit side before they enter the streak unit via its own entrance slit. That slit oriented in the direction of wavelength resolution transmits the dispersed signal of the monochromator. The transmitted photons impinge on a photosensitive screen, from which electrons are released by the photons quasi instantaneously. Thereby, the spatial and temporal order of the emitted electrons correlates with that of the photons. The accelerated electrons in the unit are then deflected between a plate capacitor by a synchronized sinusoidal voltage (at its approximately linear-slope parts) oriented perpendicular to the flight path.[1] After being amplified typically by a multi-channel plate, the electrons impinge on a phosphorous screen, from which photons are released with the achieved spatial redistribution based on their energy–time relationship. These photons are then recorded by a detector camera (CCD) of the streak system. Yet, the performance of the device depends on different experimental parameters and a generalized classification of the resolution is avoided here. Best systems with lowest temporal jitter are reported to have temporal resolution of about 0.5 ps and better. For more details on such devices, the interested reader is referred to the literature, e.g. an online product manual which also explains and depicts the working principle in [24].

To summarize, time-trace analysis with spectral resolution is strongly facilitated by this acquisition technique. In addition to lifetime studies, streak cameras can also be employed for single-photon counting with picosecond time resolution, as polariton studies regarding photon statistics have shown [25].

[1] The repetition rate of the streak camera's deflector unit is adjusted to the laser frequency of the mode-locked laser source, commonly a Titanium-Sapphire oscillator, which typically delivers fs to ps pulses at a rate of 80 MHz for time-resolved photoluminescence experiments.

Polaritons in the Spotlight

A considerable number of reports have used streak cameras for the characterization of polaritons. Early on, time-resolved experiments were sought to monitor the nonlinear polariton dynamics in microcavities by Mueller et al. [26] and the dynamics of a coherent cavity–polariton population by Bloch et al. [27]. Also, parametric amplification and strongly-interacting polariton "liquids" in semiconductor microcavities were subject of such streak camera investigations reviewed by Baumberg and Lagoudakis [28]. In general, polariton lifetime measurements—with polariton decay times on the picosecond time-scale in high-quality microcavities—can be conveniently performed with a streak-camera setup, as the observation of long-lived polariton states across the parametric threshold in an optical-parametric oscillator (OPO) scheme shows [29]. Lifetime information was also reported from measurements of "slow reflection" by Steger et al., which deduced the lifetime of the polaritons in high-quality samples from a trace in time and space over millimetre distances to be 180 ± 10 ps [30].

Furthermore, using a streak camera, Amo et al. visualized the in-plane fluid dynamics in a polariton gas [18] (see Fig. 7.4a, b), whereas Wertz et al. addressed polariton flow in a wire-like microcavity system [17]. One further example of the use of streak-camera data for the study of polariton systems is given in the report on the observation of oscillations in the dynamics of a microcavity polariton condensate formed under pulsed non-resonant excitation [31]. De Giorgi et al. attributed their observation to relaxation oscillations based on localization conditions of the condensate. By using non-resonant excitation, they demonstrated condensate injection pulses with a frequency of 0.1 THz.

This selection of different experiments performed in the literature shall highlight the possibilities regarding streak-camera measurements and provide a coarse overview. In fact, numerous more examples are out in the literature not covered by this brief summary.

7.4.2 Pump–Probe Techniques

In the strong-coupling regime, energy exchange between the microcavity's light field and the QW-exciton mode is a coherent and reversible process. This process is very fast and can be only resolved by proper (e.g. pump–probe) techniques. It is determined by the Rabi frequency, which is for III/V systems typically on the order of a few THz (in general, this frequency depends on the structure and the coupling strength), and is solely limited by the underlying damping of the coupled-oscillator system owing to dephasing as well as loss of excitons and, more noticeably, photon losses (particularly photon leakage due to the imperfect cavity).

Transient-Absorption Measurements

Transient-absorption experiments have been widely used in semiconductor spectroscopy to obtain temporal resolution on the sub-100 fs scale involving ultra-fast mode-locked lasers, such as Ti:Sapphire oscillators and their amplifiers.

In a typical pump–probe scheme, optical laser light is used to excite the sample and a spectrally-broad "white-light" pulse is shed onto the sample to probe the state of the electronic system. The temporal excitation of charge-carriers leads to a modified absorption behaviour, which decays to its original (zero-density) absorption behaviour after the relaxation of photo-induced excitations via available radiative and non-radiative recombination channels. To obtain temporal resolution, i.e. to record a transient spectrum, a delay line with commonly motorized translation stage for the probe beam is used to set the time delay between pump and probe pulses and to scan over a wide range of delay times. For every delay setting, time-integrated spectra are acquired with a common spectrometer system, whereas the acquisition time and signal-to-noise ratio are dependent on the repetition rate of the laser system, the averaging and the measured spectral intensities.

Transient-absorption experiments strongly benefit from the available ultra-short pulse durations from mode-locked lasers which are in the range of about 10–100 fs. In principle, time scales can be deduced both for population build-up and decay of excitonic or polaritonic modes. The temporal resolutions achievable are on the order of the pulse durations used. Such pump–probe experiments are not limited to the optical range of visible light. Indeed, optical-pump with THz (or IR)-probe schemes have been well used to target intra-excitonic transitions and to monitor the population build-up of incoherent excitons (see [32]).

Ultrafast optical pump–probe measurements evolved to that extent that even additional THz pulses were included to address transitions between different polariton branches via an optically-dark exciton p-state [33] or to disturb light–matter interaction on their relevant time-scales [34].

Regarding polariton systems, time-resolved studies are attractive for the observation of Rabi oscillations, which typically occur in the time frame of picoseconds for most inorganic polariton systems. Rabi oscillations in cavity–polariton systems can be monitored for instance in transient reflection measurements[2] (it resembles transient absorption, although ignoring the transmission part) using an all-optical pump–probe scheme [34]. This gives a direct measurement of exciton–polariton Rabi oscillations via the excitonic fraction (matter polarization). Also see Sect. 9.3 for the use of this technique in combination with transient THz fields.

In summary, transient spectroscopy provides a powerful tool for probing light–matter interaction in solids and to study dynamics such as non-equilibrium physics. This can be evidenced in many domains of semiconductor spectroscopy, as the interesting work of Almand-Hunter et al. [35] on the study of quantum droplets of electrons and holes, referred to as "dropletons" in quantum systems, shows.

---

[2]Due to the typically opaque substrates, many white-light spectroscopy measurements on polariton systems are performed in reflection.

Digital Holography of Photoluminescence

Rabi oscillations can also be evidenced with time-resolved off-axis digital holography of sample photoluminescence, which gives direct access to the ultra-fast processes of a strongly-coupled system via its emission [36]. This method offers a direct measurement of the coherent energy exchange between exciton and photon resonances by analysing the photonic component that leaks out of the cavity, which represents the polariton decay in the system. In addition, such technique offers a very good measurement quality, even exceeding that of transient absorption measurements as it relies on interference between the actual coherent-polariton emission and the laser pulse light. It is practically able to depict the Rabi oscillations based on emission, without the need of a probe pulse that can only retrieve the time-dependent absorbance in the excited sample. When monitoring polariton gases with μPL techniques, the holography method (see Fig. 7.10) can yield a different and yet more valuable picture, as it can give insight into the ultra-fast temporal evolution of polariton emission, e.g. under the influence of ultra-fast control pulses [36].

To detect and address Rabi oscillations in optical measurements, Dominici et al. established a digital holography photoluminescence pump–probe setup [36]. It comprises a fs-pulsed laser, the beam of which is split into an excitation path and a reference path, a delay for the reference beam, optics for beam control, and a spectrometer with high-sensitivity imaging CCD. After excitation in transmission geometry, the emitted light is collected and combined with the reference pulse. By scanning over the tunable time delay between signal and reference step-wise, an interferometric recording of the measured intensity can be obtained with temporal information. Since resonant excitation imprints the coherence of the pump pulse onto the polariton gas, the photoluminescence from the polariton mode can interfere with the reference pulse on the detector screen of the monochromator. This advanced time-resolved measurement technique was further used in a study on polarization shaping of Poincaré beams by polariton oscillations [37]. The sophisticated setup of these experiments is highlighted in Fig. 7.10 based on the extended configuration used for polarization control.

Use of Pump–Probe Measurements

While the most common use of transient spectroscopy is the study of dynamics in an electronic system after optical excitation, there are a few variations which can be highly interesting and bring a certain sophistication to the whole scheme. The examples mentioned below for an illustration of the use are a limited selection and do not represent the full scope of possibilities.

In one advanced experiment, a THz pulse was used to depopulate the upper/higher-energy polariton branch (UP/HEP) by a THz-induced transition to the excitonic $2p$ state as was predicted by theory [33]. In their work, Tomaino et al. reported THz excitation of a coherent Λ-type three-level system of exciton–polariton modes in a QW microcavity. To monitor the temporal changes in the polariton branches' population,

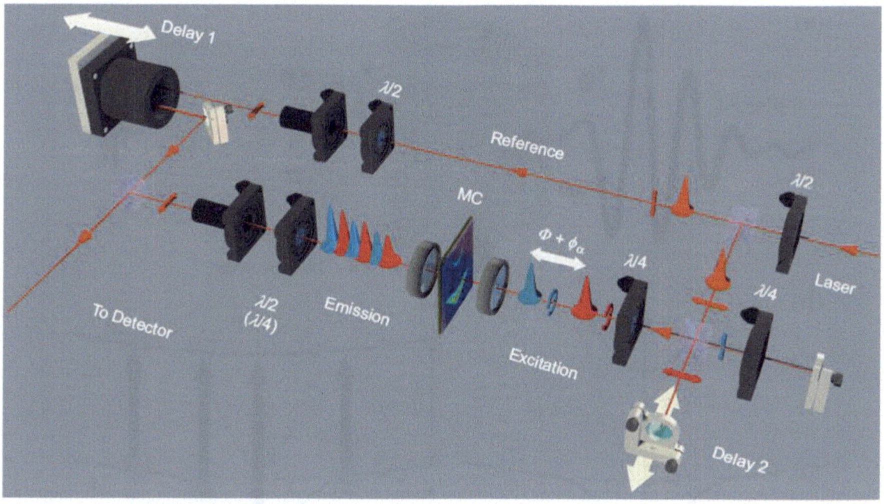

Fig. 7.10 Sketch of an experimental setup implementing multiple pulses and time-resolved off-axis digital holography, which has been employed to study ultra-fast control of Rabi oscillations in a polariton system [36] and polarization shaping of Poincaré beams by polariton oscillations [37]. The latter used the setup for counter-polarized double pulse experiments, which requires additional polarization controlling optics. The fs-pulses train is first split into a reference arm and an excitation arm (e.g. cross-polarized in this example). The latter beam is further divided into twin pulses, which are later manipulated to be counter-circularly polarized. A secondary delay path (delay 2) is used to control the double-pulse separation on time-scales of the Rabi period or optical period. For evaluation, the emission from the microcavity sample (labelled MC) is then filtered in polarization and let to interfere with the reference beam on the detector, a sensitive imaging CCD camera. Digital elaboration completes the digital holography scheme. Conveniently, the temporal evolution of the emission can be tracked with high resolution (owing to the ultra-short pulses used, with durations on the fs time-scale), by tuning the time of interference by means of a primary delay path (delay 1). Reused with permission. Reference [37] Copyright 2015, Springer Nature

an ultra-fast pump–probe scheme was employed. The system was driven weakly by 800 nm 90 fs pulses from a 1 kHz regenerative amplifier and manipulated by strong THz few-cycle pulses generated by a nonlinear crystal. Reflectance spectra were stitched together to provide energy–time spectra with sub-ps temporal resolution, showing clearly the intensity changes at zero delay time in the respective polariton branch.

Similar to the experiment performed by Kaindl et al. in 2003 on excitons [32], Ménard et al. combined a THz-probe experiment with a polariton condensation study (see Fig. 7.11a), "revealing the dark side of a bright exciton–polariton condensate"—in other words, showing the pronounced existence of an uncondensed high-energy fraction of "hot" excitons, which do not couple to the light field and thus coexist with a condensate [38]. The fact that emission from the coupled QW-microcavity system originates from the polariton dispersion is indicated in Fig. 7.11b. For the study of the uncondensed dark fraction of quasi-particles in the system, an amplifier

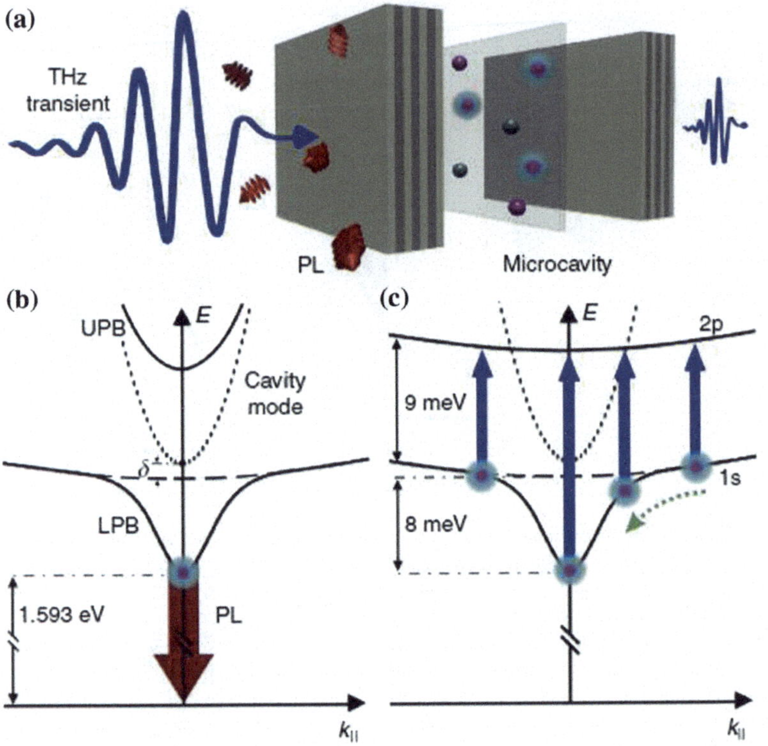

Fig. 7.11 **a** Sketch of ps-short transient fields of THz pulses passing a microcavity hosting exciton–polaritons. Due to the energy of THz photons in the meV range, intra-excitonic transitions in common III–V quantum well systems on that energy scale can be directly used to obtain information about exciton populations [32]. Such experiments are performed in an optical-pump–THz-probe scheme. **b** Schematic dispersion relation of cavity–polaritons with the two polariton branches (lower and upper one, i.e. LPB and UPB here, respectively, drawn as solid lines) and the underlying bare (uncoupled) exciton and photon dispersion (dashed and dotted, respectively) that are coupled together in the microcavity. **c** Absorption of THz photons (vertical arrows) by polaritons at different k-space vectors along their lower branch, which represents the hybrid quasi-particles originating from $1s$ excitons "dressed" with cavity photons. Thus, energies for $1s$–$2p$ hydrogen-like (intra-excitonic) transitions of the coupled system differ along the dispersion. Note that the $2p$ state does not form polaritons as it cannot be addressed by a one-photon absorption process. According to the diagram, the THz-absorption signature for ground-state polaritons, i.e. the population of a ground-state condensate, must be higher energetic (occurring at about 16–17 meV in this example) with respect to the signature for reservoir states (transition lying at about 9 meV). By studying these signatures and the system's dynamics in a pump–probe fashion regarding $1s$–$2p$ transitions, Ménard et al. were able to investigate the condensate fraction and to reveal the dark reservoir excitons, which are not accessible through the momentum-wise tight light cone. Reused with permission. Reference [38] Copyright 2014, Springer Nature

system is used for sample excitation and generation of THz pulses, which act as probe pulses set by a certain time delay with respect to the excitation pulse. Here, the THz photons can drive the excitons irrespective of their centre-of-mass momentum (i.e. their location on the dispersion) from their $1s$ to the excited states ($2p$ and so forth). More importantly, this measurement scheme allows for the discrimination of transitions from a highly-populated LP ground state (condensate mode) to the $2p$ state ($\approx$17 meV here) from that of the bare intra-excitonic transition characterized by the energy difference between the $2p$ and $1s$ states (about 9 meV in this example), as Fig. 7.11c illustrates.

In addition to the observation of Rabi oscillations in [34], a single THz-pulse protocol was proposed and demonstrated as a polarization "reset" in a transient reflection experiment on a microcavity, which was disturbed by the THz pulse at certain times. This disturbance was performed in order to probe light–matter interaction time-scales for optical manipulation and control [34].

Exploring this temporal domain of light–matter interaction thus has still the potential to unravel new applications of quantum gases and microcavity systems, such as the development of ultra-fast optical switches based on manipulation and control of coherent states, as demonstrated by recent time-resolved experiments [33, 34, 36, 37].

References

1. L. Esaki, New phenomenon in narrow Germanium p-n junctions. Phys. Rev. **109**(2), 603–604 (1958)
2. C. Schneider, A. Rahimi-Iman, N.Y. Kim, J. Fischer, I.G. Savenko, M. Amthor, M. Lermer, A. Wolf, L. Worschech, V.D. Kulakovskii, I.A. Shelykh, M. Kamp, S. Reitzenstein, A. Forchel, Y. Yamamoto, S. Höfling, An electrically pumped polariton laser. Nature **497**, 348 (2013)
3. A. Rahimi-Iman, *Nichtlineare Eekte in III/V Quantenlm-Mikroresonatoren: Von dynamischer Bose–Einstein-Kondensation hin zum elektrisch betriebenen Polariton-Laser* (Cuvillier Verlag, Göttingen, 2013)
4. M. Amthor, J. Fischer, I.G. Savenko, I.A. Shelykh, A. Chernenko, A. Rahimi-Iman, V.D. Kulakovskii, S. Reitzenstein, N.Y. Kim, M. Durnev, A.V. Kavokin, Y. Yamamoto, A. Forchel, M. Kamp, C. Schneider, S. Höfling, Exciton-polariton laser diodes, in *Proceedings of the SPIE, Nanophotonics and Micro/Nano Optics II*, ed. by Z. Zhou, K. Wada (SPIE, 2014)
5. J. Kasprzak, M. Richard, S. Kundermann, A. Baas, P. Jeambrun, J.M.J. Keeling, F.M. Marchetti, M.H. Szymańska, R. André, J.L. Staehli, V. Savona, P.B. Littlewood, B. Deveaud, L.S. Dang, Bose-Einstein condensation of exciton polaritons. Nature, **443**(7110), 409–414 (2006)
6. G. Roumpos, M. Lohse, W.H. Nitsche, J. Keeling, M.H. Szymańska, P.B. Littlewood, A. Löffler, S. Höfling, L. Worschech, A. Forchel, Y. Yamamoto, Power-law decay of the spatial correlation function in exciton-polariton condensates. Proc. Natl. Acad. Sci. USA (2012). https://doi.org/10.1073/pnas.1107970109.
7. K.G. Lagoudakis, M. Wouters, M. Richard, A. Baas, I. Carusotto, R. Andre, L.S. Dang, B. Deveaud-Plédran, Quantized vortices in an exciton-polariton condensate. Nat. Phys. **4**(9), 706–710 (2008)
8. G. Roumpos, M.D. Fraser, A. Löffler, S. Höfling, A. Forchel, Y. Yamamoto, Single vortex-antivortex pair in an exciton-polariton condensate. Nat. Phys. **7**(2), 129–133 (2011)

9. A. Amo, S. Pigeon, D. Sanvitto, V.G. Sala, R. Hivet, I. Carusotto, F. Pisanello, G. Leménager, R. Houdré, E. Giacobino, C. Ciuti, A. Bramati, Polariton superfluids reveal quantum hydrodynamic solitons. Science **332**(6034), 1167–1170 (2011)
10. L. Dominici, G. Dagvadorj, J.M. Fellows, D. Ballarini, M. De Giorgi, F.M. Marchetti, B. Piccirillo, L. Marrucci, A. Bramati, G. Gigli, M.H. Szymańska, D. Sanvitto, Vortex and half-vortex dynamics in a nonlinear spinor quantum fluid. Sci. Adv. **1**(11), e1500807 (2015)
11. T. Boulier, H. Terças, D.D. Solnyshkov, Q. Glorieux, E. Giacobino, G. Malpuech, A. Bramati, Vortex chain in a resonantly pumped polariton superfluid. Sci. Rep. **5**, 09230 (2015)
12. H. Deng, G.S. Solomon, R. Hey, K.H. Ploog, Y. Yamamoto, Spatial coherence of a polariton condensate. Phys. Rev. Lett. **99**(12), 126403 (2007)
13. C.W. Lai, N.Y. Kim, S. Utsunomiya, G. Roumpos, H. Deng, M.D. Fraser, T. Byrnes, P. Recher, N. Kumada, T. Fujisawa, Y. Yamamoto, Coherent zero-state and π-state in an exciton-polariton condensate array. Nature **450**(7169), 529–532 (2007)
14. R.I. Kaitouni, O. El Daïf, A. Baas, M. Richard, T. Paraiso, P. Lugan, T. Guillet, F. Morier-Genoud, J.D. Ganière, J.L. Staehli, V. Savona, B. Deveaud, Engineering the spatial confinement of exciton polaritons in semiconductors. Phys. Rev. B **74**(15):155311 (2006)
15. G. Roumpos, W.H. Nitsche, S. Höfling, A. Forchel, Y. Yamamoto, Gain-induced trapping of microcavity exciton polariton condensates. Phys. Rev. Lett. **104**(12), 126403 (2010)
16. E. Wertz, L. Ferrier, D.D. Solnyshkov, R. Johne, D. Sanvitto, A. Lemaître, I. Sagnes, R. Grousson, A.V. Kavokin, P. Senellart, G. Malpuech, J. Bloch, Spontaneous formation and optical manipulation of extended polariton condensates. Nat. Phys. **6**(11), 860–864 (2010)
17. E. Wertz, A. Amo, D.D. Solnyshkov, L. Ferrier, T.C.H. Liew, D. Sanvitto, P. Senellart, I. Sagnes, A. Lemaître, A.V. Kavokin, G. Malpuech, J. Bloch, Propagation and amplification dynamics of 1d polariton condensates. Phys. Rev. Lett. **109**, 216404 (2012)
18. A. Amo, D. Sanvitto, F.P. Laussy, D. Ballarini, E. del Valle, M.D. Martín, A. Lemaître, J. Bloch, D.N. Krizhanovskii, M.S. Skolnick, C. Tejedor, L. Viña, Collective fluid dynamics of a polariton condensate in a semiconductor microcavity. Nature **457**(7227), 291–295 (2009)
19. T. Gao, P.S. Eldridge, T.C.H. Liew, S.I. Tsintzos, G. Stavrinidis, G. Deligeorgis, Z. Hatzopoulos, P.G. Savvidis, Polariton condensate transistor switch. Phys. Rev. B **85**, 235102 (2012)
20. Y.J. Chen, J.D. Cain, T.K. Stanev, V.P. Dravid, N.P. Stern, Valley-polarized exciton-polaritons in a monolayer semiconductor. Nat. Photon. **11**(7), 431–435 (2017)
21. M. Richard, J. Kasprzak, R. Romestain, R. André, L.S. Dang, Spontaneous coherent phase transition of polaritons in CdTe microcavities. Phys. Rev. Lett. **94**(18), 187401 (2005)
22. D. Bajoni, E. Semenova, A. Lemaître, S. Bouchoule, E. Wertz, P. Senellart, J. Bloch, Polariton light-emitting diode in a GaAs-based microcavity. Phys. Rev. B **77**(11), 113303 (2008)
23. F. Veit, M. Aßmann, M. Bayer, A. Löffler, S. Höfling, M. Kamp, A. Forchel, Spatial dynamics of stepwise homogeneously pumped polariton condensates. Phys. Rev. B **86**(19), 195313 (2012)
24. Hamamatsu Photonics, Universal streak camera C10910 series, www.hamamatsu.com/resources/pdf/sys/SHSS0016E_C10910s.pdf
25. M. Aßmann, J.-S. Tempel, F. Veita, M. Bayer, A. Rahimi-Iman, A. Löffler, S. Höfling, S. Reitzenstein, L. Worschech, A. Forchel, From polariton condensates to highly photonic quantum degenerate states of bosonic matter. Proc. Natl. Acad. Sci. USA **108**, 1804–1809 (2011)
26. M. Müller, R. André, J. Bleuse, L.S. Dang, A. Huynh, J. Tignon, P. Roussignol, C. Delalande, Non-linear polariton dynamics in II-VI microcavities. Semicond. Sci. Technol. **18**(10), S319 (2003)
27. J. Bloch, B. Sermage, M. Perrin, P. Senellart, R. André, L.S. Dang, Monitoring the dynamics of a coherent cavity polariton population. Phys. Rev. B **71**(15), 155311 (2005)
28. J.J. Baumberg, P.G. Lagoudakis, Parametric amplification and polariton liquids in semiconductor microcavities. Phys. Status Solidi (b) **242**(11), 2210–2223 (2005)
29. D. Ballarini, D. Sanvitto, A. Amo, L. Viña, M. Wouters, I. Carusotto, A. Lemaître, J. Bloch, Observation of long-lived polariton states in semiconductor microcavities across the parametric threshold. Phys. Rev. Lett. **102**(5), 056402 (2009)
30. M. Steger, C. Gautham, D.W. Snoke, L. Pfeiffer, K. West, Slow reflection and two-photon generation of microcavity exciton-polaritons. Optica **2**(1), 1 (2015)

31. M. De Giorgi, D. Ballarini, P. Cazzato, G. Deligeorgis, S.I. Tsintzos, Z. Hatzopoulos, P.G. Savvidis, G. Gigli, F.P. Laussy, D. Sanvitto, Relaxation oscillations in the formation of a polariton condensate. Phys. Rev. Lett. **112**(11), 113602 (2014)
32. R.A. Kaindl, M.A. Carnahan, D. Hägele, R. Lövenich, D.S. Chemla, Ultrafast terahertz probes of transient conducting and insulating phases in an electron-hole gas. Nature **423**(6941), 734–738 (2003)
33. J.L. Tomaino, A.D. Jameson, Y.-S. Lee, G. Khitrova, H.M. Gibbs, A.C. Klettke, M. Kira, S.W. Koch, Terahertz excitation of a coherent Λ-type three-level system of exciton-polariton modes in a quantum-well microcavity. Phys. Rev. Lett. **108**(26), 267402 (2012)
34. A.D. Jameson, J.L. Tomaino, Y.-S. Lee, G. Khitrova, H.M. Gibbs, C.N. Böttge, A.C. Klettke, M. Kira, S.W. Koch, Direct measurement of light-matter energy exchange inside a microcavity. Optica **1**(5), 276 (2014)
35. A.E. Almand-Hunter, H. Li, S.T. Cundiff, M. Mootz, M. Kira, S.W. Koch, Quantum droplets of electrons and holes. Nature **506**(7489), 471–475 (2014)
36. L. Dominici, D. Colas, S. Donati, J.P.R. Cuartas, M. De Giorgi, D. Ballarini, G. Guirales, J.C.L. Carreño, A. Bramati, G. Gigli, E. Del Valle, F.P. Laussy, D. Sanvitto, Ultrafast control and Rabi oscillations of polaritons. Phys. Rev. Lett. **113**(22), 226401 (2014)
37. D. Colas, L. Dominici, S. Donati, A.A. Pervishko, T.C.H. Liew, I.A. Shelykh, D. Ballarini, M. De Giorgi, A. Bramati, G. Gigli, E. Del Valle, F.P. Laussy, A.V. Kavokin, D. Sanvitto, Polarization shaping of Poincaré beams by polariton oscillations. Light Sci. Appl. **4**, e350 (2015)
38. J.M. Ménard, C. Poellmann, M. Porer, U. Leierseder, E. Galopin, A. Lemaître, A. Amo, J. Bloch, R. Huber, Revealing the dark side of a bright exciton-polariton condensate. Nat. Commun. **5**, 4648 (2014)

Chapter 8
Optically-Excited Polariton Condensates

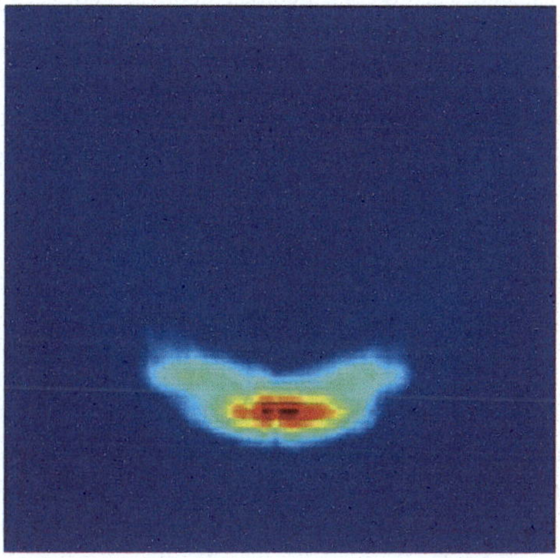

Abstract The observation of polariton condensation with the help of high-quality optical microcavities was a long sought goal in solid-state research. Optical pumping of planar microresonator structures consisting of multiple quantum wells as active region offers a common and direct path towards exciton–polariton studies. After obtaining spectral proof of polaritons in the linear regime, which describes the response at low-density excitation, the behaviour of a polariton gas at increased particle densities gets naturally into the focus, where stimulated scattering is expected to set in. The demonstration of a condensate is typically linked to signatures such as a macroscopic ground-state occupation, a change in the photon statistics and a spontaneous build-up of long-range order evidenced as spatial coherence. Indeed, to establish the link to BEC, all aspects must be carefully investigated. These topics are subject of this chapter, which concludes with examples of condensates at elevated temperature, polariton systems with special features and superfluidity studies on polariton condensates.

A. Rahimi-Iman, *Polariton Physics*, Springer Series in Optical Sciences 229,
https://doi.org/10.1007/978-3-030-39333-5_8

8.1 The Observation of Polariton Condensation

A major goal in the domain of polariton research independent of the material system has been the demonstration of Bose–Einstein condensation of polaritons. However, care has to be taken when labelling a nonlinearity in the emission behaviour and more specifically a macroscopically occupied state a BEC. Polariton condensates can be indeed different from a BEC, which exists in thermodynamic equilibrium. Nevertheless, since many properties and signatures such as spatial [1–3] and temporal [4, 5] coherence can be observed which are comparable to that of a condensate in thermal equilibrium, QW-microcavities are predestined for studies on dynamical BEC, i.e. a BEC-like scenario in a leaky photon-emitting (driven-dissipative) system in which the particle number is not preserved but constantly uphold through a refilling process from incoherent reservoir states. An optically-excited polariton gas may act as a reference system for the spectroscopic characterization and experimental demonstration of a dynamical BEC. Examples will be given on how a condensate can be identified by means of optical measurements.

8.1.1 Condensate Studies in the Literature

In addition to the aforementioned coherence-study examples, it is worth highlighting the early developments and the following wave of condensate demonstrations and characterization efforts. In this context, the experimental works of Deng et al. [4, 6, 7] and Richard et al. [8, 9] in the groups of Y. Yamamoto in Stanford, U.S.A., and L. S. Dang in Grenoble, France, respectively, deserve particular consideration. The researchers were reporting "condensation of semiconductor microcavity exciton polaritons", "quantum degenerate exciton–polaritons in thermal equilibrium" for a GaAs microcavity (Yamamoto group) and "spontaneous coherent phase transition of polaritons in CdTe microcavities" and "experimental evidence for non-equilibrium Bose condensation of exciton polaritons" for a CdTe microcavity (Dang group). In 2003, an article by Deng et al. even compared polariton lasing with photon lasing in a semiconductor microcavity [6], effectively introducing the previously anticipated polariton laser [10] experimentally to the research community. It is worth noting that early observations of bosonic final-state stimulation, macroscopic ground-state occupation and, hence, polariton lasing go back to the end of the 1990s [11, 12].

These foundational works led to the culmination in this field which was the first BEC claims by Kasprzak, Richard, Kundermann, Baas et al. [1], Deng et al. [3] and Balili et al. [13], more precisely that of a quasi-BEC or dynamical BEC. Signatures of BEC were understood to be a spectral and spatial narrowing, a peak at zero momentum in the momentum distribution, first-order coherence, and spontaneous linear polarization of the light emission [13]. Proof of enhanced spatial correlations above threshold were key to those demonstrations [1, 3, 13]. Furthermore, first temporal correlation measurements shed light on the photon statistics of polariton condensates, which are formed by interacting particles and can exhibit considerable intensity fluctuations [4, 5]. At that time, intensity noise and an increase of spatial correlations in a quantum fluid of cavity–polaritons featuring quantum degeneracy were addressed also by other work, e.g. that of Baas et al. [14]. A valuable summary of the early approaches to this condensation topic can be found in the doctoral thesis of Deng [15], and a comprehensive overview on the whole subject is given in 2010s review article of Deng et al. [16].

In the years to come, polariton condensates were intensely studied for various cavity–exciton detunings, both, in the spectral [17] as well as in the temporal domain [5, 18–20]. To reach condensation, many schemes for resonant and non-resonant excitation of polaritons were used. For example, non-resonant excitation of charge carriers with a continuous-wave (cw) laser at the first higher-energy absorption maximum of the microcavity stopband was used to generate incoherent excitons that form polaritons [21], or ns-pulsed non-resonant excitation [22]. Alternatively, quasi-resonant excitation can be pursued by pumping directly the LP branch at higher wavevectors with sub-ps [23] ($\approx$100 fs), ps [4] and ns pulses [24]. One example of cw quasi-resonant excitation near the inflection point of the LP dispersion with pump–signal–idler scheme is given in [25]. For condensate studies, typically linearly polarized laser light is chosen, since stimulated ground-state scattering would occur less efficient for excitation with circularly polarized pump light due to the injection of excitons with a single spin orientation [21].

While many reports deal with individual aspects of a condensate separately, a desirable comprehensive characterization of the pump-power-dependent transition of exciton–polaritons from a linear regime to a dynamical condensate can be found in [23], which compares spectral signatures, spatial coherence and photon statistics at different excitation densities at exactly the same experimental conditions, i.e. detuning, temperature and excitation. To provide a more or less complete characterization of a polariton condensate, the emission characteristics in the real-space and momentum-space domain were analysed. A two-point interference experiment showed interference patterns in the polariton spectrum, indicating an increase of spatial coherence in the region of the nonlinearity. Also, density-dependent changes of the polariton dispersion and distribution were studied. Furthermore, the degree of intensity fluctuations as a signature of particle–particle interactions and temporal coherence was probed by second-order temporal autocorrelation measurements. Owing to the similarities between condensation and lasing, care has to be taken when polariton systems—particularly unexplored ones—are studied with regard to nonlinearities.

Typical experiments on dynamical condensation are performed with microcavities consisting of multiple quantum wells and a $\lambda/2$ cavity. A high number of DBR pairs guarantees experimental resonator Q-factors, which are typically an estimation of a lower limit, on the order of 10,000 (see Sect. 5.3) [23]; the Q value depends significantly on the actual structure design (number of pairs, cavity length) and material system (refractive index contrast). A photon lifetime of $\tau_C \approx 7$ ps in such a microcavity can be deduced for $Q = 10000$. Via $\tau_{LP} \cong \tau_C/|C|^2$ (see (2.40) with $\tau_C \ll \tau_X$, the exciton lifetime, and $|C|^2$ the photonic Hopfield coefficient, (2.29)), a polariton lifetime of about 20 ps can be estimated for a detuning of about 3 meV. Such slightly excitonic regime with given high Q-factor of the QW-microcavity allows one to study dynamic condensation with moderate decay times. On the other hand, it can be also attractive to study highly-photonic quantum degenerate states of bosonic matter [18].

8.1.2 Optical Pumping Schemes

Continuous-wave pumping is the most common scheme for steady-state observations of polaritons in microcavity systems. Pulsed excitation with durations much longer than the exciton lifetime can be still regarded as quasi-continuous, although it must be guaranteed that rise and fall times are negligibly small and the pulse plateau is more or less constant, in order to prevent temporal averaging over different density regimes in cw acquisition.

For ultra-short-pulsed excitation on the order of picoseconds and less, a high-repetition rate facilitates time-integrated acquisition, since the dark times with respect to the bright times (signal after excitation pulse) dominate. Ultra-short pulses are commonly delivered by mode-locked Titanium:Sapphire oscillators with a duty cycle of about 80 MHz. Pulse durations typically range between 100 fs and a few ps depending on the laser configuration. Low-repetition rate mode-locked solid-state amplifiers can deliver pulses with tens of fs and less duration (and high pulse energy), however, the disadvantages overshadow the benefits for polariton excitation purposes.

While non-resonant pumping can occur basically at an arbitrary angle (commonly, normal incidence is used, taking the same path as for signal detection, see Fig. 7.1a), quasi-resonant optical pumping is typically applied under an angle, e.g. $\theta_0 \approx 50°$ ($k_\parallel \approx 7\,\mu\text{m}^{-1}$) (Fig. 7.1c). This pump scheme can lead to an increased occupation of $k = 0$ polaritons after a multi-step phonon-assisted relaxation into the region around the LP's inflection point, from which polaritons take part in the stimulated-scattering processes into the ground state [16]. Another side-pump scheme directly uses the numerical aperture of microscope objectives by displacing the laser beam from the optical axis and thereby achieving an angled irradiation on the sample through the optics (see Fig. 7.1d). The quasi-resonant excitation schemes offer an efficient injection of polaritons with finite $k_\parallel$ (see Fig. 4.4a).

Indeed, ps pulses can be sufficient to examine the decay characteristics with given spectroscopic tools with corresponding temporal resolution on the order of the excitation pulse duration. However, it has been shown that employing 150 fs pulses can be beneficial [23]. Such pulse duration is much shorter than the average polariton lifetime on the order of 10 ps in the ground state. This delta-shape excitation is advantageous when it comes to the observation of self-interaction effects in a system during the time reservoir states are populated. Furthermore, it suppresses further external perturbations of the ensemble, such as additional particle injection by pump light. It is worth noting that, for the discussed example measurement, only a fraction of $\approx$8% of the spectrally broadened pump light (with linewidth $\approx$8 meV for a 150-fs pulse) is coupled into the structure at given incidence angle owing to the spectral selectivity of the polariton mode (i.e. the resonance of that QW-microcavity system in the strong-coupling regime, with linewidth $\approx$1 meV).

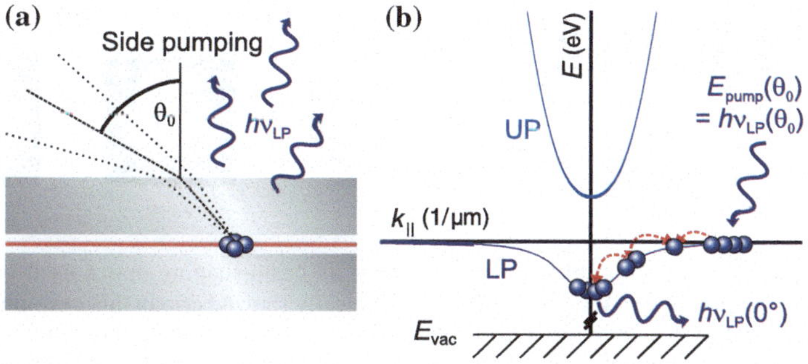

Fig. 8.1 Sketch of a side-pumping scheme for quasi-resonant excitation of polaritons in their lower-polariton (LP) branch. **a** Side pumping with a laser in resonance with reservoir states under an angle. **b** Polariton injection into reservoir states for relaxation along the LP dispersion into the quasi-particle's ground state, assisted by polariton–polariton as well as polariton–phonon scattering processes. From a polariton state, spontaneous decay takes place with photon energies corresponding to the quasi-particle state's energy under conservation of momentum. Thus, ground-state emission is perpendicular to the plane of the quantum wells. Emission at higher polariton energies will occur under a finite angle

Pumping the planar microresonator structures from the side can lead to typical spot sizes on the order of tens of micrometers. For a 75-mm planoconvex lens, the focus results in an elliptical irradiance with approximately $40 \times 30\,\mu m$ (FWHM) with Gaussian intensity distribution. A rather large size compared to the diffusion length of polaritons on the order of a few micrometers guarantees a homogeneous excitation profile of a polariton gas and prevents excitation-spot induced quantisation effects [26]. In contrast, a strong focus can be deliberately used to induce a potential landscape for spatial confinement effects [27]. The advantage of angled side-pumping is given by the particularly easy filtering of the quasi-resonantly excited photoluminescence by cross-polarization techniques, while laser back-reflection into the detection path is naturally suppressed. This allows the recording of emission profiles both in the real space as well as in the phase space (Fourier space) for a wide excitation range. Thereby, the collected laser signal can be reduced by more than six orders of magnitude below the level of polariton emission intensities. Only individual spot-like surface inhomogeneities or adsorbed debris would show local intensity spikes in the field of view due to pronounced scattering into the detection path.

In general, condensates which are pumped non-resonantly can be expected to be free of any imprinted coherence structure induced by the laser [4]. This is justified—even for side pumping into reservoir states (Fig. 8.1)—by the occurrence of incoherent scattering processes which precede the occupation of the ground state, such as parametric polariton–polariton as well as polariton–phonon scattering [6] (see Fig. 3.3b and c). In the quasi-resonant finite-k excitation into the LP branch with the given energy–momentum-space selectivity of the side-pump scheme, condensation thus occurs on the basis of spontaneous coherence build-up. In addition, a strong reduction in the generation of free charge carriers in the microcavity system can

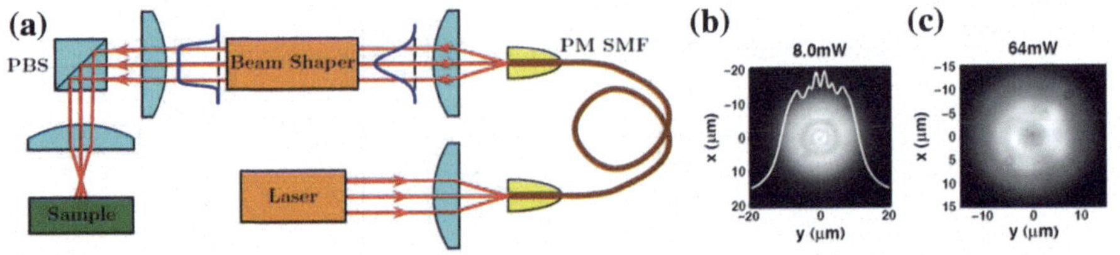

Fig. 8.2 Creation of a flat-top-like excitation profile by beam-shaping optics used prior to in-coupling of the laser light into the experiment, schematically shown here. In this example, the laser beam is transferred to the setup through a polarization-maintaining single-mode fibre (PMSMF) before beam shaping and projection onto the sample. **b** Quasi-flat-top profile of the pumping spot generated by the beam shaper measured below the condensation threshold. A white line projects the spot's cross-section along the horizontal axis onto the displayed real-space (NF) image. **c** NF image of the top-hat-pumped condensate, i.e. the emission profile recorded above threshold. Reused with permission. Reference [32] (Supplementary Information) Copyright 2011, Springer Nature

be achieved when compared to non-resonant excitation which reduces the bleaching of polaritons [28, 29]. Furthermore, the quasi-resonant excitation of polaritons suppresses the creation of hot excitons, which could cause a dephasing of polaritons through scattering processes. Also, two-photon generation of polaritons has been used in the literature [30]: The different states which were occupied using such nonlinear absorption scheme with its energy–momentum sensitivity were studied by temporally and spatially resolving the polaritons' propagation after their generation.

One should note, that since the pump landscape determines the spatial distribution of polaritons, it plays a role in many condensation studies using different excitation schemes. While many experiments are performed with Gaussian-shaped excitation spots, some experiments require beam shaping for the incident pump laser in order to obtain for instance a flat-top-like excitation profile on the sample, or in seldom cases even a donut-shaped pump landscape or other shapes. This is for example achieved by beam-shaping optics used prior to in-coupling of the laser light into the experiment, as schematically shown in Fig. 8.2, or by projection optics and spatial light modulators (not shown here).

Another example of an excitation scheme is given, when one uses a combination of optical pumping into the inflection point for resonant continuous injection of polaritons at finite k together with a pulsed idler in a triggered optical-parametric oscillator scheme for condensate fluidic studies [31], as shown in Fig. 8.3a–c. The cw-pump laser provides a steady refill of polaritons into an excited state on the LP branch, while an idler pulse creates a propagating lower-energetic signal that is constantly repopulated from the pump state. Thereby, propagation of a polariton gas can be mapped in real space, as demonstrated by Amo et al. [31].

It is important to note that the condensates achieved by different excitation conditions differ in their characteristics from each other, be it regarding their blue shift, the critical particle density, the degree of coherence, the momentum-space expansion or their thermalization dynamics. These differences occur, because condensation in those systems is a dynamical process. This process can be considered as the result of the interplay between a pump-related ground-state polariton density, dynamics at

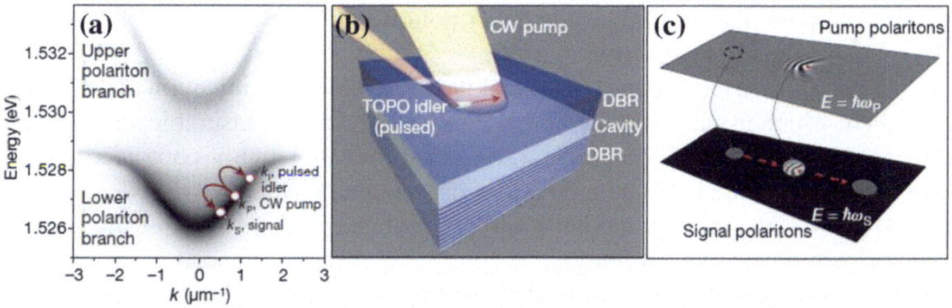

Fig. 8.3 Example of an optical-parametric oscillator (OPO) scheme: **a** Schematically indicated excitation of the continuous pump into the cavity–polariton dispersion for the formation of a polariton state at the cw pump energy, which feeds the signal polaritons at a lower energy on the LP dispersion. For a triggered optical-parametric oscillator (TOPO) scheme, the conditions $2k_P = k_S + k_I$ and $2E_P = E_S + E_I$ are fulfilled, with subscript P, I and S representing pump, idler and signal, respectively. The TOPO process is initialised by a pulse at the idler state. **b** Impinging two laser beams on the sample surface at an arbitrary direction in the plane of the cavity in an artistic drawing of the QW-microcavity system. **c** The motion of two fluids at different energy levels (gray and black plane) is schematically indicated, with pump polaritons in the upper plane and signal polaritons in the lower energy plane, which in fact are not spatially separated, but visually. The signal-polariton fluid's motion against a defect depicted by the red point becomes detectable thanks to the continuous feeding from the pump polaritons. Reused with permission. Reference [31] Copyright 2009, Springer Nature

given detuning, differently obtained spectral condensate dimensions and heat development in the system under strong laser excitation. Thus, the degree of thermalization and cooling in the polariton system determines whether the condensate regime is closer to the so-called "polariton laser" or to a quasi-BEC state in thermal equilibrium.

8.1.3 Spectral Features of Polaritons

The identification of polaritons in emission spectra is mainly facilitated by Fourier-space resolved spectroscopy (see Sect. 7.3 for details on the far-field projection technique). Indeed, the strongest signature of strong coupling is the anti-crossing behaviour of the exciton and cavity mode. However, this is not always unambiguous when multiple spectral features, broad features or small coupling strengths are exhibited. Moreover, a tuning mechanism is needed to vary the detuning in a fashion that reveals the anti-crossing nature of the modes.

In contrast to "active" tuning methods (e.g. by temperature or field strength variations), anti-crossing at a given sample position/configuration can naturally occur at finite angles for a polariton system close to zero detuning. For slightly positive, zero or negative detunings, the flattening of the dispersion towards higher $k_{\|}$ and the occurrence of an inflection point are characteristic for the coupled-system's lower-energy dispersion that resembles the lower polariton (LP) branch (correspondingly lower with respect to the exciton mode).

Only angle-resolved spectra can reveal the presence of an inflection point for the LP branch, which usually is accessible within the typically measurable angles (light cone) in the case of small detunings ($\Delta \ll \hbar\Omega$). Yet, for highly-photonic coupled modes, the difference between a polariton branch and a bare cavity mode is hardly seen. Moreover, the inflection point could occur at inaccessible in-plane momentum $k_{\parallel}$ at such large negative detunings.

On the other hand, if the detuning is already too positive, the LP branch is not very different than a bare exciton mode. Moreover, a positively detuned cavity mode will not obviously (anti-)cross the already lower-lying exciton mode. For lowest in-plane momenta, one has then to anticipate the mode repulsion of the bare resonances towards a normal-mode splitting for the coupled oscillators, supported by dispersion fit curves which indicate that the upper branch is not a bare-cavity "parabola" and the lower branch is not a bare-exciton "line" (i.e. flat dispersion).

Indeed, the position of the QW system's exciton mode(s) should be known in order to determine reliable detuning situations. In contrast, the position of the bare cavity mode for a prepared system cannot be determined precisely ad hoc. For $\Delta \ll \hbar\Omega$, the dispersion shape is governed by the coupling situation. At highly negative (or positive) detunings $|\Delta| \gg \hbar\Omega$, the observed considerably-curved dispersion can be representative of the uncoupled photon mode. The upper branch can be hardly seen in photoluminescence due to the population distribution, particularly for highly positive detunings, while in reflection spectra it might be observable. Nevertheless, access to such different and in some case extreme detunings is not readily provided for every device.

The experimental Q-factor of planar microresonator structures can be extracted after micropillar processing, since the photoluminescence from such pillar is not prone to any inhomogeneous broadening due to sample quality, a structural cavity wedge or other effects related to the planar nature of the unprocessed structures. Pillars with 4 μm for GaAs-based microcavities [33–35] have shown to offer a trade-off between too strong spatial confinement (towards 1 μm), with the corresponding negative influence of surface scattering, and too shallow confinement, with the corresponding lack of discrete microcavity modes with clear spectral separation (towards 20 μm). From pillars on wafer pieces with red (negative) detuning (Δ_-), one can extract the spectral linewidth of the emission ground state reliably when the excitonic fraction of the pillar-polariton mode $|X|^2 < 10\%$, i.e. $\left|\Delta_- = E_{C,0} - E_X\right| \gg \hbar\Omega$.

This pillar-based method is well suited for the reliable determination of the optical Q-factor in QW-microcavities. In contrast, planar samples measured in reflection would inherently be broadened by the angle dependence, unless signal for $k_{\parallel} = 0$ is carefully extracted for linewidth measurements, and by the wafer-tapering effect, unless the signal is strongly spatially filtered. In 0D optical structures such as micropillars, the measured ground-state linewidth directly represents normal incidence light ($k_{\parallel}$=0), and an experimental Q-factor automatically represents the lower limit already accounting for additional optical losses described by (5.10) in Sect. 5.3.

For spectral characterization of polariton condensates, photoluminescence experiments are typically conducted with time-integrated acquisition by an imaging monochromator in the near-field (NF) or far-field (FF) detection mode (for method details see Chap. 7). Two-dimensional FF spectra (energy over in-plane momentum) are commonly recorded for polariton condensate studies, from which line spectra can be easily extracted for a certain k value or region, i.e. "manual" k-space integration can be applied over a desired range, for evaluation purposes. Such extracted spectra are referred to in the following as $k_{\|} = 0$ (short $k = 0$) spectra and are naturally essential for the detailed analysis of ground-state emission. Momentum-integrated spectra are indicated by their integration range, e.g. $k_{\|} = (0.0 \pm 0.1)\,\mu\text{m}^{-1}$.

To observe condensation, pump-power series are performed with the aim of recording a peculiar nonlinearity in the emission behaviour. Signatures of condensation in FF spectra are mainly k-space narrowing and a macroscopic population of the ground state. Further evaluation of extracted $k = 0$ spectra typically shows an s-shape in double-logarithmic input–output curves, energy shifts and linewidth changes, which are strongly correlated with a threshold behaviour in the recorded excitation-density-dependent photoluminescence. However, the behaviour of the output can vary from structure to structure based on the material system, structure design, resonator quality or detuning situation. For a detailed discussion of polariton lasing/condensation signatures, see Sect. 4.3.

8.2 Condensation Experimentally Characterized

In the following, typical spectral signatures of polariton condensation are summarized. This includes features in real-space and momentum-space as well as spatial and temporal coherence considerations, which are of importance for quasi-BEC studies.

8.2.1 Real-Space and Momentum-Space Distribution of Condensate Emission

From historic BEC studies [36–38] it can be understood that the change in the momentum distribution with a drastic narrowing around the energy ground state—usually where the kinetic energy is zero—acts as the key signature of condensation. This strongly motivates the recording of Fourier-space images for the emitter system.

Naturally, pump-dependent experiments are at the heart of condensation studies, as the crossing of a nonlinearity threshold in connection with a phase transition has to be evidenced experimentally for a polariton gas. While most attention goes to the spectral signatures, it is worth taking a look at the NF and FF images as the phase transition to a condensate occurs (Fig. 8.4a and b, respectively).

Interestingly, the real-space image of a condensate can differ between studies. While in samples with structural imperfections or inhomogeneities, multiple bright spots within the pump area may occur as shown in various reports (e.g. in [8]),

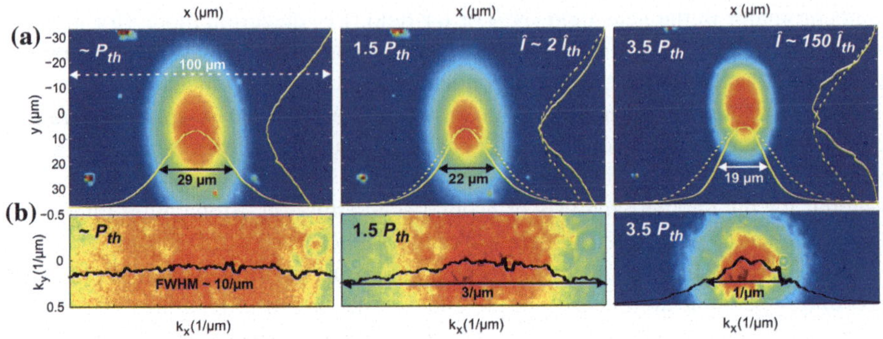

Fig. 8.4 **a** Example of a pump-power-dependent NF-imaging measurement of the sample emission across a condensation threshold, shown as normalized time-integrated false-colour images (from blue to red with increasing intensity). $\hat{I}$ labels the LP emission's peak intensity at given pump rate in relation to that of the value $\hat{I}_{\text{th}}$ at P_{th} **b** Corresponding images of the Fourier-space plane are presented for the same excitation power densities of 1.0, 1.5, and 3.5 times the threshold pump power density (P_{th}). Solid lines represent cross-sectional views with normalized intensity projected into the respective false-colour image. In the FF images, the FWHM of the momentum-space intensity distribution decreases from about 10 to 1 μm^{-1}. This clear k-space emission narrowing towards the centre of the phase space corresponds to a macroscopic population of the $k = 0$ state. In analogy, cross-sectional views in both x and y direction of the NF signal (solid lines) reveal a narrowing of the emission region when compared to the spot size before condensation, visible for P_{th} and projected as dotted lines. In this quasi-resonant side pump configuration, laser reflections were strongly reduced in a cross-polarized pump–detection scheme by orders of magnitudes below the PL intensity, while shiny spots originating from laser back scattering from surface debris/defects well indicate the position on the sample. All plots are in linear intensity scale. Adapted from [39]. Courtesy of the author

individual observations show large area smooth condensate emission profiles, e.g. in the power-dependent study of [23].

Time-integrated CCD-camera images of the polariton emission from a large-area condensate with strong attenuation of the pump light had shown for different powers around the condensation threshold, how the Gaussian-shaped NF emission profile had reduced in size within the pump-spot area and changed shape with the phase transition. Above threshold, a non-Gaussian intensity-distribution profile with about two-third of the size compared to the size in the at-threshold (linear) regime was observed, which represents the formation of a spatially localized polariton condensate (in the example above, 20 μm in expansion, measured as full-width-at-half-maximum FWHM). Moreover, its non-Gaussian shape exhibited steeper flanks. Such spatial emission profile indicates the establishment of a dynamical BEC-like condensate and can be distinguished from the behaviour of cavity lasing [6].

While a spatial shrinking of the polariton gas above the condensation threshold into a smaller spot or multiple small spots could be attributed to localization of polariton condensates in the (inhomogeneity-induced or gain-induced) potential landscape of microcavities or defect sites, it might be well a feature of the condensed system, as theoretical considerations in the literature suggest [40]. Based on strong feedback between the condensate and reservoir, instability-induced textures in space and time can occur.

Similarly recorded CCD-camera images of the emission profiles in the FF projection, i.e. in-plane-momentum resolved, had also been obtained for the same emission regimes in the aforementioned study with 20 μm condensate size. In the linear emission regime, a homogeneous k-space distribution was observed. A cross-sectional analysis indicated a Gaussian distribution with a FWHM value of $\approx$ 10 μm^{-1}. For pump densities above threshold, a reduction to about 1 μm^{-1} was observable. This narrowing of the momentum distribution represents clearly condensate formation by a macroscopic ground-state occupation at $k_{\|} = 0$ [1, 23, 24]. In addition to the occupation narrowing, an analysis of the polarization distribution in the emission profile is discussed in [24]. In the literature, condensates with position and momentum uncertainty product close to the Heisenberg limit were reported, highlighting that a minimum uncertainty wave packet of the experimentally obtained polariton condensate could provide access to interesting coherent matter–wave phenomena [3].

The here discussed signatures in NF and FF projection show a clear modification of the emission characteristics without any spectral analysis. These considerations may now be amended by a spectral characterization, which particularly serves the identification of polariton dispersions.

8.2.2 Stimulated Scattering and Macroscopic Ground-State Occupation

The polariton character can be probed spectrally by recordings of the mode dispersion of the shiny LP branch in luminescence studies with the help of FF spectroscopy. Though, this can be challenging above the condensation threshold. Compared to the photon dispersion in planar microresonator structures which features a parabolic behaviour in the investigated in-plane-momentum range, the energy dispersion of the LP mode exhibits an inflection point at specific wavevectors around resonance detuning (around $k_{\|} \approx 2\,\mu m^{-1}$ for common GaAs systems). This allows one to identify a polaritonic regime based on FF spectroscopy. However, when a strong momentum-space as well as spectral narrowing occurs, hardly any dispersion remains visible. In some cases, the dispersion nature is fully lost, while in other cases a faint dispersion signal remains detectable as background emission. Particularly, time-integrated detection of pulsed excitation may lead to an overlap of low-density and high-density signal components in the acquired spectra. Yet, this occurrence of a background dispersion may also occur in some cw schemes, whereas it does not necessarily has to be a dispersion branch belonging to the spot of the designated condensate emission.

By normalizing a measured intensity profile of a dispersion curve to its maximum and plotting the intensity profile both in linear and in logarithmic intensity scale next to each other, one can distinguish high intensity and low intensity features practically. An auxiliary virtual empty-cavity (uncoupled-cavity) resonance can be plotted on top of the emitting branch in FF spectra to better differentiate polariton modes from cavity modes, particularly, when an inflection point in the dispersion is not obtained clearly (e.g. in certain detuning situations). Such modelled cavity resonance dispersion can be artificially aligned to the ground-state in order to highlight the reduced curvature of the polariton branch from which condensation is observed. But care has to be taken that the test curvature of the uncoupled resonator mode is verified for a similar weak-coupling structure with its angle-dependent cavity mode, i.e. Fabry–Pérot filter function. This prevents errors from two possible sources: an assumed effective cavity photon mass may be inaccurately determined using an estimated effective refractive index of the cavity structure; the experimental scaling of the detectable Fourier-space (angle) scale may be imprecise. Often, the neighbouring cavity stopband fringes can be used to get a clue about the photon dispersion, or the reflectivity spectrum of bare DBRs of that structure can reveal the corresponding curvature of an uncoupled mode at those observed fringe wavelengths.

To obtain macroscopic ground-state occupation via stimulated scattering between polaritons (see Sect. 3.2), the exciton–polariton density must be experimentally ramped up and remain below the exciton-dissociation threshold. Density-dependent luminescence measurements in the FF spectroscopy mode reveal typical features of macroscopic ground-state occupation. Among them are the formation of a relaxation bottleneck with its pronounced emission at finite k [41] and an increased relaxation into lower energy states accompanied by a threshold behaviour [4, 29]. The occurrence of a relaxation bottleneck is for instance seen in Fig. 8.5a, which in the example of an optically-pumped organic microcavity system is overcome by temperature-dependent thermalization of polaritons (from (a) to (c)). The threshold marks a strong increase of ground-state occupation, which is evidenced as an intensity nonlinearity. Typically, a spectral blue shift of the intensely emitting $k = 0$ mode is observed as a consequence of polariton–polariton interaction [42] and phase-space filling [43]. This transition from a linear to a nonlinear regime marks the onset of effective stimulated scattering into the condensate. Such phase transition can be seen in FF spectra of the example measurement shown in Fig. 8.5 at room temperature (from (d) to (f)).

The observed phase transition typically comes with three pronounced "narrowings" as signatures of an increased temporal and spatial coherence: (1) a significant linewidth narrowing of the emitter (for a coherent gas in the absence of dephasing interaction processes); (2) a momentum-space narrowing (redistribution of the polariton population into a single energy state); and (3) spatial narrowing (formation of condensate areas within the excitation spot in 2D systems) [1, 6, 8, 44]. For a detailed analysis of interaction features and ground-state occupation of a dynamical condensate, one typically extracts $k = 0$ line spectra, and one determines the momentum-space and energy distributions, as discussed later.

Stimulated scattering processes have been the subject of various publications in the field of polariton research. For instance, exciton–exciton scattering dynamics and stimulated scattering into polaritons had been discussed for semiconductor microcavities in the late 1990s by Tassone and Yamamoto [45]. Shortly after, the depopulation of polaritons in a microcavity from reservoir states in a two-beam experiment had been regarded as evidence for stimulated scattering, whereas one beam in the exper-

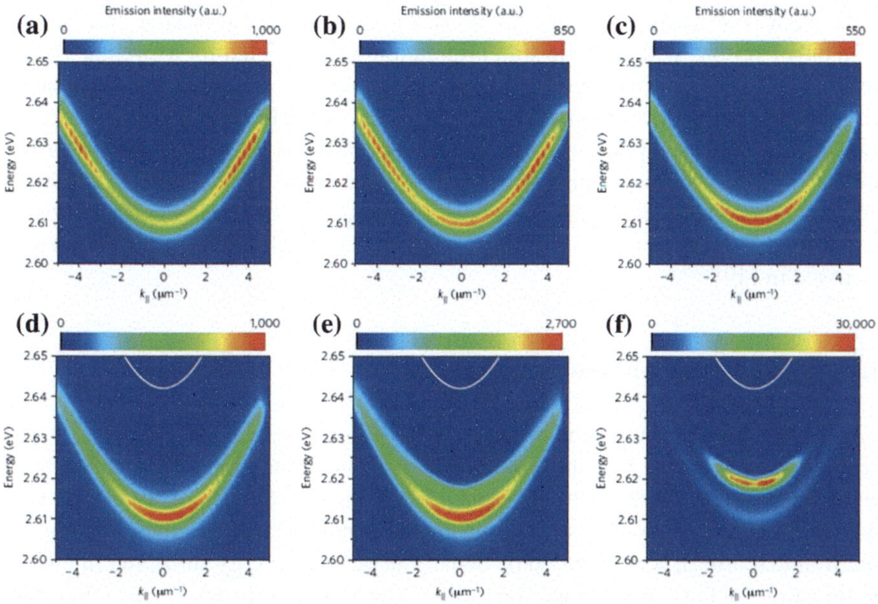

Fig. 8.5 Temperature-dependent thermalization of polaritons and power-dependent condensation: **a–c** show time-integrated Fourier-space-resolved emission spectra for different temperatures T with an excitation fluence below threshold at **a** $T = 6\,\text{K}$, **b** $T = 60\,\text{K}$ and **c** $T = 160\,\text{K}$. With increasing temperature, the clearly observed relaxation bottleneck in the lower-polariton branch is overcome due to increased relaxation efficiency, which provides Boltzmann-like energy and momentum distributions below the condensation threshold (P_{th}). **d–f** show the emergence of a polariton condensate at room temperature as a consequence of ramped up excitation density from **d** below the threshold ($P = 0.22\ P_{\text{th}}$), **e** near the threshold ($P \approx P_{\text{th}}$) to **f** above the nonlinearity threshold for a stimulated macroscopic ground-state occupation ($P = 1.7\ P_{\text{th}}$). The white line indicates the dispersion of the uncoupled cavity mode. Reused with permission. Reference [44] Copyright 2013, Springer Nature

iment was non-resonant and a second beam was resonant. By resonantly injecting polaritons into an energetically lower polariton final state, polariton emission from finite $k_{\parallel}$ vectors was quenched, from which stimulated scattering was deduced by Senellart et al. [46]. At that time, stimulated energy relaxation was also considered by Savvidis et al. when proposing an angle-resonant stimulated polariton amplifier that utilizes boson amplification [47].

Later, the role of spin dynamics in addition to the previously discussed polariton dynamics in semiconductor microcavities were highlighted considering the non-resonant excitation case [48]. Furthermore, signatures of the exciton–polariton relaxation mechanism were revealed in the polarization of the emitted light from a microcavity [21]. Therefore, spin-dependent real and momentum space spectroscopy was used to study spontaneously formed polariton condensates for a non-resonant excitation scheme. Roumpos et al. suggested that for such a system under linear-polarization pumping of excitons with excess energy, relaxation into the final state occurs after multiple phonon-scattering events and only one polariton–polariton scattering event was involved to reach the lowest energy of the LP branch. In contrast, when only excitons with a single spin orientation were injected for a circular pumping case, a bottleneck effect occurred from which inefficient relaxation was deduced. Stimulated polariton–polariton scattering and dynamical BEC of polaritons were also discussed for the case of near-resonance exciton excitation [24]. Polariton–polariton interaction such as repulsion has also been discussed from the theoretical side. Such considerations on the kinetics of polariton condensate formation in microcavity systems were provided for instance in [42]. The subject of polariton interactions is also discussed in [49].

Macroscopical Ground-State Occupation

Pump-power-dependent ground-state emission can be analyzed well in $k = 0$ line spectra. Lorentzian line fits can be used to extract integrated intensity, ground-state energy and spectral linewidth. In input–output curves, the phase transition to a condensate usually reveals itself as an intensity nonlinearity (s-shape in double-logarithmic scaling), accompanied by density-dependent energy shifts and linewidth modifications of the $k = 0$ mode. The onset of the nonlinearity defines the condensation threshold, as indicated in Fig. 8.6a. The output intensities recorded with arbitrary units are often normalized to the value at the threshold, which represents the fact that the ground-state occupation is on average one at the threshold, thereby enabling the onset of stimulated scattering [45].

Input–output curves usually show excitation density, pump fluence or normalized pump powers P/P_{th} (P_{th} is the threshold value) on the horizontal scale. When estimating absolute excitation densities based on pump powers and spot sizes, absorption, transmission and reflection of the pump light should be adequately taken into account.

Time-integrated PL under pulsed excitation usually allows one to record signatures of the linear regime in addition to the condensation emission around the threshold due to the fact that the temporal evolution of the emission pulse ranges between low and high densities (as there is no steady state) (see [50]). The linear LP branch is thus detectable in the background with unchanged energy or linewidth, representing the lower-energy dispersion according to the eigen modes of the single-particle Hamiltonian. A double-mode fit of the $k = 0$ spectra can help to separate this contribution from the condensate mode (see [23]). Similarly, one can include emitted intensity from finite k-vectors (obtained from respective line spectra) in the input–output diagram to indicate the density-dependent evolution of reservoir states, for instance to highlight depopulation effects [46].

Considering polariton occupation numbers, the condensate fraction υ_0/υ (ground-state to total polariton number or density) as a function of pump power typically shows values in excess of 0.5 above the condensation threshold as a consequence of macroscopic occupation through final-state stimulation (cf. [23, 51]). For increased pump rates, saturation effects can cause a decrease of the condensate fraction. Such saturation effects can originate from dephasing of polaritons in the ground state

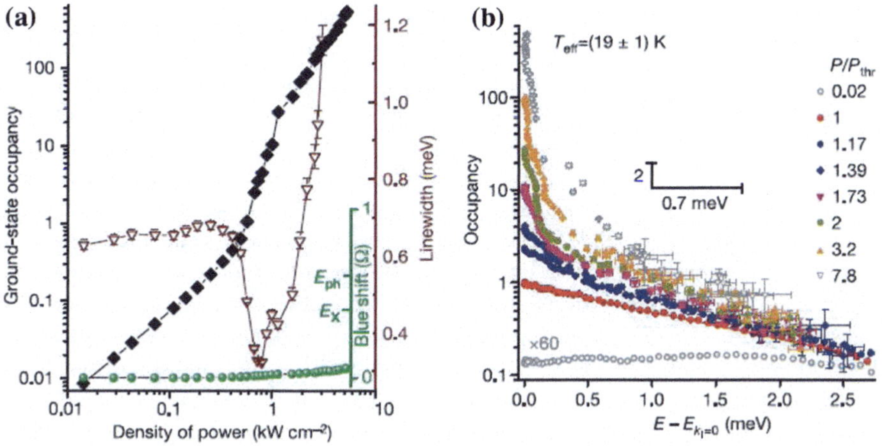

Fig. 8.6 Ground-state occupancy (solid black diamonds), i.e. $k = 0 = k_0$ emission intensity, k_0-emission energy blue-shift (solid green circles) and k_0-emission linewidth (open red triangles) as a function of the excitation power. The energy scale for the blue-shift data is given in units of the Rabi splitting (here $E_{\text{Rabi}} = \Omega = 26\,\text{meV}$). The nonlinearity in the input–output diagram is accompanied by a pronounced decrease of the linewidth which is attributed to the increase of coherence in the polariton ground state gas. However, polariton–polariton interaction is understood to widen the emission lines spectrally at increased densities. The registered emission blue-shift of the condensate with respect to the single-particle LP ground state is less than 7% of the Rabi splitting for densities up to a few times the threshold density. The positions of the uncoupled exciton (E_X) and photon mode (here E_{ph}, i.e. E_C) are marked on the scale and energetically well separated from the condensate mode, an indicator for the strong-coupling regime. **b** The polariton occupancy as a function of the energy separation from the (density-dependent) LP's ground state, i.e. the distribution function, is plotted in a semi-logarithmic scale for various excitation powers. Such occupancy information is deduced from FF emission data, whereas the radiative lifetime of polaritons along their dispersion has to be taken into account when deducing occupancies from PL intensities. Typically, a Boltzmann-like exponential decay of the distribution function is only found up to the threshold density, from which the thermal gas temperature can be deduced. In contrast, a Bose–Einstein-like distribution is obtained well above the condensation threshold, with the ground state massively occupied and the excited states saturated. Reused with permission. Reference [1] Copyright 2006, Springer Nature

through the high particle density and interactions with reservoir states [52]. In practice, the condensate fraction can be reduced well above threshold both by saturation effects through increased interactions in high-density regimes (as indicated by photon statistics) and by local heating effects in the sample through possible changes of the threshold conditions (as indicated by dispersion characteristics) [23]. One should note that for off-resonant excitation, estimations of the condensate fraction cannot be correct when only considering optical output, since the considerable dark exciton cloud (reservoir/hot excitons outside the light cone) remains undetected. However, complementary spectroscopic techniques can be used to reveal the "dark side" of exciton–polariton condensates (see [53]).

Next, the power-dependence of the emission energy is briefly discussed. The typically obtained gradual blue shifts of the condensate emission is often explained as a signature of polariton–polariton interaction and commonly used as justification for a polariton condensation claim, when the bare cavity mode is energetically distant to the emitter's energy. If the condensate's emission energy is considerably lower than the cavity mode energy $E_{C,0}$, e.g. separated on the order of half of the Rabi splitting, this can be regarded as a sign that despite high pump rates, strong coupling is preserved. Density-dependent energy shifts above threshold up to 1 meV are not unusual. The higher the particle density, the stronger the repulsive interactions of polaritons owing to their matter component. This condensate blue shift is discussed in the literature [16, 51], with repulsive fermionic exchange interaction between quantum-well excitons of the same spin orientation [54, 55], and a reduced Rabi splitting due to a pronounced phase-space filling [43]. However, local heating effects of the pump area must be also taken into consideration, which can cause energy shifts for the bare cavity and exciton modes and even affect the coupling strength.

In power-dependent measurements, a pronounced linewidth drop above a nonlinearity threshold is a measure of the first-order temporal coherence build-up of the emission. Typically, a clear broadening occurs for the LP mode in the density range of the threshold, until the expected reduction of the ground-state mode's linewidth is evidenced which indicates an increased coherence of that mode according to the classical coherence time $\tau_c \approx \hbar/\gamma$ ($\gamma = \hbar\Delta\omega$ energy linewidth). This is in good agreeement with observations of a spontaneous coherence build-up based on spatial first-order correlation function measurements ($g^{(1)}(r)$). In such measurements, the pronounced enhancement of interference patterns (indicative of increased spatial coherence) clearly correlates with the phase transition around the threshold excitation density [2, 3], which is accompanied by a typical linewidth reduction shortly above threshold (hinting at a rise in temporal coherence). This is supported by theoretical considerations as well [56]. At higher excitation densities, the obtained condensate mode often features a gradual linewidth broadening. A mode broadening around the threshold and well above the threshold can be attributed to particle-density dependent interaction effects in the system [57], whereas such signatures can also be caused by a temporally-integrated acquisition of spectra from pulsed systems which undergo several density regimes within the evolution of the emission pulse.

Indeed, individual signatures of a dynamical and self-interacting exciton–polariton system may bear strong similarities to the features of a photon laser and this is often a major concern for the identification of polariton lasing [58, 59]. The often justified criticism about ambiguities shows the difficulties regarding the identification and proof of a polariton condensate. However, the evidence of a preserved polaritonic dispersion and the existence of a matter component above a nonlinearity threshold for the emitting mode can be used to rule out a weak-coupling regime and therewith the occurrence of conventional photon lasing.

Particularly for positive detunings, aforementioned ground-state-energy blueshifts represent changes of the rather excitonic polariton system within a strong-coupling regime unambiguously, whereas for negative detunings, a density-dependent condensate shift is sometimes not well distinguishable from the breaking up of strong coupling—the exclusion of photon lasing from the cavity may be ambiguous. Thus, this case deserves special attention. It is known that cavity lasing can pin itself at the gain maximum (conventional laser frequency pulling) and the photon laser does not necessarily need to occur at the bare cavity mode energy.

The occurrence of a well visible cavity dispersion in FF spectra is a clear sign of the weak-coupling regime (see [50], which visualizes dispersion for cavity lasing). The preservation of a polariton dispersion in the background with inflection point [23], while the condensate energy off-set is little, hints at condensate emission, i.e. lasing in the strong-coupling regime, or in other words polariton lasing. An example, where the polaritonic dispersion situation and effective mass alter above the threshold as a function of pump power while being in the strong-coupling regime is shown in [23]. Such changes can be on the one hand understood as a consequence of density-dependent interaction effects [6, 52] and the reduction of the Rabi splitting due to higher particle densities [28, 29, 43]. On the other hand, it could be caused by induced heat or refractive index changes due to high exciton densities in the quantum well [58]. Both effects (i.e. temperature and index change) would shift the cavity mode slightly red and affect the detuning situation, and therewith the effective polariton mass.

Generally, a very good understanding of the detuning situation and the energies of exciton and photon resonances is required to rule out a transition to photon lasing. An educated guess of the detuning situation and a mere reliance on coupled-oscillator dispersion modelling with detuning and coupling strength as fit parameter might lead to wrong conclusions. In fact, a double threshold in the input–output characteristics for $k = 0$ emission can be an indicator of two distinct light–matter coupling regimes. Typically, different emission properties can be revealed when a double-threshold behaviour is obtained [32, 50, 60], as discussed by Tempel et al. Particularly, emission transients and photon statistics (coherence properties) exhibited distinguishable features above the first and the second threshold attributed to polariton condensation and cavity lasing, respectively [50], but also the response to a magnetic field differs [60]. Moreover, vortices can occur in the excitation-density range between the two thresholds which are attributed to superfluidity behaviour in (interacting, coherent, bosonic) polariton gases [31, 32, 61–71] that are absent in the photonic regime.

For photonic polariton regimes (or quantized polariton modes without dispersion character), the matter component needs to be verified well above threshold to fully exclude conventional cavity lasing, i.e. lasing in the weak-coupling regime. This can be done through interaction with external fields (see Chap. 9) (cf. [60, 72]. Particularly, for highly-photonic lower-polaritons, this discrimination between strong-coupling and weak-coupling lasing is hardly possible unless a detailed characterization can show clear differences between the two regimes. A study on polariton condensates and highly-photonic quantum-degenerate states of bosonic matter presented an overview on different detuning regimes and the characteristics of their condensation features [18]. In their work, Assmann et al. discuss the mode features above threshold by comparison to homogeneous-equilibrium Bogoliubov-mode and diffusive Goldstone-mode dispersions. The observation of Bogoliubov excitations in exciton–polariton condensates was first reported in 2008 by Utsunomiya et al. [51].

Given the fact that polariton (quasi-)BEC is a non-equilibrium process of a degenerate polariton gas in self-equilibrium, which is not in equilibrium with the baths it is coupled to, its characteristics must deviate from that of atomic BEC which features a thermodynamic phase transition. In this context, it is further shown by Assmann et al. that key signatures of Bose–Einstein condensation can even be observed in highly-photonic quantum-degenerate non-equilibrium systems that do not fulfill the aforementioned self-equilibrium condition of dynamic polariton BEC [18].

The macroscopic ground-state occupation can be well evidenced in FF spectra above threshold, in which the energy and momentum distribution for the emission are strongly narrowed down. The power-dependent momentum-space distribution of polaritons offers an important tool to probe the development and formation of a condensate and its thermodynamic properties. Such an analysis is summarized and compared with numerical simulations in [16]. If the intensities recorded in time-integrated FF spectra are weighted with estimated k-space dependent lifetimes $1/\tau_{LP}(k) = |C|^2(k)/\tau_C + |X|^2(k)/\tau_X \approx |C|^2(k)/\tau_C$ (usually $\tau_X \gg \tau_C$) according to their Hopfield coefficients (see Sect. 2.3.2), occupation numbers $N(k_\parallel)$ can be estimated for polaritons across the LP branch.

To account for the whole $k_\parallel$ space, a radial-dependent factor is considered, since FF spectra originate from a slice through the $k_\parallel$ plane along k_x or k_y. Thus, integrating over the half circle (assuming radial symmetry in the k-space) adds up for each radius $\pm k_\parallel$ the population in the "hidden" states (signal not acquired through the entrance slit); i.e., if the full circumference is $2\pi|k_\parallel|$, a factor of $\pi|k_\parallel|$ corrects the estimated population $N(k_\parallel)$ based on recorded FF-spectra intensities. For comparability in power-dependent studies, the occupation curve can be normalized to the $k = 0$ value of the respective curve (1 representing maximum occupancy) or to the $k = 0$ value of the curve at P_{th} (representing unity in the occupation number of the ground state). In such density-dependent chart, a clear transition can be found in the estimated distribution of polaritons from higher occupation of reservoir states to a macroscopically-occupied ground state. For the condensed gas, $\approx 68\%$ of the emitters would lie within one standard deviation of a normal distribution if such a distribution was used to represent its k-space profile. Deng et al. showed in very early work that $N(k)$ is better represented by a Bose–Einstein (BE) distribution than a Maxwell–Boltzmann (MB) distribution above threshold [6, 7].

In addition to the momentum-space distribution, the analysis of the density-dependent energy distribution reveals a transition to a condensate with a change from an MB distribution at low densities to a distribution above threshold resembling a BE distribution [1, 73] (see Fig. 8.6b). The energetic polariton distribution $N(E)$ with respect to the ground-state energy E_0 for various increasing excitation densities can be analysed when considering k-space integrated lifetime-corrected intensities $I(E - E_0; k_\parallel = 0 \pm |k_{\text{max}}|\,\mu\text{m}^{-1})$. Again, the correction factor for the unmeasured k-states of the whole $k_\parallel$ plane in FF spectroscopy needs to be accounted for (assuming radial symmetry). For better comparability, curves are commonly normalized to their E_0 value, or to the E_0 value for the curve measured at P_{th}, and plotted in semilogarithmical scale.

Typically, above threshold, a pronounced spectral narrowing takes place regarding the whole measured momentum space which directly translates into a narrowing of the occupation distribution. Well below threshold, the distribution of polaritons can be modelled by an MB distribution with a certain gas temperature, which is usually higher than the lattice temperature and which represent the thermalized nature of the polaritons among themselves. However, at elevated pump rates above the condensate threshold, the distribution cannot be matched with an MB curve but with a BE-like distribution. An MB distribution would simply not feature a macroscopic occupation in the ground state. For a quasi-BEC the increased degree of thermalization of the condensed and bosonic gas explains the distribution observed. A transition to a Bose gas in quasi-thermal equilibrium is evidenced, as shown by Deng et al. [7]. At higher densities, dephasing of polaritons through interaction with reservoir states may cause a reduction of the ground-state occupation (see [23], $P > 3.5P_{\mathrm{th}}$ in that experiment).

The occupancy as a function of relative energy for a quasi-BEC of exciton–polaritons reported by Kasprzak et al. 2006 is shown in Fig. 8.6b for different densities from below to well-above condensation threshold. Similar occupation functions for an electrically-driven polariton laser have been shown in Fig. 4.5 of Chap. 4. In the three different excitation-density-dependent regimes characterized by a double-threshold behaviour, i.e. polariton LED operation, dynamic condensation (polariton lasing) and conventional laser operation (weak-coupling laser, VCSEL), respectively, different distribution functions can be evidenced.

Indeed, polariton condensates with their macroscopic population in one energy state and the stimulated scattering processes involved are not only attractive due to their coherent light output and the superfluid nature of their quasi-particle cloud, but also for their polarization-sensitive phenomena [21, 24, 74–76] and their magneto-optical properties [77–80].

8.2.3 *Link to BEC via Spatial Coherence Measurements*

A spontaneous coherence build-up in a Bose gas is considered strong evidence for condensate formation due to the rise in spatial correlations. Thus, the link to Bose–Einstein condensation is commonly established for polariton condensates through spatial coherence measurements, allowing to refer to the condensed gas as a quasi-BEC or dynamical BEC. Typically, two methods are found in the literature in the context of polariton condensate characterization. The Young's double-slit experiment and Michelson interferometry.

Young's Double-Slit Experiment

An analysis of spatial coherence as indicator for a condensation threshold can be achieved with double-slit interference measurements with spatial [3] or energy and momentum resolution [2]. By reading out the excitation-density dependent interference pattern in the spatial domain, the increase of a spatial coherence length can be obtained by an analysis of the fringes and their visibility [3]. For this investigation, correlation between two spatially separated spots (emitters) on the surface area of the overall light source (emitter) is achieved by a projected spatial source separation $r = r'/Z$ on the sample. Therefore, a double slit with slit separation r' is placed into a NF projection plane [2, 3]. $Z = f_{RL}/f_{obj}$ corresponds to the optical magnification of the real-space image in the NF plane of the setup, with f_{RL} and f_{obj} the focal lengths of the relay lens and microscope objective, respectively. When the slit is mapped by the optical projection onto the sample surface, its effective slit width s seen by the emitters is usually less than the intrinsic coherence length ($\approx$1 µm) of a single polaritonic quasi-particle (at wavelength 0.8 µm) [3].

The intensity modulation (normalized to the intensity maximum) in the momentum-space projection measured in such interference experiment can be expressed by a term proportional to $|g^{(1)}(r)| * \cos(2r\, k_{\parallel})$ with Gaussian envelope function, whereas $r = |r_1 - r_2|$ is the slit separation, and correspondingly r_1 and r_2 are the two separated spots on the centred sample surface, positioned via the double

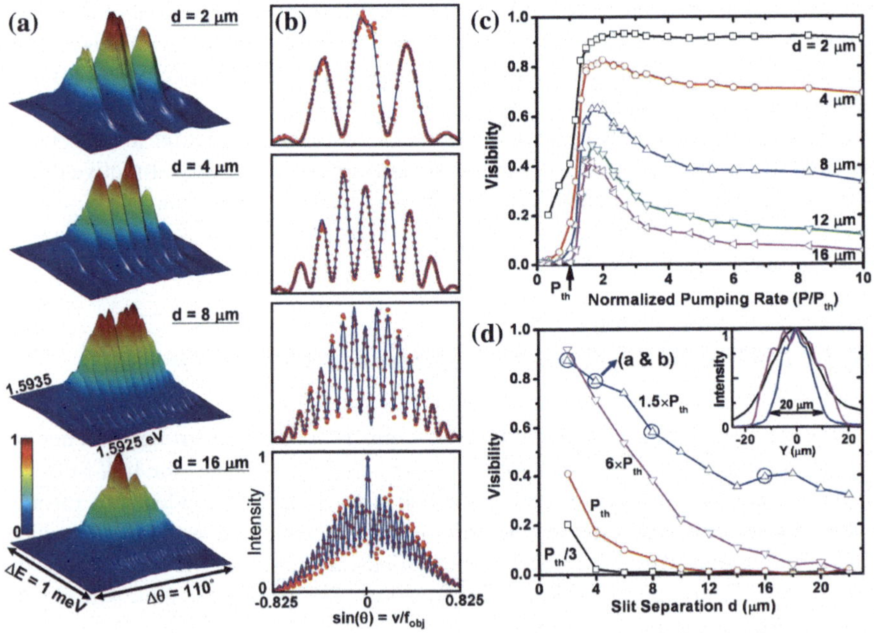

Fig. 8.7 Spatial coherence measured by Young's double-slit interference measurement. **a** Interference in the E–θ plane with cross-sectional fringe patterns along the angle axis (here v is the coordinate in the plane of observation) shown in (**b**). From top to bottom, the slit separation (here d) was varied for an excitation density of $1.5\,P_{th}$. **c** Density-dependence of the measured visibility for different slit separations. **d** Visibilities as a function of the slit separation d for representative excitation densities given in units of P_{th}. Below threshold, the spatial correlations are short ranged, while above threshold, the coherence length is considerably higher. Encircled values correspond to the plots shown in (**a**) and (**b**). Reused with permission. Reference [2] (Supplementary Information) Copyright 2007, Springer Nature

slit in the NF plane symmetrically at $\pm r'/2$ from the optical axis. The 1st-order spatial correlation function $g^{(1)}$ represents the degree of spatial coherence through its relationship to the visibility V (contrast of the fringes) $|g^{(1)}| = V$. For a given slit separation, a transition from low to high visibilities can be measured for polariton gases as a function of pump power. In [2], it is further mentioned that the quasi-monochromatic condition has to be satisfied, i.e. optical path difference being much less than the longitudinal coherence length, $r \sin(\theta) \ll c/\Delta\nu$, where $\Delta\nu$ is the spectral resolution. In this centred NF projection of the double slit, the two slits produce equal intensities across the plane of observation for a spatially homogeneous emission spot which overlap and create the interference pattern on the detection plane. Further details are found in [2, 3] and their supplementary information. Figure 8.7 shows results of such a projection scheme in a representative fashion with interference patterns resolved in the E–θ space.

When emission takes place from many k states with a wide distribution of polaritons among their energy states well below a condensation threshold, hardly any interference fringes can be observed. Above threshold, the increased monochromaticity of the emission drastically improves the visibility of fringes, which is also dependent on the spatial separation of the slits. By increasing r' or r, the spatial separation of emitters is increased. Thus, correlations are expected to decrease for a given coherence length d_c. By analysing the slit-separation dependence of the visibility $g^{(1)}(r)$, the spatial coherence length in the emitting gas can be extracted from the 1/e drop of the visibility. It was show that d_c increases significantly (from a few to several μm) above threshold for elevated polariton densities [2, 3, 23].

The increase of the measured coherence length d_c as a consequence of stronger excitation is a clear indication of a spontaneous coherence build up of the polariton system above threshold in a partially thermalized, spatially expanded condensate [2, 3, 23]. However, due to the short polariton lifetime in pulsed dynamical condensates, only a short interaction time on the ps time scale is provided for the gas which is on the order of the thermalization time of approximately 1–10 ps [16]. This hampers the formation of a long-range order and therewith spatial coherence on the length scale of the condensate expansion in such systems. Moreover, at densities well above the threshold, saturation of the visibility can set in or even a reduction of the coherence length can be observed due to interactions in the gas. The slit-separation dependence of the double-slit interference for above threshold emission is shown for an example from the literature in Fig. 8.7a, with cross-sectional fringe patterns along the angle axis (here v is the coordinate in the plane of observation) shown in (b). A density-dependence of the measured visibility for different slit separations is summarized in Fig. 8.7c, whereas an overview of the separation-dependence is provided in (d) for representative excitation densities; encircled values in (d) correspond to the plots shown in (a) and (b) [2].

In general, for very large separations r, only well above-threshold measurements allow the extraction of a visibility value due to the enhanced spatial coherence. For very low separations on the order of the light wavelength and the intrinsic coherence length, the visibility can reach maximal values of nearly 1 as a consequence of self-interference and should show hardly any density dependence for monochromatic light. This gives no information about the system's spontaneous coherence build-up. In the case when polychromatic LP emission is shed on the slit and fringes are mapped in the spatial domain, the increase of visibility for very low slit separations merely demonstrates the transition to a monochromatic LP emission regime.

Typically, instead of observing interference fringes for a monochromatic coherent light source on a screen behind a Young's double slit, the FF spectrum will be considered for polariton signal passing such double slit inserted into an intermediate NF plane of the optical setup. Correspondingly, interference patterns occur in the FF spectrum that can be conveniently recorded for different excitation densities below, around and above threshold.

An example of interference patterns in the FF spectrum is given in [23]. Such FF configuration allows resolving the Fourier space (k_x–k_y) in the imaging mode as well as obtaining energy resolution (k_x–E) in the spectroscopy mode. For pump laser light back-reflected from the sample and passing the double slit, fringe visibilities in excess of 90% can be expected for a spatial separation $r \approx 6\,\mu\text{m}$ owing to the extended spatial correlations of the so-to-say emitter. In contrast, for polaritons in the linear emission regime, no interference fringes are obtainable. This changes above threshold, when interference patterns appear in FF spectra on the same energy level as the ground state. For a large-area pulsed-emission condensate, a coherence length of 4 μm was extracted by this method. For clarity, a comparison with bare dispersion measurements can reveal which features in the FF spectrum at finite $k_{\|}$ originate from the double-slit measurements. In addition, one can vary r for a given pump rate and observe a specific location of the interference maxima in the direction of $k_{\|}$ according to the separation in real space [2]. This can provide a clear discrimination between fringe signal and polaritonic emission at higher k states.

This straight-forward spatial-coherence characterization method can easily help to identify the coherently emitting mode in FF spectra, since interference fringes originating from a mono-mode-emitting system above threshold are distributed in the k direction and can be directly attributed to it, whereas the energy distribution of signal is simultaneously provided. This is an advantage of spectrally-resolved measurements compared to interference detection from a double-slit experiment in the bare FF-projection plane. This becomes obvious in the case of multiple coherent modes at different energy levels, when every spectrally-resolved coherent mode can be analysed without superposition of their interference patterns. For example, simultaneously recorded lasing modes can be characterized in parallel, for instance when polariton condensation and photon lasing occur in the same time-integrated spectrum of the sample emission under pulsed excitation. Similarly, competing laser modes of multi-mode emitters that are energetically separated can be probed.

Another example of energy-resolved double-slit fringes (but not in the FF spectroscopy mode) can be found for a polariton wire system, which gives rise to multiple emitter modes due to polariton trapping [81]. For the study of macroscopic off-diagonal long-range order above threshold, the visibility at different energies was evaluated as a function of slit separation and excitation power for the double-slit interference pattern measured at infinity as projected onto the entrance slit of the spectrometer. In one configuration of the study in [81], the slit was placed onto the polariton wire symmetrically onto the excitation spot, and in another configuration, it was placed onto the wire on one side of the excitation spot, which provided additional information about spatial correlations in wire systems away from the pump spot.

Michelson-Interferometric Experiment

Michelson interferometry has a long history and was initially applied for astronomy [82]. Later, the rise of coherent light sources made the employment of this interferometry technique attractive for light-source characterization. Currently, Michelson interferometry is an indispensable tool when it comes to coherence length measurements, as it correlates temporal electric fields spatially by the first-order spatial correlation function $g^{(1)}(\mathbf{r}, t)$.

Nowadays, many quasi-BEC reports for polaritonic systems that are obtained in microcavities demonstrate a spontaneous coherence build-up in support of their condensation claims. In 2006, Kasprzak et al. used a Michelson-interferometer setup in connection with their condensate characterization and showed that at low excitation densities, the polaritons exhibit short-range correlations as expected, with a correlation length determined by the thermal de-Broglie wavelength (representing the particle size). However, above threshold a significant increase of visibility fringes and a spatial spread of correlations were obtained as expected for a condensed phase [1]. While complete coherence within the condensate is expected for an ideal gas of condensed bosons [83, 84], practical polariton condensates show a finite expansion of the long-range order within the condensate area. Follow-up polariton-condensate studies on different material systems or structures typically employ the same method to claim a BEC-like regime [44] and to demonstrate long-range order (e.g. [85, 86]).

Using the first-order spatial correlation function [87]

$$g^{(1)}(\mathbf{r}_1, t_1; \mathbf{r}_2, t_2) = \frac{\left\langle a_1^\dagger a_2 \right\rangle}{\sqrt{\left\langle a_1^\dagger a_1 \right\rangle \left\langle a_2^\dagger a_2 \right\rangle}}, \tag{8.1}$$

whereas $a_i^\dagger$ and a_i are the field's creation and annihilation operators at space–time point $(\mathbf{r}_i, t_i)$, the correlations of two separated spots r_1 and r_2 can be analyzed. One can measure this $g^{(1)}$ by overlapping the real-space image of a polariton emission spot with its reflected version in a Michelson-interferometer setup. Changing one interferometer arm's length yields a relative phase shift between the two split beams in the setup. Since the two recombined beams project the mirrored images of the emitter on a camera, each pixel shows a sinusoidal modulation of intensity by the arm's movement according to the phase relation at that point. From such data, one can extract the phase difference between the overlapped images pixel-wise and also determine the fringe visibility, i.e. the first-order correlation function, for each point. Using a prism (or retro-reflector) instead of a plane mirror in the moving arm, the reflected image from the emitter spot is axis-mirrored (point-mirrored) along the axis (centre) of the prism (retro-reflector). For condensate emission, this leads to correlation measurements between spots across the excitation spot over one mirror axis depending on the orientation of the prism, that is point (x, y) overlaps either with $(-x, y)$ or with $(x, -y)$ on the screen/camera. As described by Roumpos et al. [87], this allows to measure

$$g^{(1)}(x, -x; \tau) \equiv \langle g^{(1)}(x, y, t + \tau; -x, y, \tau) \rangle_t , \tag{8.2}$$

or

$$g^{(1)}(y, -y; \tau) \equiv \langle g^{(1)}(x, y, t + \tau; x, -y, \tau) \rangle_t , \tag{8.3}$$

whereas in this equations $\langle\rangle_t$ denotes the time average.

Position changes of the prism lead to a delay τ in one arm. By scanning through different delays around $\tau = 0$, the modulation of intensity can be recorded for each pixel based on the (temporal) coherence of the emission. For the extraction of the phase map, a sine function is fitted to the trace at each pixel (across the whole spot). Based on the spectral width of the emitter, the intensity modulation as a function of the temporal delay (that corresponds to path length difference) features a Gaussian envelope centred at zero delay, as $\Delta E = \hbar/\tau$ in the Fourier limit.

For the spatial distribution of visibilities which is mapped by this method, typically only interference data at $\tau = 0$ is of interest. Thus, the time argument is not always mentioned explicitly. An example of the short-distance decay behaviour of the visibility is given in [87], while single vortex–antivortex pairs in an exciton–polariton condensate were detected using the same technique in [32]. With this Michelson-interferometry technique, Roumpos et al. identified for increased condensate sizes a convergence of the spot-separation-dependent visibility to a power-law behaviour [87]. An example of the setup is provided here in Fig. 8.8 after [32].

8.2.4 Photon Statistics

The photon statistics of an emitter can be characterized by temporal intensity autocorrelation experiments. For instance, the second-order temporal correlation function $g^{(2)}(\tau)$ serves as a means of identification of a transition from a thermal emission regime to a coherent output of light in a laser system. This is evidenced by a drop of the correlation function from 2 to 1 for simultaneously recorded photons.

The first-order temporal correlation function

$$g^{(1)}(\tau) = \frac{\langle E^*(t)E(t + \tau) \rangle}{\langle |E(t)|^2 \rangle} \tag{8.4}$$

represents the correlation between fields and is used to obtain the temporal coherence length of a light source. It quantifies the fluctuations of the electric field over time. A complex electric field $E(t)$ corresponds to an intensity $I(t) = E^*(t)E(t) = |E(t)|^2$.

In contrast, the second-order correlation function

$$g^{(2)}(\tau) = \frac{\langle E^*(t)E^*(t+\tau)E(t)E(t+\tau)\rangle}{\langle E^*(t)E(t)\rangle^2} = \frac{\langle I(t)I(t+\tau)\rangle}{\langle I(t)\rangle^2} \tag{8.5}$$

represents the correlation of intensities and is used to obtain temporal photon statistics. It quantifies the intensity fluctuations and is sensitive to the photon number. This correlation function can be used to identify the probability of detecting two photons at the same time.

The second-order autocorrelation function is particularly relevant at zero time delay ($\tau = 0$) where the correlation between photons is highest, whereas for increased time intervals between photons their correlation decays.

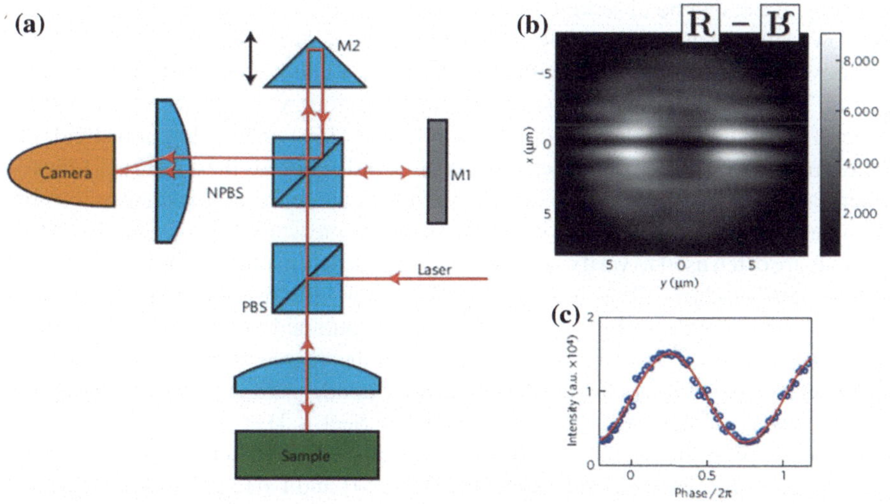

Fig. 8.8 **a** Sketch of a Michelson interferometer. This experimental setup enables the measurement of phase and fringe visibility maps. Mirror (M1) and right-angle prism (M2) create the reflection of the original image along one axis, depending on the prism's orientation. Using a microscope configuration, the two real-space images of a polariton condensate are projected onto the camera for interference pattern read-out. Here, polarizing and non-polarizing cube beam splitters (PBS and NPBS, respectively) were employed. **b** Typical interference pattern measured above the polariton condensation threshold. A schematic using the letters R and flipped R indicate the orientation of the two overlapping images. **c** Normalized intensity trace on one of the camera pixels as a function of the phase delay, set by the displacement of one interferometer arm (blue circles) with a fit to a sine function (red line). Reused with permission. Reference [32]. Copyright 2011, Springer Nature

$$g^{(2)}(\tau = 0) = \frac{\langle : \hat{n}^2 : \rangle}{\langle \hat{n} \rangle^2}, \tag{8.6}$$

with $\hat{n}$ the photon counting operator. Colons indicate normal ordering of the photon field operators [18]. The experimental value $G^{(2)}(0)$ represents a time-averaged $g^{(2)}(0)$ for every pulse [4].

Below the condensation threshold, polaritons occupy a wide range of states in the k space. For this regime, thermal emission is expected, which is represented by $g^{(2)}(0) = 2$. For increasing delay τ, the value approaches the value for random photon numbers $g^{(2)}(|\tau| \gg 0) = 1$. In order to resolve the so-called thermal "bunching" feature at $\tau = 0$ in photon statistics measurements, a high temporal resolution is needed. With standard Hanbury–Brown-and-Twiss (HBT) setups [88] based on Si single-photon detectors, the temporal resolution on the sub-100-ps scale can typically be worse than the relevant coherence time of the emitter. Thus, the bunching effect can only be resolved around the nonlinearity threshold, when the temporal coherence of the emitter significantly increases [89, 90]. Only if the coherence time is close to the temporal resolution of the HBT setup or comparable to the pulse duration, the bunching effect can be observed. In fact, advanced single-photon detection systems can nowadays reach temporal resolutions in the lower ps range.

Above laser threshold, a transition to $g^{(2)}(0) = 1$ can be resolved for a coherent emitter. This leads to a characteristic bunching behaviour with a maximum detectable value of $g^{(2)}(0)$ around the threshold pump power [35], since below threshold, the finite temporal resolution of the detectors significantly exceeds the coherence times of a thermal emitter. In the transition range when the coherence time increases towards a coherent state above threshold, bunching becomes measurable [90]. Standard HBT methods are often used to verify laser operation for different photonic microstructured systems, but similarly they are used to probe the polariton-laser nonlinearity, as is shown as an example in Fig. 8.9 for an optically-pumped polariton-laser structure [91]. For condensate emission, $g^{(2)}(0)$ can reveal the degree of phase coherence in a polariton gas and indicate the purity of the Bose condensate in terms of uncoupling from the reservoir states (cf. [92], also see below Sect. 8.3).

However, preliminary investigations on the second-order temporal autocorrelation function of polariton condensates for GaAs [4] and CdTe [5] showed different results which were not readily understood at that time. The differences can be partially explained by the differences in the excitation scheme. GaAs polaritons that were quasi-resonantly excited from the side exhibited strong intensity fluctuations above the nonlinearity threshold, as pioneering work by Deng et al. in 2002 showed in the form of slowly decreasing bunching towards higher pump powers above threshold. Later HBT measurements with comparable pump scheme by Horikiri et al. in 2010 [19] ($\approx$ps-pulsed excitation) and Rahimi-Iman et al. in 2012 [23] ($\approx$100-fs-pulsed excitation) for similar GaAs structures also showed such strong fluctuations for polariton condensates, which can be attributed to polariton–polariton interactions and the large spatial expansion of the condensate. In contrast, CdTe polaritons investigated by Kasprzak et al. in 2008, which were non-resonantly excited by a

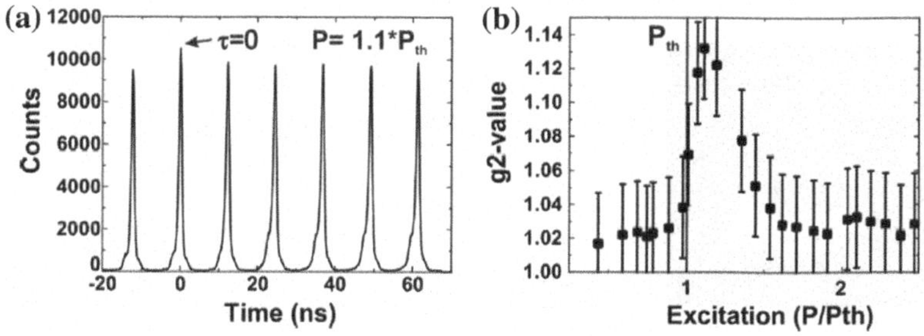

Fig. 8.9 **a** Histogram recorded with an HBT setup for ground-state emission from an optically-pumped polariton gas slightly above the nonlinearity threshold. The $g^{(2)}(\tau)$ measurement clearly shows notably increased photon counts for the peak at $\tau = 0$, i.e. "bunching", which is a signature of a non-coherent state resolvable in this transition regime. **b** The power-dependence obtained for the $g^{(2)}(0)$ value shows a clear increase around the threshold with a drop towards unity for higher pump densities, representing an increase in phase coherence for emission out of the polariton condensate. Reused with permission. Reference [91] Copyright 2014, Optical Society of America

cw laser, apparently featured a high degree of coherence shortly above the condensation threshold [5]. Far above threshold, the $g^{(2)}(0)$ value increased again which was attributed to interactions within the polariton condensate and scatterings with non-condensed excitons and polaritons. For comparison with existing results in the literature at that time, also pulsed measurements were performed in [5].

Systematic power-dependent investigations were performed by second-order and third-order temporal correlation experiments with polaritons using a comparable side-pump scheme with GaAs samples by Assmann et al. [18]. In addition, their work considered different detunings and excitation polarisations. In contrast to HBT measurements, the results were obtained using a unique streak camera read-out technique developed for photon counting, which enables picosecond time resolution [18]. A clear dependence on the cavity–exciton detuning Δ and on the excitation laser polarization was identified. Nevertheless, third-order correlations were also obtained with an avalanche-photodetector-based (APD) HBT-like measurement scheme by Horikiri et al. [19].

One experiment on spinor condensates with GaAs microcavities even revealed different behaviour for Zeeman-split condensates of polaritons in an external magnetic field up to 5 T [80]: with increasing field, opposing polarizations approached each others $G^{(2)}(0)$ with noticable intensity fluctuations, whereas at 0 T, one of the two circularly-polarized condensate signals featured a high degree of coherence with near unity value. Chernenko et al. also reported in [80], that one of the two coexisting circularly-polarized condensates quickly approached $G^{(2)}(0) = 1$ above threshold for 5 T, whereas the counterpart still exhibited noticeable intensity fluctuations. The reported correlation experiments were performed using Fourier-space resolved spectroscopy, which allows for a strong energy and momentum resolution (for light passing the imaging monochromator) and, thus, enables photon-statistics measurements for ground-state emission.

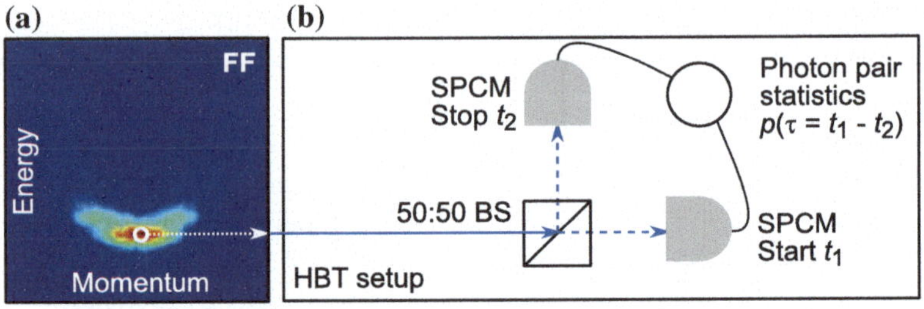

Fig. 8.10 **a** Fourier-space-resolved emission spectrum with strong ground-state occupation of a polariton branch. Here, a white circle indicates the fibre-core aperture in the FF-projection plane behind the monochromator in the exit plane used for coupling into a fibre-coupled HBT setup. The working principle of an HBT configuration is sketched in (**b**), showing two single-photon counting modules (SPCMs), a beam splitter (BS) and a correlation unit (circle) which together enable recordings of photon pair statistics

Commonly, a fibre-coupled HBT setup with fast APDs directly behind a monochromator output is used to record the autocorrelation histogram. The aperture of the fibre itself defined by the core can act as a strong frequency- and momentum-selective element in the FF-spectrum projection plane behind the imaging monochromator, facilitating mono-mode detection. A sketch of the FF-acquisition configuration for time-averaged second-order correlation $G^{(2)}(\tau)$ is shown in Fig. 8.10. If no special means are applied to filter the signal spatially, the results in FF configuration correspond to spatially-integrated signals within the excitation spot. In the literature, a few reports employed such setup with strong E–k filtering capability (on the order of $\delta E \approx 150\,\mu\text{eV}$ and $\delta k \approx 0.02\,\mu\text{m}^{-1}$ due to the fibre core) in the FF $G^{(2)}$ configuration [23, 80, 91]. For such measurements, individual fine adjustment of the grating position for each pump-density step are required for a $G^{(2)}(\tau)$ power series, when the fibre position is fixed in the exit-slit plane with respect to the E and $k_{\|}$ direction. In fact, $k = 0$ $G^{(2)}(\tau)$ and $g^{(2)}(\tau)$ (for pulsed and cw emission, respectively) were also recorded using FF filtering ($\delta\theta \approx 1$ deg) for the CdTe system by Kasprzak et al. [5]. One should note that the signal maximum in the k plane can vary between experiments. For energy adjustments, an attenuated sharp laser line can be used to maximize wavelength-sensitive photon counting. Comparable methods or a strong narrow ground-state signal can be used for an initial k space adjustment.

Even time-resolution-limited HBT systems can show measurable bunching values for polariton condensates in the region of the pump threshold for signal selected at $k_{\|} = 0$. As Deng et al. pointed out, the photonic fraction of polaritons in the ground state is higher and, thus, the loss rates are higher than for the excitonic reservoir states. This renders signal pulse durations comparable to the coherence time of condensed polaritons owing to the stimulated ground-state scattering [4]. Thus, the time-integrated function $G^{(2)}(0)$ well represents $g^{(2)}(0)$.

As discussed in [23] based on previous studies, intensity fluctuations above threshold were generally assumed to be a sign of either dephasing effects owing to a strong interaction between polaritons in the condensed state [5, 18, 19] or an incomplete thermalization process within the lifetime of condensed particles, whereas scattering is efficient enough to achieve macroscopic ground-state occupation on that time scale [4, 93]. While some results indicate a thermalized, weakly interacting system that can reach a high level of coherence well above threshold [18, 50], other reports on condensate $G^{(2)}$ remain in a regime of non-equilibrium condensation even for elevated pump rates [4, 23]. Such differences can be related to the detuning, spot size, spatial or momentum filtering, or pump conditions, but also to the sample quality. In those cases with remaining intensity fluctuations, the photon statistics clearly indicate that the coherence properties of the investigated condensate differs from that of a conventional photon laser, since above threshold no abrupt transition from bunching to the coherent state occurs. These intensity fluctuations that are evidenced for a partially coherent system were attributed to polariton–polariton scattering by Schwendimann et al. [52]. It was explained that a strong noise component is added to the photon statistics when interactions prevent the condensate from full occupation, i.e. when the condensate fraction υ_0/υ is reduced. Thereby, calculated autocorrelation values for $k = 0$ emission from polaritons above threshold can reproduce the experimentally observed power dependence for non-equilibrium condensates, when interactions between the ground-state particles and particles of different k (ground-state–reservoir interactions) are considered.

Particularly, for large scale condensates with tens of μm in-plane expansion within a Gaussian-profile pump spot and with low spatial coherence length (on the order of few μm), the condensate's measured degree of coherence can be reduced when spatially integrating ground-state emission owing to repulsive polariton interactions in the whole non-equilibrium condensate [23]. In contrast, when condensate size (or signal-collection area) and coherence length are comparable, higher degrees of coherence for ground-state emission can be obtained from FF $G^{(2)}$ experiments. Moreover, polariton-laser structures in the micropillar geometry with their strongly-reduced availability of states at finite k values feature an enhanced temporal coherence due to a suppression of unfavourable interaction effects [91]. Amthor et al. clearly observed a drop of the $g^{(2)}(0)$ function towards 1 with increasing optical pump power, indicating coherent emission in the polariton-laser regime. Similarly, alternative polariton structures with strong mode selection feature shot-noise-limited intensity stability in the coherent state [92], delivering coherent polariton lasers.

8.3 Special Condensate Features

Indeed, polaritons have already been well studied in many regards and there may be plenty of works not adequately covered in this chapter. Nevertheless, the world of polariton condensate studies is moving forward with high pace. Still intriguing features come up in late studies, e.g. those involving the dynamics of exciton–polaritons [53, 94], such as the role of dark states or relaxation oscillations, and the behaviour of quantum fluids based on condensed polaritons [31, 63, 65–67, 69–71, 95, 96], with the many demonstrations of superfluidity and spontaneous vortex formation. Furthermore, p-wave, d-wave or Dirac-point condensates in lattices [2, 97, 98] and condensates in excited states in patterned structures [99] were explored. For an electrically-pumped system, optical probe pulses were used to study the effect of repulsion by locally injecting carriers [100], whereas for an optically-pumped system, a spontaneous real-space collapse at high densities has been investigated and the role of a collective polaron effect in the self-trapping phenomenon for a polariton fluid evaluated [96] (see Fig. 8.11). Moreover, further systems have been proposed with which condensation studies can be performed, such as structures that give rise to so-called dark polaritons when exploiting quantum tunnelling experiments, in which direct and indirect excitonic states were coupled with cavity photons using double-QW structures: Cristofolini et al. suggested that dark polaritons formed in them could offer new possibilities for electromagnetically induced transparency, room-temperature condensation, and adiabatic photon-to-electron transfer [101].

The relationship of polaritons with terahertz (THz) radiation in terms of the Rabi-oscillation frequency being on the order of THz radiation in common polariton systems has also triggered research activities. Some theoretical works propose the realization of bosonic lasers, e.g. bosonic cascade lasers [102], and novel polariton-based THz emitters [103, 104]. The generation of coherent THz radiation in non-linear polariton regimes [105] has become attractive for the development of novel THz sources. In this regard, important physics still need to be studied carefully to enable the exploitation of many-particle effects in novel device concepts and quantum information schemes.

Thus, the exploration of this domain of light–matter interaction has still the potential of unravelling new applications of quantum gases and microcavity systems. Based on the development of ultrafast optical switches using coherent control [106, 107], electronic bias [91] or operation bistability [108–110], polarization shaping [111] and cooling of quantum gases [112], novel schemes can be envisaged for optoelectronics based on polaritons. Experimental realizations of all-optical polariton devices based on transistors [113], routers [114], resonant-tunneling diodes [115] and interferometers [116] have further promoted the idea of *polaritonics*. Even *qubits* (quantum bits) based on polariton Rabi oscillators have been proposed [117].

Many interesting phenomena, among others in the time domain, have been reported for polaritons, such as the manipulation of coherent states [106, 107] and the evidence of relaxation oscillations [94], to name but a few. Some of the experiments use the common pump–probe scheme for transient-absorption measurements, others resolve the PL in time using ultrafast APDs, streak cameras, or phase-sensitive imaging in a digital holography scheme. For polaritons undergoing a condensation effect, time-resolved investigations (mainly concerning lifetimes and dynamics) using PL spectroscopy were carried out by the polariton research community typically employing ultrafast streak cameras (examples are given in Sect. 7.4). In the low-density polariton regime, Dominici et al. have demonstrated the first direct ultrafast imaging (using a linear technique) of the polariton sub-picosecond oscillations with both temporal and spatial resolution (see [106]). The authors successfully implemented a coherent control technique in a double pulse experiment based upon a delay line. Thereby, optical control of the quantum state of the polariton system

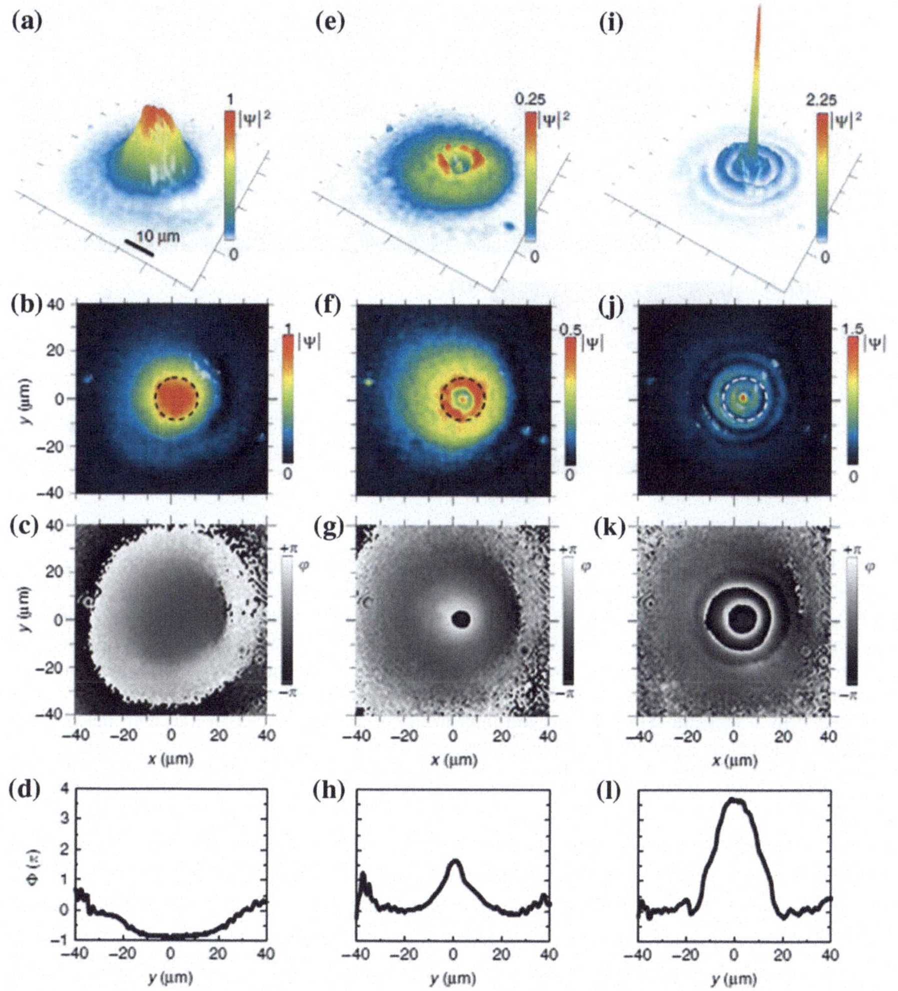

Fig. 8.11 First row: density maps of the in-plane polariton fluid in a planar microcavity structure displayed three-dimensionally as spatial contour profile for an area spanning 80 times 80 μm, for three different temporal moments (data arranged in columns for this figure), i.e. about 0 ps (**a**), 2.8 ps (**e**) and 10.4 ps (**i**). These three temporal moments correspond to injection-pulse arrival, ignition of the dynamical peak and, following a shock wave formation at early times, the evolution of a central peak as the system's long-lived state, respectively, surrounded by concentric intensity fringes. A supplementary animation is provided in [96]. Second row: 2D contour plots of the amplitude for the corresponding times. The dashed circles indicate the size of the initial pump spot in FWHM. Third row: corresponding phase maps. Fourth row: unwrapped phase profiles along the radius. These profiles indicate a phase curvature that leads to the development of an opposite flow of the polaritons toward the centre. The peak that was formed reached a localization below the resolution limit of 2 μm, and was 10 times sharper than the initial Gaussian laser exciation spot (18.5 μm). Reused with permission. Reference [96] (CC BY license) Copyright 2015, Springer Nature

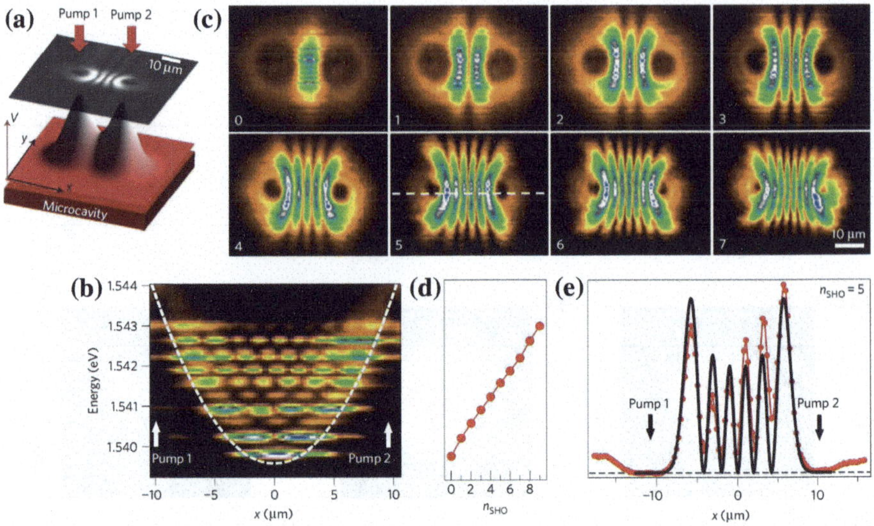

Fig. 8.12 **a** Sketch of the experiment with two spatially-separated pump spots. **b** Energy-resolved intensity distribution as a function of the real-space coordinate representing the axis from one pump spot to the other, along which a simple harmonic oscillator (SHO) potential is induced. Polariton condensates are formed on various harmonic-oscillator levels, with their individual spatially mapped emission profiles plotted energy selective in (**c**). **d** Mode energies as a function of the quantum number (n_{SHO}). **e** An extracted image cross-section from (**c**), marked by the dashed line, is well reproduced by a Hermite–Gaussian wave-function fit curve $\psi^{n=5}_{SHO}(x)$. Reused with permission. Reference [118] Copyright 2012, Springer Nature

in relation to their exciton–photon content with phase degree of freedom became feasible. Moreover, it allowed photonic switching into a desired state on the Bloch sphere. Further refinements to the experiment allowed to control the polarization degree which enabled the implementation of a record speed polarization scrambling on the Poincaré sphere of polarizations, on the scale of Trad/sec (see [111]).

8.3.1 Polaritons at Their Extremes

Ongoing investigations target exciton–polariton systems at extreme conditions. These conditions include high temperatures, strong confinement, suppressed particle interactions, extreme excitation intensities, extremely long polariton lifetimes, and extremely low particle numbers, which allow to investigate quantum effects at the few particle level.

In fact, trapping of polaritons has been addressed by several technological or experimental means (see [13, 26, 51, 58, 119, 120], to name but a few). Remarkably, it has been also shown that trapped condensates can form between two spatially-separated cw-pump spots on the sample, by exploiting self-repulsion of polaritons to modify the quasi-particles' own potential, the induced potential landscape along the line between the two injection points resembles a simple harmonic oscillator (SHO) potential [118]. By temporal and spatial interference measurements, Tosi et al. revealed that the condensate wave-packets arise from non-equilibrium solitons. Quantum oscillator wave-functions were probed in real space and the oscillations at tunable THz frequencies studied. Figure 8.12 shows a sketch of the experiment (a) with measurement data from the polariton condensates across the harmonic potential states (b); the emission was spatially mapped for different energies in (c)—with their mode energies plotted as a function of the quantum number in (d). Furthermore, it is shown in (e) how the cross-section of the measured intensity distribution can be fitted with a Hermite–Gaussian model for the corresponding quantum number.

Strikingly, on the quantum level, the path to quantum polaritonics have been entered by recent experiments, which demonstrated quantum entanglement using polaritons. Remarkably, Cuevas et al. demonstrated entanglement of a polariton with a single photon [121]. By swapping the entanglement between a polariton and an external photon from a photon pair generated by parametric down-conversion, the first experimental demonstration of a truly-quantum manifestation of cavity–polaritons was shown. Furthermore, perturbation of the entangled state with a rarefied polariton condensate revealed how interactions manifest at a single polariton level. Such experiments can be regarded as the first application of quantum spectroscopy.

In the past years, microcavity samples with polariton lifetime orders of magnitude longer than before became available. It is understood that energy relaxation effects appear to be very important in these samples. This requires the development of an updated theoretical description. Many experiments have been so far limited by the lifetime of cavity–polaritons, which hardly exceeded several or several tens of picoseconds. However, samples with polariton lifetime of several hundreds of picoseconds have been demonstrated recently due to a number of major developments and improvements in the production technology of microcavities designed for polariton research. For instance, it was shown that motion of long-living cavity–polariton (with lifetimes of about 180 ps deduced) due to the wedge-shaped cavity is well described by the equations of motion for a moving mass under a constant force which resembles a parabolic trajectory [30]. Another example shows, that in an exceptionally high-Q microcavity structure, condensates of thermalized polaritons can be created, as experimental work of Caputo et al. addresses [122]. Within one small pump spot, two condensates in the regime of long-lifetime polaritons were reported, whereas a low-energy condensate was effectively separated from the hot excitonic reservoir. Caputo et al. also showed that correlation function measurements are in good agreement with the Berezinskii–Kosterlitz–Thouless (BKT) theory of topological phase transitions (see [122]). Yet, formation of such condensate has not been entirely understood.

One of the other aforementioned extreme regimes concerning polariton studies is given with respect to excitation densities. Since in common microcavities, a two-threshold behaviour is not unusual due to the occurrence of conventional photon lasing above a second threshold, i.e. when strong pumping is applied and the coupling between cavity photons and excitons is lost due to bleaching, the polariton experiments are usually confined to excitation regimes close to the first threshold. This relatively low excitation regime of condensates far away from the second threshold is widely characterized, whereas the regime of high excitation density still lacks full understanding. If investigations similar to that of Dominici et al. [96] were performed with intense and ultrashort laser pulses, the physics of polariton gases could be probed further with eyes towards the exploration of the dynamics and characteristics of such quantum gases. An alternative high-density scenario in experimental work shows, how the strong coupling is maintained until the photoluminescence smoothly moves from the LP energy to the cavity mode energy [123].

Another possible extreme regarding condensate features can be given by the obtainable sizes. Recently, macroscopic two-dimensional polariton condensates were reported [124]. Based on a ballistically-expanding polariton flow that relaxes and condenses over a large area outside of the excitation spot, a condensate width on the order of 100 μm lateral dimensionality free from the presence of an exciton reservoir was obtained.

Ultimately, the elevated operation temperatures, at which polariton condensation can be obtained, have been in the focus of many years of polariton research, as a sort of extreme for BEC studies. Using different material systems as platform for the investigation of strong coupling at elevated temperatures, optically-pumped polariton lasing (condensation) has been claimed up to room temperature since the end of the 2000s for microcavities based on GaN [125–128], ZnO [129–135], perovskite material [136] and organic semiconductor molecules (polymers) [44, 85, 137]. Some of the studied polariton lasers at room temperature even where obtained using electrical current injection schemes [138, 139].

In fact, one could consider the observation or acquisition time for polariton studies another scenario were an extreme can be reached, e.g. by looking at the dynamics on a single-condensate level. As a driven–dissipative condensate, i.e. subject to pumping and decay, further understanding of the condensate's nature can be obtained by a unique single-shot optical-excitation scheme, which does not lead to time-averaging over multiple condensate realizations. For a high-quality inorganic polariton system, single condensates were analyzed with regard to formation and interaction effects by Estrecho et al., who showed that condensation was strongly influenced by an incoherent reservoir [140]. It was further revealed that reservoir depletion, referred to as spatial hole burning, was critical for the transition to the ground state. Interestingly, the study also could show the difference between photon-like and exciton-like polaritons: the former exhibited strong shot-to-shot fluctuations and density filamentation due to an effective self-focusing associated with reservoir depletion, the latter established smoother spatial density distributions and became second-order coherent [140].

8.3.2 Coherent Polariton Lasers

The device concept behind a polariton-laser diode has been built on the idea that the threshold for condensation leading to coherent emission from the ground-state bath of polaritons can be orders of magnitude lower than that of conventional lasing in similar structures (see Chap. 4). This justified the approach of harnessing the phase coherence of polaritons in a BEC-like state for the generation of a coherent light beam in the strong-coupling regime of an emitter–cavity system. However, as the discussion about intensity fluctuations and the low degree of phase coherence evidenced in second-order temporal autocorrelation studies shows, common condensate systems do not act as good-enough sources of coherent radiation.

A coherent state, i.e. the quantum-mechanical electro-magnetic field description of the coherent state referred to as the *Glauber* state, is defined by a high intensity stability. The photon statistics of a Glauber state is given by a Poisson distribution peaking at the expectation value of the photon number in a mode N. While a sub-Poissonian distribution is characteristic for *non-classical light*, a super-Poissonian distribution is common for thermal light sources. Correspondingly, those two distinct states have narrower or wider photon number distributions than the coherent state. For polariton condensates, a transition to a coherent state is expected. In contrast, often observed remaining intensity noise above the condensation threshold can be attributed to strong interactions between condensed polaritons and excited states of the system, e.g. polaritons and excitons in reservoir states (see above discussion and [14, 52, 53, 141]), or even coupling to the phonon bath of the host crystal lattice in the sense of a collective *polaron* formation [96] could be thought of as a possible force behind the reduction of correlations.

In 2016, Kim et al. demonstrated a polariton laser with shot-noise-limited intensity stability, as expected from a fully coherent state [92]: To obtain a high intensity stability, an optical cavity with high mode selectivity was used in order to enforce single-mode lasing, to suppress condensate depletion, and to establish gain saturation. This study has shown that the absence of intensity fluctuations can give rise to a phase-coherent ground-state emission with near unity $g^{(2)}(0)$. From all the above discussion regarding second-order temporal autocorrelation measurements, it also becomes clear that the intensity noise remains an issue and perfect coherence such as for a conventional laser is not necessarily obtained for most experimentally achieved polariton condensates. It was understood to be mainly the case owing to the interactions between polaritons in the ground state with the uncondensed fraction remaining in reservoir states. However, as the degree of coherence of condensate emission matters for polariton laser applications, single-mode operation had been proposed as one promising path towards coherent light sources based on polariton condensates, or matter(-wave) lasers.

In conclusion, one can say that obtaining a coherent polariton laser clearly requires a good control of polariton–polariton interactions, i.e. high stimulated-scattering but low ground-state-depletion rates, and sufficient decoupling from reservoir states, taking into account self-interaction energies with respect to the laser bandwidth.

Single-mode polariton lasers recently demonstrated in the group of Deng [92] based on an optimized microcavity concept and previously demonstrated micropillar polariton lasers with near-unity second-order temporal autocorrelation function open up the possibility to envision coherent exciton–polariton devices. Further discussions on such devices can be found in [142].

8.3.3 *Superfluidity and Vortices in Condensates*

Naturally, the phase transition to a condensate opens an important playground for studies on quantum fluids (and superfluidity as a whole subject), in which vortices, backjets and further exotic properties can be analyzed. Superfluidity is a manifestation of the collective behaviour typical of BEC [143] and is typically observed for Bose fluids at ultra-low (cryogenic) temperatures.

In fact, superfluidity can occur without the presence of a (true) BEC, that is macroscopic homogeneous phase coherence in the real space with infinite and constant condensate wave-function. Since it is the Kosterlitz–Thouless (KT) transition that takes place in two dimensions for which formation of a BEC in the strict sense is forbidden, spatial phase coherence between two points in space (i.e. for finite-size condensates) can give rise to superfluidity, which needs a phase-coherent path for its particles to move dissipation-less (e.g. in a polariton system) [144]. In this context of a KT transition with its own critical temperature, formation of vortex pairs plays an important role for the creation of an expanded polariton condensate with local phase order and establishment of a superfluid in a 2D system. Similar to superfluidity of atoms, below the KT temperature, a normal fluid and a superfluid coexist. Only at 0 K, all vortices vanish and true long-range order can be obtained. In practical systems, local Bose condensation (condensate droplets formation, a system without superfluidity, referred to as *Bose glass*) can occur before the KT transition (delocalization of the condensate needed, achieved by condensate droplet percolation, facilitated by the vanishing of free vortices) at low densities, whereas these condensate droplets become larger when the critical density for a KT phase is reached. Moreover, localization of condensates can become more prominent for exciton–polariton systems at elevated temperatures due to inhomogeneous exciton broadening, whereas disorder potentials can generally play a role in quasi-condensate formation in excitonic (exciton–polariton) systems, as discussed in the literature (see for instance [144]).

For non-equilibrium condensates, such as those formed of exciton–polaritons, a variety of many-body effects related to condensates and superfluidity have been reported in the literature, such as the formation of persistent currents [65], excitation of quantized vortices [61, 65, 68] and—as anticipated from a superfluid—collective motion in the sense of frictionless flow across obstacles, a suppression of Rayleigh scattering of polaritons, formation of hydrodynamic solitons and splitting of fluids by obstacles on the size of the fluid's wave packet [63, 66, 67]. These results were mainly obtained in inorganic semiconductor microcavities, but the spontaneous formation of vortices has also been observed in organic polariton systems [44, 85, 145, 146].

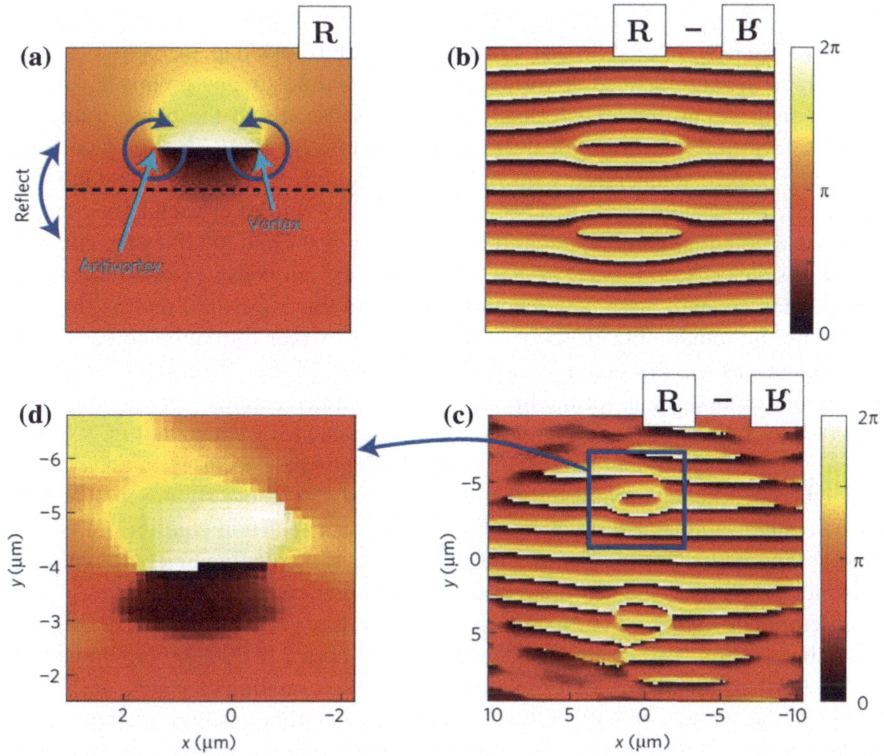

Fig. 8.13 **a** Expected phase map for a polariton condensate with single vortex–antivortex pair. The direction of the phase increase around the vortex and antivortex is indicated by the counter-clockwise and clockwise arrow, respectively, in this example. The letter R indicates the real-space orientation for the NF map. **b** By mirroring the sample image along the horizontal (dashed) line and projecting it in the Michelson apparatus ontop of its original (see R and upside-down R), the experimentally measured phase map of an interference experiment had been simulated. To emphasize the features, a global phase slope along the vertical direction was added. **c** For a polariton fluid at pump densities nearly three-times above the condensation threshold, an experimentally measured phase map clearly featured a pinned vortex–antivortex pair. The position of the obtained double-dislocation pattern is indicated by a blue box and an expanded view of it is shown in (**d**). In this close-up view, the global slope was subtracted to reveal the actual phase map of a single vortex–antivortex pair. Reused with permission. Reference [32] Copyright 2011, Springer Nature

Recently, even room-temperature superfluidity has been reported by the demonstration of a transition from supersonic to superfluid flow in an exciton–polariton system based on the robust Frenkel excitons [146], whereas aforementioned studies were typically limited by the weaker binding energies of Wannier–Mott excitons, which give rise to cavity–polaritons in most common microcavity systems. Corresponding to the phase dislocations highlighted in the phase map in the supersonic regime, a vortex–antivortex pair is observed, whereas in the superfluid regime, the features of supersonic flow, i.e. the elastic scattering ring, the interference pattern in front of the defect, the dark cone in the wake of the defect and the vortex–antivortex pair, are according to the expected behaviour of a frictionless flow suppressed, as was shown by Lerario et al. in 2017 [146] (see Fig. 1.10 in Chap. 1).

Owing to the significance of vortices in polariton fluids, many groups have been studying their occurrence. Historically, it was discussed that the phase transition to a BEC is prohibited in an ideal 2D system. While for a two-dimensional Bose liquid at non-zero temperatures no ordering of infinite range can be established [147, 148], a superfluid phase had been predicted [149–151]. For weakly interacting Bosons in a potential landscape that breaks the symmetry and modifies the constant density of states (DOS) in a 2D system, bound vortex–antivortex pairs dominate the thermodynamics and phase coherence properties in this superfluid regime. In such a scenario, the remarkable effect of vortex–antivortex pair formation screens local phase defects, as was discussed for polariton systems [32, 62]. With an experimental technique, Roumpos et al. for the first time demonstrated a single vortex–antivortex pair in a two-dimensional exciton–polariton condensate [32], shown in Fig. 8.13 in a comparison between model and experiment. A pinned pair occuring in phase maps obtained from Michelson interferometric measurements is clearly observable well above a condensation threshold. It was further found for such non-equilibrium system by modelling with an open-dissipative Gross–Pitaevskii equation that pair evolution is distinctly different when compared to that in atomic condensates [152].

A plethora of studies on vortices exist in the literature [32, 61, 62, 65, 68, 70], and even half-quantum vortices in a condensate were reported in 2009 by Lagoudakis et al. [64]. Furthermore, the role of supercurrents on the formation of vortex–antivortex pairs [71], the vortex and half-vortex dynamics in a nonlinear spinor quantum fluid [95] and vortex chains [153] were studied. All these studies reveal the rich phenomenology linked to non-equilibrium condensate observations and physics. It shall be noted that half-vortices cannot appear in the NF pattern of a conventional laser for fundamental reasons [154].

The understanding, control and use of superfluidity in polariton systems may also affect the development of future (room-temperature) polariton devices, in which the polaritons can be moved in suitable waveguide structures without resistance for information processing and become screened from obstacles and scattering processes, rendering them more robust and persistive in transport schemes. Some devices and applications for polaritons have been recently summarized for instance in [155].

References

1. J. Kasprzak, M. Richard, S. Kundermann, A. Baas, P. Jeambrun, J.M.J. Keeling, F.M. Marchetti, M.H. Szymańska, R. André, J.L. Staehli, V. Savona, P.B. Littlewood, B. Deveaud, L.S. Dang. Bose-Einstein condensation of exciton polaritons. Nature **443**(7110), 409–414 (2006)
2. C.W. Lai, N.Y. Kim, S. Utsunomiya, G. Roumpos, H. Deng, M.D. Fraser, T. Byrnes, P. Recher, N. Kumada, T. Fujisawa, Y. Yamamoto, Coherent zero-state and π-state in an exciton-polariton condensate array. Nature **450**(7169), 529–532 (2007)

3. H. Deng, G.S. Solomon, R. Hey, K.H. Ploog, Y. Yamamoto, Spatial coherence of a polariton condensate. Phys. Rev. Lett. **99**(12), 126403 (2007)
4. H. Deng, G. Weihs, C. Santori, J. Bloch, Y. Yamamoto, Condensation of semiconductor microcavity exciton polaritons. Science **298**(5591), 199–202 (2002)
5. J. Kasprzak, M. Richard, A. Baas, B. Deveaud, R. Andre, J.-P. Poizat, L.S. Dang, Second-order time correlations within a polariton Bose-Einstein condensate in a CdTe microcavity. Phys. Rev. Lett. **100**(6), 067402 (2008)
6. H. Deng, G. Weihs, D. Snoke, J. Bloch, Y. Yamamoto, Polariton lasing vs. photon lasing in a semiconductor microcavity. Proc. Natl. Acad. Sci. USA **100**(26), 15318–15323 (2003)
7. H. Deng, D. Press, S. Götzinger, G.S. Solomon, R. Hey, K.H. Ploog, Y. Yamamoto, Quantum degenerate exciton-polaritons in thermal equilibrium. Phys. Rev. Lett. **97**(14), 146402 (2006)
8. M. Richard, J. Kasprzak, R. André, R. Romestain, L.S. Dang, G. Malpuech, A. Kavokin, Experimental evidence for nonequilibrium Bose condensation of exciton polaritons. Phys. Rev. B **72**(20), 201301 (2005)
9. M. Richard, J. Kasprzak, R. Romestain, R. André, L.S. Dang, Spontaneous coherent phase transition of polaritons in CdTe microcavities. Phys. Rev. Lett. **94**(18), 187401 (2005)
10. A. Imamoglu, R.J. Ram, S. Pau, Y. Yamamoto, Nonequilibrium condensates and lasers without inversion: exciton-polariton lasers. Phys. Rev. A **53**(6), 4250 (1996)
11. L.S. Dang, D. Heger, R. André, F. Bœuf, R. Romestain, Stimulation of polariton photoluminescence in semiconductor microcavity. Phys. Rev. Lett. **81**(18), 3920 (1998)
12. R.M. Stevenson, V.N. Astratov, M.S. Skolnick, D.M. Whittaker, M. Emam-Ismail, A.I. Tartakovskii, P.G. Savvidis, J.J. Baumberg, J.S. Roberts, Continuous wave observation of massive polariton redistribution by stimulated scattering in semiconductor microcavities. Phys. Rev. Lett. **85**(17), 3680–3683 (2000)
13. R. Balili, V. Hartwell, D. Snoke, L. Pfeiffer, K. West, Bose-Einstein condensation of microcavity polaritons in a trap. Science **316**(5827), 1007–1010 (2007)
14. A. Baas, J.-P. Karr, M. Romanelli, A. Bramati, E. Giacobino, Quantum degeneracy of microcavity polaritons. Phys. Rev. Lett. **96**(17), 176401 (2006)
15. H. Deng, Dynamic condensation of semiconductor microcavity polaritons. Ph.D. thesis, Department of Applied Physics, Stanford University (2006)
16. H. Deng, H. Haug, Y. Yamamoto, Exciton-polariton Bose-Einstein condensation. Rev. Mod. Phys. **82**(2), 1489 (2010)
17. J. Kasprzak, D.D. Solnyshkov, R. Andre, L.S. Dang, G. Malpuech, Formation of an exciton polariton condensate: thermodynamic versus kinetic regimes. Phys. Rev. Lett. **101**(14), 146404 (2008)
18. M. Aßmann, J.-S. Tempel, F. Veita, M. Bayer, A. Rahimi-Iman, A. Löffler, S. Höfling, S. Reitzenstein, L. Worschech, A. Forchel, From polariton condensates to highly photonic quantum degenerate states of bosonic matter. Proc. Natl. Acad. Sci. USA **108**, 1804–1809 (2011)
19. T. Horikiri, P. Schwendimann, A. Quattropani, S. Höfling, A. Forchel, Y. Yamamoto, Higher order coherence of exciton-polariton condensates. Phys. Rev. B **81**(3), 033307 (2010)
20. E. Kammann, H. Ohadi, M. Maragkou, A V. Kavokin, P.G. Lagoudakis, Crossover from photon to exciton-polariton lasing. New J. Phys. **14**(10), 105003 (2012)
21. G. Roumpos, C.-W. Lai, T.C.H. Liew, Y.G. Rubo, A.V. Kavokin, Y. Yamamoto, Signature of the microcavity exciton-polariton relaxation mechanism in the polarization of emitted light. Phys. Rev. B **79**(19), 195310 (2009)
22. A.V. Larionov, V.D. Kulakovskii, S. Höfling, C. Schneider, L. Worschech, A. Forchel, Polarized nonequilibrium Bose-Einstein condensates of spinor exciton polaritons in a magnetic field. Phys. Rev. Lett. **105**(25), 256401 (2010)
23. A. Rahimi-Iman, A.V. Chernenko, J. Fischer, S. Brodbeck, M. Amthor, C. Schneider, A. Forchel, S. Höfling, S. Reitzenstein, M. Kamp, Coherence signatures and density-dependent interaction in a dynamical exciton-polariton condensate. Phys. Rev. B **86**(15), 155308 (2012)
24. A. Demenev, S. Gavrilov, V. Kulakovskii, Stimulated polariton-polariton scattering and dynamic Bose-Einstein condensation of polaritons in GaAs microcavities under excitation near the exciton resonance. JETP Lett. **95**(1), 38–43 (2012)

25. P. Walker, T.C.H. Liew, D. Sarkar, M. Durska, A.P.D. Love, M.S. Skolnick, J.S. Roberts, I.A. Shelykh, A.V. Kavokin, D.N. Krizhanovskii, Suppression of Zeeman splitting of the energy levels of exciton-polariton condensates in semiconductor microcavities in an external magnetic field. Phys. Rev. Lett. **106**(25), 257401 (2011)
26. G. Roumpos, W.H. Nitsche, S. Höfling, A. Forchel, Y. Yamamoto, Gain-induced trapping of microcavity exciton polariton condensates. Phys. Rev. Lett. **104**(12), 126403 (2010)
27. E. Wertz, A. Amo, D.D. Solnyshkov, L. Ferrier, T.C.H. Liew, D. Sanvitto, P. Senellart, I. Sagnes, A. Lemaître, A.V. Kavokin, G. Malpuech, J. Bloch, Propagation and amplification dynamics of 1d polariton condensates. Phys. Rev. Lett. **109**, 216404 (2012)
28. R. Houdré, J.L. Gibernon, P. Pellandini, R.P. Stanley, U. Oesterle, C. Weisbuch, J. O'Gorman, B. Roycroft, M. Ilegems, Saturation of the strong-coupling regime in a semiconductor microcavity: free-carrier bleaching of cavity polaritons. Phys. Rev. B **52**(11), 7810 (1995)
29. J. Bloch, B. Sermage, M. Perrin, P. Senellart, R. André, L.S. Dang, Monitoring the dynamics of a coherent cavity polariton population. Phys. Rev. B **71**(15), 155311 (2005)
30. M. Steger, C. Gautham, D.W. Snoke, L. Pfeiffer, K. West, Slow reflection and two-photon generation of microcavity exciton-polaritons. Optica **2**(1), 1 (2015)
31. A. Amo, D. Sanvitto, F.P. Laussy, D. Ballarini, E. del Valle, M.D. Martín, A. Lemaître, J. Bloch, D.N. Krizhanovskii, M.S. Skolnick, C. Tejedor, L. Viña, Collective fluid dynamics of a polariton condensate in a semiconductor microcavity. Nature **457**(7227), 291–295 (2009)
32. G. Roumpos, M.D. Fraser, A. Löffler, S. Höfling, A. Forchel, Y. Yamamoto, Single vortex-antivortex pair in an exciton-polariton condensate. Nat. Phys. **7**(2), 129–133 (2011)
33. S. Reitzenstein, C. Hofmann, A. Gorbunov, M. Strauss, S.H. Kwon, C. Schneider, A. Löffler, S. Höfling, M. Kamp, A. Forchel, AlAs/GaAs micropillar cavities with quality factors exceeding 150.000. Appl. Phys. Lett. **90**(25), 251109 (2007)
34. C. Böckler, S. Reitzenstein, C. Kistner, R. Debusmann, A. Löffler, T. Kida, S. Höfling, A. Forchel, L. Grenouillet, J. Claudon, J.M. Gérard, Electrically driven high-Q quantum dot-micropillar cavities. Appl. Phys. Lett. **92**(9), 091107 (2008)
35. S. Reitzenstein, T. Heindel, C. Kistner, A. Rahimi-Iman, C. Schneider, S. Höfling, A. Forchel, Low threshold electrically pumped quantum dot-micropillar lasers. Appl. Phys. Lett. **93**(6), 061104 (2008)
36. M.H. Anderson, J.R. Ensher, M.R. Matthews, C.E. Wieman, E.A. Cornell, Observation of Bose-Einstein condensation in a dilute atomic vapor. Science **269**(5221), 198–201 (1995)
37. K.B. Davis, M.O. Mewes, M.R. Andrews, N.J. van Druten, D.S. Durfee, D.M. Kurn, W. Ketterle, Bose-Einstein condensation in a gas of sodium atoms. Phys. Rev. Lett. **75**(22), 3969 (1995)
38. C.C. Bradley, C.A. Sackett, J.J. Tollett, R.G. Hulet, Evidence of Bose-Einstein condensation in an atomic gas with attractive interactions. Phys. Rev. Lett. **75**(9), 1687 (1995)
39. A. Rahimi-Iman. *Nichtlineare Eekte in III/V Quantenlm-Mikroresonatoren: Von dynamischer Bose–Einstein-Kondensation hin zum elektrisch betriebenen Polariton-Laser* (Cuvillier Verlag, Göttingen, 2013)
40. T.C.H. Liew, O.A. Egorov, M. Matuszewski, O. Kyriienko, X. Ma, E.A. Ostrovskaya, Instability-induced formation and nonequilibrium dynamics of phase defects in polariton condensates. Phys. Rev. B **91**(8), 085413 (2015)
41. F. Tassone, C. Piermarocchi, V. Savona, A. Quattropani, and P. Schwendimann. Bottleneck effects in the relaxation and photoluminescence of microcavity polaritons. Phys. Rev. B **56**(12), 7554 (1997)
42. Y.G. Rubo, Kinetics of the polariton condensate formation in a microcavity. Phys. Stat. Sol. (a) **201**(4), 641–645 (2004)
43. S. Schmitt-Rink, D.S. Chemla, D.A.B. Miller, Theory of transient excitonic optical nonlinearities in semiconductor quantum-well structures. Phys. Rev. B **32**(10), 6601 (1985)
44. J.D. Plumhof, T. Stöferle, L. Mai, U. Scherf, R.F. Mahrt, Room-temperature Bose–Einstein condensation of cavity exciton-polaritons in a polymer. Nat. Mater. **13**, 247 (2014). Published online Dec 2013

45. F. Tassone, Y. Yamamoto, Exciton-exciton scattering dynamics in a semiconductor microcavity and stimulated scattering into polaritons. Phys. Rev. B **59**(16), 10830 (1999)
46. P. Senellart, J. Bloch, B. Sermage, J.Y. Marzin, Microcavity polariton depopulation as evidence for stimulated scattering. Phys. Rev. B **62**(24), R16263 (2000)
47. P.G. Savvidis, J.J. Baumberg, R.M. Stevenson, M.S. Skolnick, D.M. Whittaker, J.S. Roberts, Angle-resonant stimulated polariton amplifier. Phys. Rev. Lett. **84**(7), 1547 (2000)
48. M.D. Martin, G. Aichmayr, A. Amo, D. Ballarini, Polariton and spin dynamics in semiconductor microcavities under non-resonant excitation. J. Phys.: Condens. Matter **19**(29), 295204 (2007)
49. B. Deveaud, Polariton interactions in semiconductor microcavities. Comptes Rendus Phys. **17**(8), 874–892 (2016)
50. J.-S. Tempel, F. Veit, M. Aßmann, L.E. Kreilkamp, A. Rahimi-Iman, A. Löffler, S. Höfling, S. Reitzenstein, L. Worschech, A. Forchel, M. Bayer, Characterization of two-threshold behavior of the emission from a GaAs microcavity. Phys. Rev. B **85**(7), 075318 (2012)
51. S. Utsunomiya, L. Tian, G. Roumpos, C.W. Lai, N. Kumada, T. Fujisawa, M. Kuwata-Gonokami, A. Löffler, S. Höfling, A. Forchel, Y. Yamamoto, Observation of Bogoliubov excitations in exciton-polariton condensates. Nat. Phys. **4**(9), 700–705 (2008)
52. P. Schwendimann, A. Quattropani, Statistics of the polariton condensate. Phys. Rev. B **77**(8), 085317 (2008)
53. J.M. Ménard, C. Poellmann, M. Porer, U. Leierseder, E. Galopin, A. Lemaître, A. Amo, J. Bloch, R. Huber, Revealing the dark side of a bright exciton-polariton condensate. Nat. Commun. **5**, 4648 (2014)
54. C. Ciuti, V. Savona, C. Piermarocchi, A. Quattropani, P. Schwendimann, Role of the exchange of carriers in elastic exciton-exciton scattering in quantum wells. Phys. Rev. B **58**(12), 7926–7933 (1998)
55. G. Rochat, C. Ciuti, V. Savona, C. Piermarocchi, A. Quattropani, P. Schwendimann, Excitonic Bloch equations for a two-dimensional system of interacting excitons. Phys. Rev. B **61**(20), 13856–13862 (2000)
56. T.D. Doan, H. Thien Cao, D.B. Tran Thoai, H. Haug, Coherence of condensed microcavity polaritons calculated within Boltzmann-Master equations. Phys. Rev. B **78**(20), 205306 (2008)
57. D. Bajoni, P. Senellart, E. Wertz, I. Sagnes, A. Miard, A. Lemaître, J. Bloch, Polariton laser using single micropillar GaAs-GaAlAs semiconductor cavities. Phys. Rev. Lett. **100**(4), 047401 (2008)
58. D. Bajoni, P. Senellart, A. Lemaître, J. Bloch, Photon lasing in GaAs microcavity: similarities with a polariton condensate. Phys. Rev. B **76**(20), 201305(R) (2007)
59. B. Deveaud-Plédran, On the condensation of polaritons. J. Opt. Soc. Am. B **29**(2), A138–A145 (2012)
60. C. Schneider, A. Rahimi-Iman, N.Y. Kim, J. Fischer, I.G. Savenko, M. Amthor, M. Lermer, A. Wolf, L. Worschech, V.D. Kulakovskii, I.A. Shelykh, M. Kamp, S. Reitzenstein, A. Forchel, Y. Yamamoto, S. Hoefling, An electrically pumped polariton laser. Nature **497**, 348 (2013)
61. K.G. Lagoudakis, M. Wouters, M. Richard, A. Baas, I. Carusotto, R. Andre, L.S. Dang, B. Deveaud-Plédran, Quantized vortices in an exciton-polariton condensate. Nat. Phys. **4**(9), 706–710 (2008)
62. M.D. Fraser, G. Roumpos, Y. Yamamoto, Vortex-antivortex pair dynamics in an exciton-polariton condensate. New J. Phys. **11**, 113048 (2009)
63. A. Amo, J. Lefrere, S. Pigeon, C. Adrados, C. Ciuti, I. Carusotto, R. Houdré, E. Giacobino, A. Bramati, Superfluidity of polaritons in semiconductor microcavities. Nat. Phys. **5**(11), 805–810 (2009)
64. K.G. Lagoudakis, T. Ostatnický, A.V. Kavokin, Y.G. Rubo, R. André, B. Deveaud-Plédran, Observation of half-quantum vortices in an exciton-polariton condensate. Science **326**(5955), 974–976 (2009)
65. D. Sanvitto, F.M. Marchetti, M.H. Szymanska, G. Tosi, M. Baudisch, F.P. Laussy, D.N. Krizhanovskii, M.S. Skolnick, L. Marrucci, A. Lemaître, J. Bloch, C. Tejedor, L. Viña, Persistent currents and quantized vortices in a polariton superfluid. Nat. Phys. **6**(7), 527–533 (2010)

66. A. Amo, S. Pigeon, D. Sanvitto, V.G. Sala, R. Hivet, I. Carusotto, F. Pisanello, G. Lemé-nager, R. Houdré, E. Giacobino, C. Ciuti, A. Bramati, Polariton superfluids reveal quantum hydrodynamic solitons. Science **332**(6034), 1167–1170 (2011)
67. D. Sanvitto, S. Pigeon, A. Amo, D. Ballarini, M. De Giorgi, I. Carusotto, R. Hivet, F. Pisanello, V.G. Sala, P.S.S. Guimaraes, R. Houdré, E. Giacobino, C. Ciuti, A. Bramati, G. Gigli, All-optical control of the quantum flow of a polariton condensate. Nat. Phot. **5**, 610–614 (2011)
68. G. Nardin, G. Grosso, Y. Leger, B. Pietka, F. Morier-Genoud, B. Deveaud-Plédran, Hydrodynamic nucleation of quantized vortex pairs in a polariton quantum fluid. Nat. Phys. **7**(8), 635–641 (2011)
69. G. Tosi, D. Sanvitto, M. Baudisch, E. Karimi, B. Piccirillo, L. Marrucci, A. Lemaître, J. Bloch, L. Viña, Vortex stability and permanent flow in nonequilibrium polariton condensates. J. Appl. Phys. **109**, 102406 (2011)
70. G. Tosi, F.M. Marchetti, D. Sanvitto, C. Antón, M.H. Szymańska, A. Berceanu, C. Tejedor, L. Marrucci, A. Lemaître, J. Bloch, L. Viña, Onset and dynamics of vortex-antivortex pairs in polariton optical parametric oscillator superfluids. Phys. Rev. Lett. **107**(3), 036401 (2011)
71. C. Antón, G. Tosi, M.D. Martén, L. Viña, A. Lemaître, J. Bloch, Role of supercurrents on vortices formation in polariton condensates. Opt. Express **20**(15), 16366–16373 (2012)
72. A. Rahimi-Iman, C. Schneider, J. Fischer, S. Holzinger, M. Amthor, S. Höfling, S. Reitzenstein, L. Worschech, M. Kamp, A. Forchel, Zeeman splitting and diamagnetic shift of spatially confined quantum-well exciton polaritons in an external magnetic field. Phys. Rev. B **84**(16), 165325 (2011)
73. T.D. Doan, H. Thien Cao, D.B. Tran Thoai, H. Haug, Relaxation kinetics for temperature and degeneracy of microcavity polaritons. Solid State Commun. **145**, 48–51 (2008)
74. D. Read, T.C.H. Liew, Y.G. Rubo, A.V. Kavokin, Stochastic polarization formation in exciton-polariton Bose-Einstein condensates. Phys. Rev. B **80**(19), 195309 (2009)
75. N. Na, Y. Yamamoto, Massive parallel generation of indistinguishable single photons via the polaritonic superfluid to Mott-insulator quantum phase transition. New J. Phys. **12**, 123001 (2010)
76. I.A. Shelykh, A.V. Kavokin, Y.G. Rubo, Polariton polarization-sensitive phenomena in planar semiconductor microcavities. Semicond. Sci. Technol. **25**(1), 013001 (2010)
77. Y.G. Rubo, A.V. Kavokin, I.A. Shelykh, Suppression of superfluidity of exciton-polaritons by magnetic field. Phys. Lett. A **358**(3), 227–230 (2006)
78. T.C.H. Liew, Y.G. Rubo, I.A. Shelykh, A.V. Kavokin, Suppression of Zeeman splitting and polarization steps in localized exciton-polariton condensates. Phys. Rev. B **77**(12), 125339 (2008)
79. J. Fischer, S. Brodbeck, A.V. Chernenko, I. Lederer, A. Rahimi-Iman, M. Amthor, V.D. Kulakovskii, L. Worschech, M. Kamp, M. Durnev, C. Schneider, A.V. Kavokin, S. Höfling, Anomalies of a nonequilibrium spinor polariton condensate in a magnetic field. Phys. Rev. Lett. **112**, 093902 (2014)
80. A.V. Chernenko, A. Rahimi-Iman, J. Fischer, M. Amthor, C. Schneider, S. Reitzenstein, A. Forchel, S. Höfling, Polariton condensate coherence in planar microcavities in a magnetic field. Semiconductors **50**(12), 1609–1613 (2016)
81. E. Wertz, L. Ferrier, D.D. Solnyshkov, R. Johne, D. Sanvitto, A. Lemaître, I. Sagnes, R. Grousson, A.V. Kavokin, P. Senellart, G. Malpuech, J. Bloch, Spontaneous formation and optical manipulation of extended polariton condensates. Nat. Phys. **6**(11), 860–864 (2010)
82. A.A. Michelson, E.W. Morley, On the relative motion of the Earth and the luminiferous ether. Am. J. Sci. **s3–34**(203), 333–345 (1887)
83. A.J. Leggett, Bose-Einstein condensation in the alkali gases: some fundamental concepts. Rev. Mod. Phys. **73**(2), 307 (2001)
84. L.P. Pitaevskii, S. Stringari, *Bose-Einstein Condensation* (Oxford University Press, Oxford, 2003)
85. K.S. Daskalakis, S.A. Maier, R. Murray, S. Kéna-Cohen, Nonlinear interactions in an organic polariton condensate. Nat. Mater. **13**(3), 271–278 (2014)

86. J. Fischer, I.G. Savenko, M.D. Fraser, S. Holzinger, S. Brodbeck, M. Kamp, I.A. Shelykh, C. Schneider, S. Höfling, Spatial coherence properties of one dimensional exciton-polariton condensates. Phys. Rev. Lett. **113**(20), 203902 (2014)
87. G. Roumpos, M. Lohse, W.H. Nitsche, J. Keeling, M.H. Szymańska, P.B. Littlewood, A. Löffler, S. Höfling, L. Worschech, A. Forchel, Y. Yamamoto, Power-law decay of the spatial correlation function in exciton-polariton condensates. Proc. Natl. Acad. Sci. USA (2012). https://doi.org/10.1073/pnas.1107970109.
88. R. Hanbury Brown, R.Q. Twiss, Correlation between photons in two coherent beams of light. Nature **177**, 27–29 (1956)
89. S. Strauf, K. Hennessy, M.T. Rakher, Y.-S. Choi, A. Badolato, L.C. Andreani, E.L. Hu, P.M. Petroff, D. Bouwmeester, Self-tuned quantum dot gain in photonic crystal lasers. Phys. Rev. Lett. **96**(12), 127404 (2006)
90. S.M. Ulrich, C. Gies, S. Ates, J. Wiersig, S. Reitzenstein, C. Hofmann, A. Löffler, A. Forchel, F. Jahnke, P. Michler. Photon statistics of semiconductor microcavity lasers. Phys. Rev. Lett. **98**(4), 043906 (2007)
91. M. Amthor, S. Weißenseel, J. Fischer, M. Kamp, C. Schneider, S. Höfling, Electro-optical switching between polariton and cavity lasing in an InGaAs quantum well microcavity. Opt. Express **22**(25), 31146–31153 (2014)
92. S. Kim, B. Zhang, Z. Wang, J. Fischer, S. Brodbeck, M. Kamp, C. Schneider, S. Höfling, H. Deng, Coherent polariton laser. Phys. Rev. X **6**(1), 011026 (2016)
93. N.D. Vy, H.T. Cao, D.B.T. Thoai, H. Haug, Time dependence of the ground-state population statistics of condensed microcavity polaritons. Phys. Rev. B, **80**(19), 195306 (2009)
94. M. De Giorgi, D. Ballarini, P. Cazzato, G. Deligeorgis, S.I. Tsintzos, Z. Hatzopoulos, P.G. Savvidis, G. Gigli, F.P. Laussy, D. Sanvitto, Relaxation oscillations in the formation of a polariton condensate. Phys. Rev. Lett. **112**(11), 113602 (2014)
95. L. Dominici, G. Dagvadorj, J.M. Fellows, D. Ballarini, M. De Giorgi, F.M. Marchetti, B. Piccirillo, L. Marrucci, A. Bramati, G. Gigli, M.H. Szymañska, D. Sanvitto, Vortex and half-vortex dynamics in a nonlinear spinor quantum fluid. Sci. Adv. **1**(11), e1500807 (2015)
96. L. Dominici, M. Petrov, M. Matuszewski, D. Ballarini, M. De Giorgi, D. Colas, E. Cancellieri, B.S. Fernández, A. Bramati, G. Gigli, A. Kavokin, F. Laussy, D. Sanvitto, Real-space collapse of a polariton condensate. Nat. Commun. **6**, 8993 (2015)
97. N.Y. Kim, K. Kusudo, C. Wu, N. Masumoto, A. Löffler, S. Höfling, N. Kumada, L. Worschech, A. Forchel, Y. Yamamoto. Dynamical d-wave condensation of exciton-polaritons in a two-dimensional square-lattice potential. Nat. Phys. **7**(9), 681–686 (2011)
98. N.Y. Kim, K. Kusudo, A. Löffler, S. Höfling, A. Forchel, Y. Yamamoto, Exciton-polariton condensates near the Dirac point in a triangular lattice. New J. Phys. **15**, 035032 (2013)
99. K. Winkler, J. Fischer, A. Schade, M. Amthor, R. Dall, J. Geßler, M. Emmerling, E.A. Ostrovskaya, M. Kamp, C. Schneider, S. Höfling, A polariton condensate in a photonic crystal potential landscape. New J. Phys. **17**, 023001 (2015)
100. M. Klaas, S. Mandal, T.C.H. Liew, M. Amthor, S. Klembt, L. Worschech, C. Schneider, S. Höfling, Optical probing of the Coulomb interactions of an electrically pumped polariton condensate. Appl. Phys. Lett. **110**(15), 151103 (2017)
101. P. Cristofolini, G. Christmann, S.I. Tsintzos, G. Deligeorgis, G. Konstantinidis, Z. Hatzopoulos, P.G. Savvidis, J.J. Baumberg, Coupling quantum tunneling with cavity photons. Science **336**(6082), 704–707 (2012)
102. T.C.H. Liew, M.M. Glazov, K.V. Kavokin, I.A. Shelykh, M.A. Kaliteevski, A.V. Kavokin, Proposal for a bosonic cascade laser. Phys. Rev. Lett. **110**(4), 047402 (2013)
103. I.G. Savenko, I.A. Shelykh, M.A. Kaliteevski, Nonlinear terahertz emission in semiconductor microcavities. Phys. Rev. Lett. **107**(2), 027401 (2011)
104. S. Huppert, O. Lafont, E. Baudin, J. Tignon, R. Ferreira, Terahertz emission from multiple-microcavity exciton-polariton lasers. Phys. Rev. B **90**(24), 241302(R) (2014)
105. K.V. Kavokin, M.A. Kaliteevski, R.A. Abram, A.V. Kavokin, S. Sharkova, I.A. Shelykh, Stimulated emission of terahertz radiation by exciton-polariton lasers. Appl. Phys. Lett. **97**(20), 201111–3 (2010)

106. L. Dominici, D. Colas, S. Donati, J.P. Restrepo Cuartas, M. De Giorgi, D. Ballarini, G. Guirales, J.C. López Carreño, A. Bramati, G. Gigli, E. Del Valle, F.P. Laussy, D. Sanvitto, Ultrafast control and Rabi oscillations of polaritons. Phys. Rev. Lett. **113**(22), 226401 (2014)
107. A.D. Jameson, J.L. Tomaino, Y.-S. Lee, G. Khitrova, H.M. Gibbs, C.N. Böttge, A.C. Klettke, M. Kira, S.W. Koch, Direct measurement of light-matter energy exchange inside a microcavity. Optica **1**(5), 276 (2014)
108. S.S. Gavrilov, A.S. Brichkin, A.A. Demenev, A.A. Dorodnyy, S.I. Novikov, V.D. Kulakovskii, S.G. Tikhodeev, N.A. Gippius, Bistability and nonequilibrium transitions in the system of cavity polaritons under nanosecond-long resonant excitation. Phys. Rev. B **85**, 075319 (2012)
109. M. Amthor, T.C.H. Liew, C. Metzger, S. Brodbeck, L. Worschech, M. Kamp, I.A. Shelykh, A.V. Kavokin, C. Schneider, S. Höfling, Optical bistability in electrically driven polariton condensates. Phys. Rev. B **91**, 081404 (2015)
110. M. Klaas, H. Sigurdsson, T.C.H. Liew, S. Klembt, M. Amthor, F. Hartmann, L. Worschech, C. Schneider, S. Höfling, Electrical and optical switching in the bistable regime of an electrically injected polariton laser. Phys. Rev. B **96**(4), 041301 (2017)
111. D. Colas, L. Dominici, S. Donati, A.A. Pervishko, T.C.H. Liew, I.A. Shelykh, D. Ballarini, M. De Giorgi, A. Bramati, G. Gigli, E. Del Valle, F.P. Laussy, A.V. Kavokin, D. Sanvitto, Polarization shaping of Poincaré beams by polariton oscillations. Light Sci. Appl. **4**, e350 (2015)
112. S. Klembt, E. Durupt, S. Datta, T. Klein, A. Baas, Y. Léger, C. Kruse, D. Hommel, A. Minguzzi, M. Richard, Exciton-polariton gas as a nonequilibrium coolant. Phys. Rev. Lett. **114**(18), 186403 (2015)
113. D. Ballarini, M.D. Giorgi, E. Cancellieri, R. Houdré, E. Giacobino, R. Cingolani, A. Bramati, G. Gigli, D. Sanvitto, All-optical polariton transistor. Nat. Commun. **4**, 1778 (2013)
114. F. Marsault, H.S. Nguyen, D. Tanese, A. Lemaître, E. Galopin, I. Sagnes, A. Amo, J. Bloch, Single-wavelength, all-optical switching based on exciton-polaritons. Appl. Phys. Lett **107**, 5263 (2015)
115. H.S. Nguyen, D. Vishnevsky, C. Sturm, D. Tanese, D. Solnyshkov, E. Galopin, A. Lemaître, I. Sagnes, A. Amo, G. Malpuech, J. Bloch, Realization of a double-barrier resonant tunneling diode for cavity polaritons. Phys. Rev. Lett. **110**, 236601 (2013)
116. C. Sturm, D. Tanese, H. Nguyen, H. Flayac, E. Galopin, A. Lemaître, A. Amo, G. Malpuech, J. Bloch, All-optical phase modulation in a cavity-polariton Mach-Zehnder interferometer. Nat. Commun. **5**, 3278 (2014)
117. S.S. Demirchyan, I. Yu Chestnov, A.P. Alodjants, M.M. Glazov, A.V. Kavokin, Qubits based on polariton Rabi oscillators. Phys. Rev. Lett. **112**(19), 196403 (2014)
118. G. Tosi, G. Christmann, N.G. Berloff, P. Tsotsis, T. Gao, Z. Hatzopoulos, P.G. Savvidis, J.J. Baumberg, Sculpting oscillators with light within a nonlinear quantum fluid. Nat. Phys. (2012)
119. O. El Daïf, A. Baas, T. Guillet, J.-P. Brantut, R. Idrissi Kaitouni, J.L. Staehli, F. Morier-Genoud, B. Deveaud, Polariton quantum boxes in semiconductor microcavities. Appl. Phys. Lett. **88**(6), 061105–3 (2006)
120. E.A. Cerda-Méndez, D. Krizhanovskii, K. Biermann, K. Guda, R. Bradley, R. Hey, P.V. Santos, M.S. Skolnick, One dimensional confinement of microcavity polaritons using non-piezoelectric surface acoustic waves. Phys. E: Low-Dimens. Syst. Nanostructures **42**(10), 2548–2551 (2010)
121. Á. Cuevas, J.C.L. Carreño, B. Silva, M. De Giorgi, D.G. Suárez-Forero, C.S. Muñoz, A. Fieramosca, F. Cardano, L. Marrucci, V. Tasco, G. Biasiol, E. Del Valle, L. Dominici, D. Ballarini, G. Gigli, P. Mataloni, F.P. Laussy, F. Sciarrino, D. Sanvitto, First observation of the quantized exciton-polariton field and effect of interactions on a single polariton. Sci. Adv. **4**(4), eaao6814 (2018)
122. D. Caputo, D. Ballarini, G. Dagvadorj, C.S. Muñoz, M. De Giorgi, L. Dominici, K. West, L.N. Pfeiffer, G. Gigli, F.P. Laussy, M.H. Szymanska, D.Sanvitto, Topological order and thermal equilibrium in polariton condensates. Nat. Mater. **17**(2), 145–151 (2018)

123. N. Ishida, T. Byrnes, T. Horikiri, F. Nori, Y. Yamamoto, Photoluminescence of high-density exciton-polariton condensates. Phys. Rev. B **90**(24), 241304 (2014)
124. D. Ballarini, D. Caputo, C.S. Muñoz, M. De Giorgi, L. Dominici, M.H. Szymanska, K. West, L.N. Pfeiffer, G. Gigli, F.P. Laussy, D. Sanvitto, Macroscopic two-dimensional polariton condensates. Phys. Rev. Lett. **118**(21), 215301 (2017)
125. S. Christopoulos, G.B.H. von Högersthal, A.J.D. Grundy, P.G. Lagoudakis, A.V. Kavokin, J.J. Baumberg, G. Christmann, R. Butté, E. Feltin, J.-F. Carlin, N. Grandjean, Room-temperature polariton lasing in semiconductor microcavities. Phys. Rev. Lett. **98**(12), 126405 (2007)
126. G. Christmann, R. Butté, E. Feltin, J.-F. Carlin, N. Grandjean, Room temperature polariton lasing in a GaN/AlGaN multiple quantum well microcavity. Appl. Phys. Lett. **93**(5), 051102 (2008)
127. J.J. Baumberg, A.V. Kavokin, S. Christopoulos, A.J.D. Grundy, R. Butté, G. Christmann, D.D. Solnyshkov, G. Malpuech, G.B.H. von Högersthal, E. Feltin, J.-F. Carlin, N. Grandjean, Spontaneous polarization buildup in a room-temperature polariton laser. Phys. Rev. Lett. **101**(13), 136409 (2008)
128. A. Das, J. Heo, M. Jankowski, W. Guo, L. Zhang, H. Deng, P. Bhattacharya, Room temperature ultralow threshold GaN nanowire polariton laser. Phys. Rev. Lett. **107**(6), 066405 (2011)
129. L. Sun, S. Sun, H. Dong, W. Xie, M. Richard, L. Zhou, L.S. Dang, X. Shen, Z. Chen, Room temperature one-dimensional polariton condensate in a ZnO microwire (2010). arXiv:1007.4686v1 (unpublished)
130. T. Guillet, M. Mexis, J. Levrat, G. Rossbach, C. Brimont, T. Bretagnon, B. Gil, R. Butté, N. Grandjean, L. Orosz, F. Réveret, J. Leymarie, J. Zúñiga-Pérez, M. Leroux, F. Semond, S. Bouchoule, Polariton lasing in a hybrid bulk ZnO microcavity. Appl. Phys. Lett. **99**(16), 161104 (2011)
131. C. Sturm, H. Hilmer, R. Schmidt-Grund, M. Grundmann, Exciton-polaritons in a ZnO-based microcavity: polarization dependence and nonlinear occupation. New J. Phys. **13**(3), 033014 (2011)
132. A. Das, J. Heo, A. Bayraktaroglu, W. Guo, T.-K. Ng, J. Phillips, B.S. Ooi, P. Bhattacharya, Room temperature strong coupling effects from single ZnO nanowire microcavity. Opt. Express **20**(11), 11830–11837 (2012)
133. W. Xie, H. Dong, S. Zhang, L. Sun, W. Zhou, Y. Ling, L. Jian, X. Shen, Z. Chen, Room-temperature polariton parametric scattering driven by a one-dimensional polariton condensate. Phys. Rev. Lett. **108**(16), 166401 (2012)
134. Q. Duan, X. Dan, W. Liu, L. Jian, L. Zhang, J. Wang, Y. Wang, G. Jie, H. Tao, W. Xie, X. Shen, Z. Chen, Polariton lasing of quasi-whispering gallery modes in a ZnO microwire. Appl. Phys. Lett. **103**(2), 022103 (2013)
135. F. Li, L. Orosz, O. Kamoun, S. Bouchoule, C. Brimont, P. Disseix, T. Guillet, X. Lafosse, M. Leroux, J. Leymarie, M. Mexis, M. Mihailovic, G. Patriarche, F. Réveret, D. Solnyshkov, J. Zuniga-Perez, G. Malpuech. From excitonic to photonic polariton condensate in a ZnO-based microcavity. Phys. Rev. Lett. **110**(19), 196406 (2013)
136. R. Su, C. Diederichs, J. Wang, T.C.H. Liew, J. Zhao, S. Liu, W. Xu, Z. Chen, Q. Xiong. Room-temperature polariton lasing in all-inorganic perovskite nanoplatelets. Nano Lett. **17**(6), 3982–3988 (2017)
137. S. Kéna-Cohen, S.R. Forrest, Room-temperature polariton lasing in an organic single-crystal microcavity. Nat. Phot. **4**(6), 371–375 (2010)
138. P. Bhattacharya, T. Frost, S. Deshpande, M.Z. Baten, A. Hazari, A. Das, Room temperature electrically injected polariton laser. Phys. Rev. Lett. **112**, 236802 (2014)
139. A. Bhattacharya, M.Z. Baten, I. Iorsh, T. Frost, A. Kavokin, P. Bhattacharya. Output polarization characteristics of a GaN microcavity diode polariton laser. Phys. Rev. B **94**(3), 5203 (2016)
140. E. Estrecho, T. Gao, N. Bobrovska, M.D. Fraser, M. Steger, L. Pfeiffer, K. West, T.C.H. Liew, M. Matuszewski, D.W. Snoke, A.G. Truscott, E.A. Ostrovskaya, Single-shot condensation of exciton polaritons and the hole burning effect. Nat. Commun. **9**(1), 2944 (2018)

141. M. Glauser, R. Butté, Relative intensity noise and emission linewidth of polariton laser diodes. Phys. Rev. B **88**(11), 115305 (2013)
142. M.D. Fraser, Coherent exciton-polariton devices. Semicond. Sci. Technol. **32**(9), 093003 (2017)
143. R. Onofrio, C. Raman, J.M. Vogels, J.R. Abo-Shaeer, A.P. Chikkatur, W. Ketterle, Observation of superfluid flow in a Bose-Einstein condensed gas. Phys. Rev. Lett. **85**(11), 2228–2231 (2000)
144. A.V. Kavokin, J.J. Baumberg, G. Malpuech, F.P. Laussy, *Microcavities*, vol. 1 (Oxford University Press, Oxford, 2017)
145. K.S. Daskalakis, S.A. Maier, S. Kéna-Cohen, Spatial coherence and stability in a disordered organic polariton condensate. Phys. Rev. Lett. **115**, 035301 (2015)
146. G. Lerario, A. Fieramosca, F. Barachati, D. Ballarini, K.S. Daskalakis, L. Dominici, M. De Giorgi, S.A. Maier, G. Gigli, S. Kéna-Cohen, D. Sanvitto, Room-temperature superfluidity in a polariton condensate. Nat. Phys. **13**, 837 (2017)
147. P.C. Hohenberg. Existence of long-range order in one and two dimensions. Phys. Rev. **158**(2), 383 (1967)
148. N.D. Mermin, H. Wagner, Absence of ferromagnetism or antiferromagnetism in one- or two-dimensional isotropic Heisenberg models. Phys. Rev. Lett. **17**(22), 1133 (1966)
149. V.L. Berezinskii, Destruction of long-range order in one-dimensional and two-dimensional systems possessing a continuous symmetry group. II. Quantum systems. J. Exp. Theor. Phys. **34**, 610 (1972)
150. J.M. Kosterlitz, D.J. Thouless, Ordering, metastability and phase transitions in two-dimensional systems. J. Phys. C: Solid State Phys. **6**(7), 1181 (1973)
151. A. Posazhennikova, Colloquium: weakly interacting, dilute Bose gases in 2D. Rev. Mod. Phys. **78**(4), 1111–1134 (2006)
152. T.W. Neely, E.C. Samson, A.S. Bradley, M.J. Davis, B.P. Anderson, Observation of vortex dipoles in an oblate Bose-Einstein condensate. Phys. Rev. Lett. **104**(16), 160401 (2010)
153. T. Boulier, H. Terças, D.D. Solnyshkov, Q. Glorieux, E. Giacobino, G. Malpuech, A. Bramati, Vortex chain in a resonantly pumped polariton superfluid. Sci. Rep. **5**, 09230 (2015)
154. V.B. Timoffeev, Bose condensation of exciton polaritons in microcavities. Fizika i Tekhnika Poluprovodnikov **46**(7), 865–883 (2012). engl. translation in Semiconductors **46**(7), S. 843, Springer (2012)
155. D. Sanvitto, S. Kéna-Cohen, The road towards polaritonic device. Nat. Mater. **15**, 1061 (2016)

Chapter 9
Polaritons in External Fields

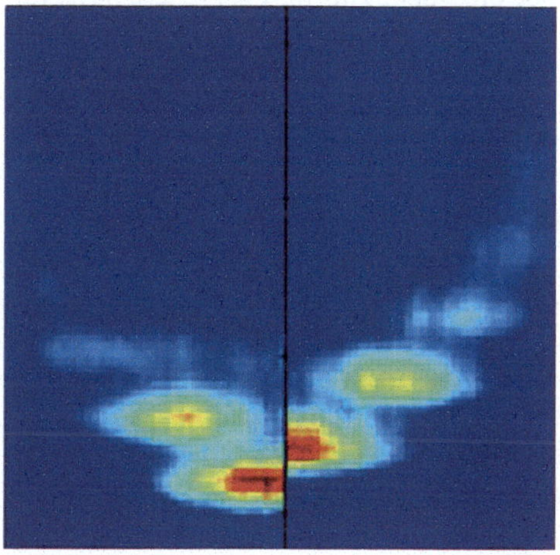

Abstract External fields such as magnetic or electric fields are widely used to manipulate and control quantum emitters in semiconductor structures based on their magnetic or electronic response. Since coupling of the light field to an emitter is governed by a few factors, such as resonance conditions, the oscillator strength of electric dipoles, and selection rules, external fields with their effects on the spatial wave-function overlap, the dipole orientation, spin degeneracy or resonance energies directly affect light–matter interaction. Some properties can be for instance electrically manipulated through band structure tilts and fields using tunable external bias, whereas the magnetic moment and spin properties of excitations in matter can be addressed with variable magnetic fields. Typically, static fields are employed, whereas transient fields can be used as a tool for additional control of the quantum state on ultrafast time scales. Many examples of polariton experiments clearly show that the rich polariton physics in optical microcavities can be further studied when exposing the exciton–polaritons to external fields. Prominent effects of external fields on excitons and, consequently, on polaritons will be summarized in the following.

A. Rahimi-Iman, *Polariton Physics*, Springer Series in Optical Sciences 229,
https://doi.org/10.1007/978-3-030-39333-5_9

9.1 Effects of External Fields on Quantum-Well Excitons

Exposing polaritons to external electric and magnetic fields can add very exciting possibilities to the standard experiments concerning polariton physics. Naturally, the impact of external fields on polariton resonances, particle dynamics and device properties is not only of concern for fundamental studies, but also for practical polariton structures and future quantum optical devices based on *polaritonics*. Indeed, the coverage of all phenomena related to external fields in this section is out of the scope. Here, a brief overview on the influence of external fields primarily in the linear polariton regime shall be provided. This may serve as a basis for later considerations and as a profound background for the dealing with electro-optical and magneto-optical studies in the literature.

9.1.1 *Electro-Optical Tuning*

Electric fields are known to have a significant influence on semiconductor band structures and are heavily utilized in electrical semiconductor devices such as diodes, transistors and many other functional heterostructures. Many textbooks cover fundamental descriptions of such effects for quantum structures and semiconductor heterostructures in general. Thus, a good overview concerning electro-optic effects can be obtained for instance through [1–3].

Franz–Keldysh Effect

For an intrinsic semiconductor in an external field, charge carriers are expected to experience a drift force. By redistribution of charges in the structure, a potential build-up as a consequence of the applied bias voltage can be obtained. This leads to a tilt in the band structure in the direction of applied bias, which modifies the absorption edge of the semiconductor, well-known as the Franz–Keldysh effect: As the wave-functions in both the valence and conduction band have evanescent tails inside the energy band gap with certain overlap, an energy red shift of the bulk excitons is obtained as a function of the field strength. Here, spatial overlap of the wave-functions is dependent on the magnitude of the field and decreases with increasing fields. Besides, exciton dissociation is facilitated with increasing fields, broadening and smearing out the excitonic resonance already at moderate fields.

Quantum-Confined Stark Effect

Similarly, an energy shift can be evidenced for excitons in quantum-well structures, if an external field is applied in growth direction, i.e. perpendicular to the QW plane.

However, the effect differs in the case of QWs compared to the bulk case, in that larger electric fields can be applied without exciton dissociation and the energy shift is larger due to a significant change in the potential well of the QW. At the same time, in a tilted potential under an acting field (e.g. built-in or external bias) electron and hole wave-functions within the well have ground-state maxima at opposite sides, as can be seen in Fig. 9.1b, in contrast to the unbiased (flat-band) symmetrical case (a). Since, electron–hole overlap and their correlations are not vanished, excitons with most of their oscillator strength are preserved in the structure. With increasing electric field, the exciton resonance is increasingly red-shifted with respect to its unbiased level, and simultaneously weakened with the amount of electron–hole wave-function separation. This effect is referred to as the quantum-confined Stark effect (QCSE).

The QCSE is a widely used effect in order to tune excitonic resonances in quantum structures such as quantum wells and—in the 0D limit—in quantum dot systems [4], for which often little tuning parameters are given after growth. While temperature tuning of resonances in microcavity systems does not leave the cavity mode untouched, electro-optical tuning offers a unique tool to alter the resonance detuning situation without raising the temperature, if no other tuning mechanisms are available (such as pressure, thickness variation, etc.).

The QCSE is inevitably experienced in quantum structures which are embedded in diode junctions—commonly of p-i-n type for photonic structures—since a change of bias naturally leads to a modification of the band tilt. The QCSE is readily obtained for polariton diode structures at zero external fields, as the band structure of the QWs is already tilted (considered in Fig. 9.1b). At forward bias, the intrinsic bias is reduced until flat-band conditions are obtained (considered in Fig. 9.1a), at which the QCSE is completely neutralized, while charge carriers flood the active region of the device from opposite sides of the p-i-n structure.

Indeed, a local charge carrier accumulation and the asymmetric filling of multiple QWs, i.e. an asymmetric electron–hole distribution across a structure, can also lead to band edge deformation along the growth direction, so that the designed scenario is usually not anymore existent. At reverse bias, the quantum-well band edges are further tilted and the QCSE-related energy shift increased, as shown in Fig. 9.1c. This provides for example a unique tool for reversible electro-optical fine-tuning of excitonic resonances for cQED experiments [4, 5], (ultra-)fast optical switches based on (transient) electric fields at high frequencies [6], and the transition between strong coupling and weak coupling in polariton diodes on GHz timescales [7, 8], to name but a few. In this context, also strong electric fields from terahertz (THz) pulses become appealing candidates for transient manipulation of the band edge tilt.

In contrast to a perpendicular field orientation, an electric field parallel to the QW plane causes the excitons to experience a situation similar to unconfined excitons, i.e. broadening and smearing out of the excitonic resonance as in the bulk system. Such an in-plane field could be employed as an on–off switch for strong coupling, whereas excitons suffer a drastical decrease in oscillator strength due to charge separation until at stronger fields excitons are dissociated or do not form (see [9]). Remarkably, exciton features are conserved up to field strengths of 10^5 V/cm in the QCSE case, while in bulk GaAs excitons disappear at two orders of magnitude smaller field strength [1]. Nevertheless, in-plane fields could serve another useful purpose, irrespective of

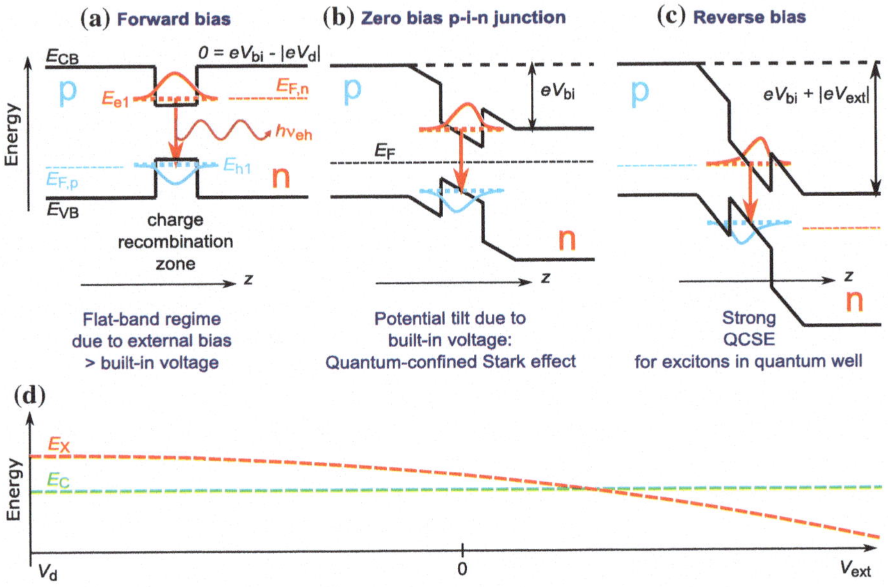

Fig. 9.1 Sketch of the quantum-confined Stark effect (QCSE) experienced by QW excitons in a p-i-n structure under external bias. To resemble the common situation given in polariton LEDs, the representative QW is placed in an intrinsic semiconductor environment, whereas the outer regions (e.g. DBR structure) are doped for sufficient charge-carrier injection into the active region. $E_{\mathrm{CB/VB}}$ denote conduction/valence band edges, E_{F} and $E_{\mathrm{F,n/p}}$ denote the Fermi level across the junction and within the doped regions, respectively. For simplicity, the *imref* (quasi-Fermi levels inside the charge-recombination zone) are not displayed. The energy levels are only indicative, with gap and transition energies not in true scale. **a** Flat-band condition for the double heterostructure's band structure (in growth direction z), with lowest-lying energy states for the confined electron and hole $E_{\mathrm{e1/h1}}$ indicated (dotted lines) together with corresponding particle's wave-function (solid curve) inside the quantum-potential structure. The flat-band configuration is also representative for the band structure in an undoped QW-microcavity system, which is free of a built-in voltage V_{bi}. Here, the external forward bias applied to the p-i-n diode compensates the built-in potential at its characteristic diode voltage V_{d} (for $V_{\mathrm{forward}} > V_{\mathrm{d}}$, the system remains in flat-band configuration, with the additional voltages increasing currents). The optical transition $E_{\mathrm{e1}} - E_{\mathrm{h1}} = h\nu_{\mathrm{eh}}$ is indicative of that of the QW exciton's excitation energy. However, no considerations of the excitonic binding energy are made here (in the single-particle band-structure diagram) to represent the actual optical band gap of the system in this schematic drawing which would be determined by the lowest transition energy (corresponding to the exciton's 1s state). **b** Potential of a QW LED without bias. The built-in voltage already causes a QCSE for the QW transition, represented by a red shift in the resonance. Here, wave-function and energy modifications are schematically indicated. **c** The QCSE is increased under an increased reverse bias. **d** Sketch of the energies of the exciton and cavity modes as a function of the external bias

the negative impact on the oscillator strength. It could be employed for the identification of a Mott transition in etched planar microcavities with moderate and small lateral dimensions, for instance during operation of a polariton laser. For laterally contacted QW microcavities with a sufficiently small bias, the in-plane conductivity of an electron–hole plasma in the QW structure should become measurable (at least upon an offset of background conductivity) as soon as the exciton Mott density is surpassed. Whereas, the neutral exciton gas in a polariton system should reveal an insulating behaviour. In case of resonant or quasi-resonant polariton injection by optical pumping techniques, the amount of free electrons in the structure could be kept at a minimum.

9.1.2 Coupling to Strong Transient Electric Fields

In the domain of light–matter interaction studies, coupling experiments are not restricted to a certain frequency range or confinement scenarios. Similar to Rabi oscillations in systems employing atomic transitions and electro-magnetic fields (cf. [10]), other systems with an optically-sensitive transition between two energy states acting as a quasi-two-level system can be driven into the strong-coupling regime. Therefore, the intensity of the spectrally and spatially resonant electro-magnetic mode must be sufficient, as the coupling strength via the dipole transition-matrix element M ((2.19) in Sect. 2.3.1) is proportional to the electric field (its component in dipole orientation) and thus to the number of photons in the mode. While optical cavities are used for local field enhancement through confinement, the optical mode interacting with an optical transition can also be a free-space mode. One remarkable aspect of quantum electrodynamics is, in fact, that the coupling of atomic energy levels with the 3D electro-magnetic vacuum field (vacuum fluctuations) occurs (see *Lamb* shift).

In the work of Ewers et al. in 2012, an excitonic system in QW structures was directly coupled to photons from THz pulses that were irradiated onto the sample. This is possible, since excitons as hydrogen-like quasi-particles feature intra-excitonic transitions that are dipole allowed for optical transitions and are typically on the order of half or less the excitonic binding energy. In a time-resolved pump-probe experiment, the spectral signatures of the exciton mode clearly indicated a Rabi splitting due to the strong coupling between the $1s$–$2p$ transition of the excitons (matter excitation) with energy difference $\hbar\omega_{21}$ and the approximately 4 meV light field of the incident THz pulse at zero delay time. This characteristic behaviour in the spectral domain was only seen for strong THz pulses with electric fields above a few kV/cm. For higher fields, the splitting observed for the $1s$ exciton resonance saturated at values around 0.6 meV. Theoretical modelling in that report also indicated the corresponding behaviour in the time domain.

Exciton Rabi splitting was previously also studied in 2010 by Wagner et al., who showed a strongly-coupled effective two-level system driven with THz field strengths of up to 10 kV/cm: an Autler–Townes splitting on resonance ($\omega_{\mathrm{THz}} = \omega_{21}$) of about 0.6 times the transition energy was observed when the THz beam was tuned near the $1s$–$2p$ transition of the heavy-hole (hh) exciton [11], whereas the original levels split by the Rabi frequency Ω as obtained in NIR spectra of the 1s-hh resonance.

Rabi flopping (oscillations in the time domain) have been also obtained for QW systems experimentally using a time-resolved optical four-wave mixing technique, whereas the excitons were disturbed by an additional THz pulse—a third pulse in the experiment in addition to the two optical pulses. In their experiment, Drexler et al. employed an ultrafast regenerative amplifier for the generation of fs-pulses and the THz pulse [12]. For strong fields, the population transfer between intra-excitonic states became irreversible due to an ionization of excitons, as transients of the four-wave mixing signal showed.

From these example experiments, it becomes understandable that transient electric fields can be utilized to manipulate and control polariton states in microcavities, as has been pursued by a number of experiments summarized in Sect. 9.3.

9.1.3 Magneto-Optics with Excitons

Magneto-optical spectroscopy offers a unique access to excitonic properties in external fields. The technique of exciton manipulation and characterization in external magnetic fields had been widely used well before the occurrence of cavity–polaritons. However, the plethora of experimental studies concerning magneto-excitons and the observed effects for excitons and electronic systems in semiconductors shall not be the subject of this section. Instead, a very brief summary of the basics will be given in order to promote a fundamental understanding of effects relevant to the study of polaritons in external magnetic fields—or magneto-polaritons in analogy to magneto-excitons.

Interaction of polaritons with an external magnetic field is particularly useful as a means of identification concerning strong coupling of light and matter. Due to the fact that the excitonic component of polaritons—composite quasi-particles—is affected by a magnetic field, one can distinguish polaritonic modes easily from purely photonic ones, which are not influenced by the external field. This is of utter importance in strongly confining 0D photonic box structures for which no characteristic polariton dispersion is obtained. Instead, discrete energy states occur in polaritonic quantum boxes with strong confinement which are hardly distinguishable from photonic modes of the corresponding resonator structure (see for example [13–15]).

In addition, in the applied magnetic field, the polaritons' spin degeneracy is lifted and a spinor system is obtained, the dynamics of which are governed by the spin components of their particles [16–19]. With the spin degeneracy of polaritons lifted due to their matter component, a polariton lasing mode can be unambiguously distinguished from its photonic lasing counterpart, providing a powerful tool for the characterization of electrically-driven polariton-laser devices [20] (see Chap. 4).

Diamagnetic Shift and Zeeman Splitting of Excitons

The influence of an external magnetic field on excitons in quantum wells can be summarized as follows. Accordingly, this influence is directly transferred onto the

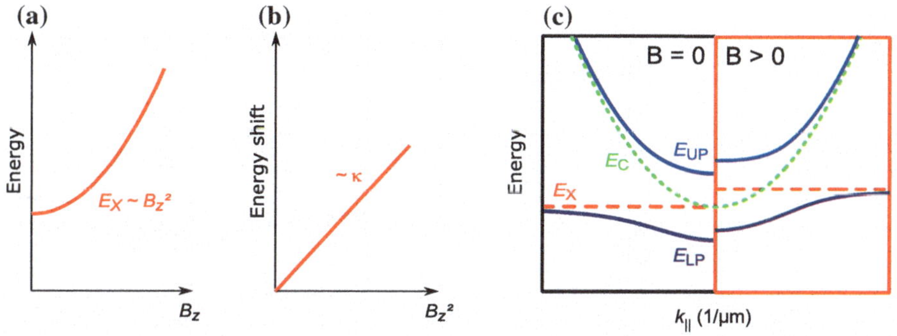

Fig. 9.2 Schematic drawing of the exciton diamagnetic shift in a magnetic field perpendicular to the QW plane. **a** Energy as a function of B (here in z orientation). **b** Relative exciton resonance as a function of B^2, i.e. diamagnetic shift. **c** Diagram of the E–k dispersion for the case of zero field at resonance detuning $\Delta = 0$ (left) in comparison to the situation at elevated fields, i.e. $B > 0$ (right). The dotted and dashed lines represent cavity and exciton resonance, respectively, whereas the solid lines show polariton branches for the given settings. Here, the changes of the Rabi splitting and the Zeeman splitting are neglected for simplicity

polariton in correspondence to the polariton's excitonic fraction, represented by the (excitonic) Hopfield coefficient $|X_{k_\|}|^2$ which is determined by (2.30). Here, an external field in Faraday configuration shall be considered: In such a field (magnetic flux $\mathbf{B} = (0, 0, B_z) \perp k_\|$), the effect on the electron–hole pairs in QWs leads to a diamagnetic blue shift of the exciton resonance according to the following equation in the low-field limit (i.e. where the induced energy shift is much less than the exciton binding energy) [21, 22]:

$$\delta E_{\mathrm{dia,X}} = \kappa_{\mathrm{X}} B^2, \tag{9.1}$$

whereas $\kappa_X = \langle \rho^2 \rangle e^2 / 8m_r$ denotes the diamagnetic coefficient (a measure of the effects of confinement, and used for the estimation of exciton binding energies; more precisely said, it is a measure of the in-plane electron–hole separation) with elementary charge e, in-plane reduced mass m_r and 2D exciton expansion ρ (here the expectation value of the exciton's squared radius is with respect to the zero-field eigen-state; the diamagnetic coefficient contains information about the zero-field properties of the exciton) [22]; in bulk, *Landau* levels are found for electrons and holes. Consequently, polaritons formed from QW excitons experience a magnetic-field-dependent blue shift. However, the exciton shift occurs while the cavity resonance remains fixed. Thus, this directly leads to a change in the spectral detuning Δ of cavity–polaritons, with the detuning altered towards a more negative detuning than in the case of $B = 0$ which affects the Hopfield coefficient $|X_{k_\|}|^2$. Thereby, the photonic fraction $|C_{k_\|}|^2$, the polariton's effective mass and its life time are modified. Thus, the magnetic field serves as a means of resonance tuning. This is schematically presented in Fig. 9.2.

In addition to the diamagnetic shift, the magnetic field couples to the total angular momentum of the excitonic spin states and gives rise to a Zeeman splitting of the

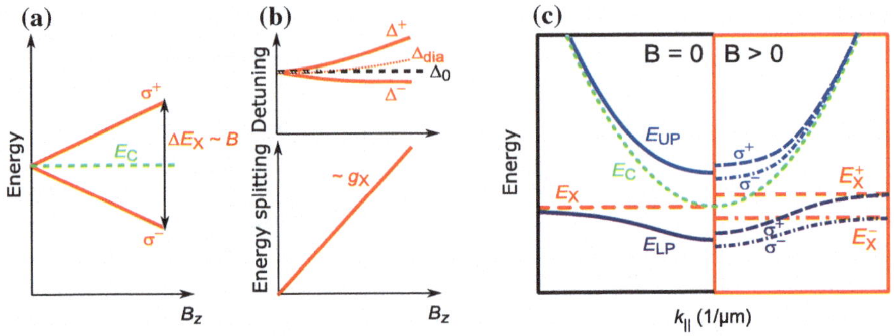

Fig. 9.3 Schematic drawing of the exciton Zeeman splitting in a magnetic field perpendicular to the QW plane. **a** Energy as a function of B (here in z orientation) for the opposite spin states, corresponding to emission under left and right circular polarization ($\sigma^{\pm}$). **b** Modified cavity–exciton detuning $\Delta(B)$ (solid lines) taking into account a diamagnetic-shift-induced detuning change (dotted curve) and the Zeeman shifts for the spin-degeneracy-lifted modes in comparison to the detuning at zero fields (dashed horizontal line). The Zeeman splitting as a function of B is shown in the lower graph. **c** Diagram of the E–k dispersion for the case of zero field at resonance detuning $\Delta = 0$ (left) in comparison to the situation at elevated fields, i.e. $B > 0$ (right). The dotted and dashed lines represent cavity and exciton resonance, respectively, whereas the solid lines show polariton branches for the given settings (dash-dotted and long-dashed for the two distinct polarizations $\sigma^{\pm}$). Here, the changes of the Rabi splitting and the diamagnetic shift are neglected for simplicity

previously degenerate energy states. This can be evidenced as an energy shift $\delta E_{\mathrm{ZS,X}}$ of excitonic resonances, or in other words as a splitting of the exciton resonance in two states with energy separation $\Delta E_{\mathrm{ZS,X}}$:

$$\delta E_{\mathrm{ZS,X}} = \pm\frac{1}{2} g_{\mathrm{X}} \mu_{\mathrm{B}} B, \tag{9.2}$$

$$\Delta E_{\mathrm{ZS,X}} = g_{\mathrm{X}} \mu_{\mathrm{B}} B, \tag{9.3}$$

whereas μ_{B} and g_{X} are the Bohr magneton and the electronic g-factor of the QW excitons, respectively [23]. Accordingly, the mode splittings of opposite-spin polaritons result from the strong couplings of the left-circularly and right-circularly polarized cavity mode with the respective states of the spin-resolved excitons with above given energy splitting. In correspondence to the underlying spin-state of the coupled emitter (states $|X^{+}\rangle$ and $|X^{-}\rangle$), radiative decay of magneto-polaritons results in the emission of σ^{+} and σ^{-} polarized photons, respectively [24]. Again, the effect of an external field on polaritons is dependent on the exciton fraction $|X|^2$ in strongly-coupled systems [15, 25]. The exciton's Zeeman splitting is sketched in Fig. 9.3, whereas modifications of the detuning situation as a result of diamagnetic and Zeeman shifts are indicated in (b).

Increase of the Exciton Oscillator Strength

Interestingly, the light–matter coupled system in such an external magnetic field also experiences an increase of the Rabi splitting. This is caused by a reduction of the excitonic wave-function due to magnetic confinement (cf. Appendix D in [26]). Indeed, studies of the binding energy of excitons, which is linked to their oscillator strength, in an external magnetic field date back to the 1980s, see for instance [27].

With increasing magnetic flux, the decrease of the exciton Bohr radius leads to a magnetic-field dependent increase of the exciton oscillator strength $f(B)$, whereby the coupling strength $\hbar\Omega \propto \sqrt{f(B)}$ increases correspondingly. The oscillator strength of the magneto-exciton can be given as a function of the Bohr radius as follows [26] (cf. supplementary information in [20]):

$$f(B) = \sqrt{\frac{2}{\pi}} \frac{1}{a_{\mathrm{B}}^2(B)} \mathrm{e}^{\left(-\frac{r}{a_{\mathrm{B}}}\right)}. \tag{9.4}$$

This allows one to approximate the Rabi splitting $E_{\mathrm{Rabi}} = 2\hbar\Omega$ in the magnetic field.

$$E_{\mathrm{Rabi}}(B) = E_{\mathrm{Rabi}}(0)\sqrt{\frac{f(B)}{f(0)}} \approx E_{\mathrm{Rabi}}(0)\frac{a_{\mathrm{B}}(0)}{a_{\mathrm{B}}(B)}, \tag{9.5}$$

whereas $E_{\mathrm{Rabi}}(0)$ and $a_{\mathrm{B}}(0)$ denote respective parameters in the absence of an external field B. The magnetic-field dependency of the exciton Bohr radius $a_{\mathrm{B}}(B)$ can be obtained from a variation solution of the Schroedinger equation, by minimization of the mean exciton energy in the QWs with respect to the Bohr radius, resulting in a transcendent equation [26]:

$$\frac{\hbar^2}{8m_{\mathrm{r}}}\left[1 - 6\left(\frac{a_{\mathrm{B}}(B)}{2L_{\mathrm{B}}}\right)^4\right] = \int_0^{\infty} r \cdot V(r)\left(1 - \frac{r}{a_{\mathrm{B}}(B)}\right)\mathrm{e}^{\left(-\frac{2r}{a_{\mathrm{B}}(B)}\right)}\mathrm{d}r, \tag{9.6}$$

with magnetic length $L_{\mathrm{B}} = \sqrt{\hbar/eB}$, spatial coordinate r of the exciton's centre of mass, the reduced effective mass m_{r}, and the Coulomb potential

$$V(r) = \frac{e^2}{4\pi\epsilon_{\mathrm{r}}\epsilon_0}\int_{-L_z/2}^{L_z/2}\int_{-L_z/2}^{L_z/2} \frac{\phi_{\mathrm{e}}^2(z_{\mathrm{e}})\phi_{\mathrm{h}}^2(z_{\mathrm{h}})}{\sqrt{r^2 + (z_{\mathrm{e}} - z_{\mathrm{h}})^2}}\mathrm{d}z_{\mathrm{e}}\mathrm{d}z_{\mathrm{h}} \tag{9.7}$$

between the confined electrons and holes in the quantum well $z_{\mathrm{e/h}}$, respectively. Those confined particles' wave-functions $\phi_{\mathrm{e/h}}$ can be considered as the solution of an infinitely high one-dimensional potential well (provided that a strong enough particle confinement in the quantum well with thickness L_z can be assumed).

This has significant implications for practical polariton systems, as the coupling strength increases with increasing magnetic field which is considered beneficial for the observation of polariton condensation in a system that is prone to polariton break-up at higher particle densities. For example, in InGaAs QWs with 15 % indium content and a thickness of 7 nm, a moderate increase of the ratio between the magnetic-field dependent Bohr radii $a_B(0)/a_B(B) = 10.5/9.3 = 1.13$ (a_B in nm) can be achieved at 5 T (see supplementary information of [20]).

9.2 Magneto-Polaritons in Microcavity Systems

A magnetic-field dependent polariton diamagnetic shift and Zeeman splitting can easily be anticipated given the strong cavity–exciton coupling in QW-microcavities. Indeed, magneto-optical effects in polaritonic systems have been thoroughly studied, particularly with respect to an increase in the oscillator strength and Rabi splitting with increasing magnetic field [24, 28–31] and regarding the Zeeman splitting of polariton modes [24, 25]. Magneto-optical studies have also been carried out on exciton–polariton condensates (e.g. in [17, 18]) and quantum-dot polaritonic systems, for which a considerable diamagnetic shift and a flux-dependent decrease of the oscillator strength was found [32, 33]. From another perspective approached, magneto-optical response has become a tool rather than a feature, as it was utilized to identify polariton emission from strongly-confining potential traps [15, 34] and to verify the strong-coupling regime in polariton lasers [20, 35].

9.2.1 *Manipulating the Excitonic Component of Polaritons*

An example of the magnetic-field effect on a polariton system is given in Fig. 9.4 which shows, both, the mode energy shift (a) and splitting (b) as a result of an increased magnetic field in Faraday geometry, i.e. perpendicular to the QW plane, whereas the typical changes of the Rabi splitting and the excitonic fraction $|X|^2$ are indicated in (c).

The response of magneto-excitons according to (9.1) and (9.3) weighted with magnetic-field-dependent Hopfield coefficients (due to the modified coupling strength, (9.5)) can approximate the coupled-system's response already well [34, 36]:

$$\delta E_{\text{dia,X}}\left(\left|X(\Delta, B)_{k_\parallel}\right|^2\right) = \left|X(\Delta, B)_{k_\parallel}\right|^2 \kappa_X B^2 = \kappa_{\text{eff}} B^2, \tag{9.8}$$

$$\Delta E_{\text{ZS,XP}}\left(\left|X(\Delta, B)_{k_\parallel}\right|^2\right) = \left|X(\Delta, B)_{k_\parallel}\right|^2 g_X \mu_B B = g_{\text{eff}} \mu_B B, \tag{9.9}$$

whereas $|X(\Delta, B)_{k_\parallel}|^2$ is the excitonic fraction $|X|^2$ (see (2.30) in Sect. 2.3.1) at corresponding cavity–exciton detuning Δ for given $k_\parallel$ and external field strength B. Note that $\kappa_{\text{eff}} = |X(\Delta, B)_{k_\parallel}|^2 \kappa_X$ and $g_{\text{eff}} = |X(\Delta, B)_{k_\parallel}|^2 g_X$ represent an effective diamagnetic coefficient and effective Bohr magneton, respectively, for the polariton state (at given $k_\parallel$ for a specific detuning). Simplified Figs. 9.2c and 9.3c indicate the actual impact on the polariton system for both effects individually, also neglecting Rabi splitting changes, and thus they do not represent the combination of effects as is described below.

From the beginning, the idea was to treat magneto-polaritons based on the coupling of the magneto-excitons with the circularly-polarized light state. However, for simplicity, the Hopfield modelling approach was often preferred for data evaluation, e.g. when dealing with spinor condensates [36]. Nevertheless, to be rigorous, one has to use a coupled oscillator model taking into account magneto-excitons.

Assuming that the different excitonic spin states $|+\rangle$ and $|-\rangle$ are independently coupled to the counter-polarized light states with photon spin projection + and − owing to conservation rules, a very accurate description of the spinor-polariton modes' energy (dispersion) can be obtained when taking into consideration both the increased Rabi splitting and E_X as a function of the field. In matrix representation, the stationary coupled system's eigen-value equation with diagonalized Hamiltonian $\hat{H}_{xp}$ and eigen-energy E_{xp} can be expressed as follows with a slight modification from (2.26) of Sect. 2.3.1:

$$\hat{H}_{xp} \left|\Psi^{\pm}\right\rangle = \begin{pmatrix} E_C & \hbar\Omega(B) \\ \hbar\Omega(B) & E_{X\pm}(B) \end{pmatrix} \begin{pmatrix} C \\ X_{\pm}(B) \end{pmatrix} = E_{xp\pm} \left|\Psi^{\pm}\right\rangle, \tag{9.10}$$

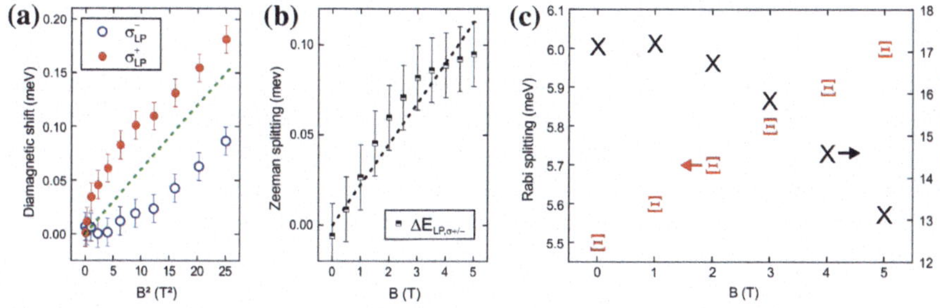

Fig. 9.4 Polariton ground-state diamagnetic shift (**a**) and Zeeman splitting (**b**) in a magnetic field perpendicular to the QW plane, measured for the electroluminescence of a polariton device. Here, the expected behaviour as described for excitons translates well to the behaviour of polaritons, i.e., for (**a**) it is linear with B^2, whereas for (**b**) it is linear with B. Dashed lines represent an approximation of the trends, using a diamagnetic coefficient and Bohr magneton value for the QW system weighted by a constant exciton fraction as effective values for the polariton mode. As can be seen, with increasing fields, a deviation of the shifts and splittings occurs which is attributed to the detuning and oscillator strength changes. The diamagnetic shift in (**a**) is spin resolved and superposed by the opposite Zeeman shifts for the respective counter-circularly-polarized modes. **c** Diagram of the magnetic-field dependent Rabi splitting (left scale, symbol Ξ) and the overall excitonic fraction deduced from the coupled-oscillator model (right scale, symbol X). The occurring blue shift of the exciton resonance of about 1.7 meV (and therewith the detuning change) is reliably extracted from dispersion fits to magnetic-field-dependent polariton emission in the linear regime at low densities. For this example, a total LP shift of about 0.15–0.16 meV up to 5 T is compatible with the low excitonic fraction of the LP ground state, whereas the Rabi splitting increase is understandable for the QW material owing to a predicted B-dependent decrease in the exciton Bohr radius, which at 0 T amounts to $\approx 1.13\, a_B(5\,\mathrm{T})$ (see Sect. 9.1.3 above). Adapted from [35]. Courtesy of the author

with $|\Psi^{\pm}\rangle$ the spin states of the polariton system, $E_{xp\pm}$ the eigen-energies from which left/right-circularly polarized emission is obtained ($\pm$ corresponding after polariton decay to $\sigma^{\pm}$, the photon polarization), C and $X_{\pm}(B)$ the corresponding amplitude for each component (cavity and exciton, respectively) of the linear mixing product. In the case of the exciton, the magnetic field dependence is taken into account, i.e. $|X|^2(B)$. $E_{X\pm}(B)$ denotes the exciton's resonance for a given field strength B.

Accordingly, one can derive the $k_{\|}$-dependent and B-dependent eigen-energies of the coupled system by the solution of the linear equation (2.25) of the diagonalized $\hat{H}_{xp}$ (modified to take into account B-field effects, see eigen-value equation (9.10)). For the spinor LP and UP branches, one obtains an expression such as

$$E_{LP,\pm}(k_{\|}, B) = \frac{1}{2}\left[E_C(k_{\|}) + E_{X\pm}(k_{\|}, B) - \sqrt{(2\hbar\Omega(B))^2 + (\Delta_{\pm}(k_{\|}, B))^2}\right], \tag{9.11}$$

and correspondingly

$$E_{UP,\pm}(k_{\|}, B) = \frac{1}{2}\left[E_C(k_{\|}) + E_{X\pm}(k_{\|}, B) + \sqrt{(2\hbar\Omega(B))^2 + (\Delta_{\pm}(k_{\|}, B))^2}\right], \tag{9.12}$$

with $\Delta_{\pm}(k_{\|}, B) = E_C(k_{\|}) - E_{X\pm}(k_{\|}, B)$ denoting the detuning for the given exciton energy

$$E_{X\pm}(k_{\|}, B) = E_X(k_{\|}, B = 0) + \delta E_X(B) + \delta E_{ZS,X\pm}(B). \tag{9.13}$$

Here, it is assumed that exciton diamagnetic shift $\delta E_X(B) = \kappa_X B^2$ and exciton Zeeman shift $\delta E_{ZS,X\pm}(B) = \pm\frac{1}{2} g_X \mu_B B$ have no specific $k_{\|}$ dependence, whereas the shallow E–k dispersion of the exciton mode due to its large effective mass is taken into account in $E_X(k_{\|}, B = 0)$ for the sake of completeness.

Accordingly, the Zeeman splitting for LP/UP states at a specific $k_{\|}$ can be expressed by

$$\Delta E_{ZS,LP/UP}(k_{\|}, B) = E_{LP/UP+}(k_{\|}, B) - E_{LP/UP-}(k_{\|}, B). \tag{9.14}$$

This was already elaborated on in [34], explaining that, in fact, if calculating the system's energy eigen-states with both magnetic-field dependent parameters $E_X(B)$ and $E_{Rabi}(B)$, the polaritons' diamagnetic shift, which is particularly the result of a detuning change with increasing B, can be well estimated for given B. Taking into account spin-split $E_X(B)$, the same can be done for the Zeeman splitting, it was further stated in [34].

In this context, further theoretical and experimental studies on the magnetic field tuning of exciton–polariton modes in semiconductor microcavities have recently shed light on the polaritons' response in external magnetic fields, catching up and revisiting this concept [37].